SERVICES IN GLOBAL VALUE CHAINS

Manufacturing-Related Services

SERVICES IN GLOBAL VALUE CHAINS

Manufacturing-Related Services

Editors

Patrick Low
Asia Global Institute, University of Hong Kong

Gloria O Pasadilla
APEC-Policy Support Unit, Singapore

World Scientific

NEW JERSEY • LONDON • SINGAPORE • BEIJING • SHANGHAI • HONG KONG • TAIPEI • CHENNAI • TOKYO

Published by

World Scientific Publishing Co. Pte. Ltd.
5 Toh Tuck Link, Singapore 596224
USA office: 27 Warren Street, Suite 401-402, Hackensack, NJ 07601
UK office: 57 Shelton Street, Covent Garden, London WC2H 9HE

British Library Cataloguing-in-Publication Data
A catalogue record for this book is available from the British Library.

This publication is based on APEC Policy Support Unit project on Services in Global Value Chains: Manufacturing-Related Services for the APEC Committee on Trade and Investment in 2015, Publication Number APEC#215-SE-01.15.

SERVICES IN GLOBAL VALUE CHAINS
Manufacturing-Related Services

ISBN 978-981-3141-45-2

Desk Editor: Sandhya Venkatesh

Typeset by Stallion Press
Email: enquiries@stallionpress.com

Printed in Singapore

CONTENTS

LIST OF FIGURES

LIST OF TABLES

LIST OF BOXES

LIST OF ABBREVIATIONS

AEC	ASEAN Economic Community
AOI/AXI	Automatic Optical Inspection/Automatic X-ray Inspection
APEC	Asia-Pacific Economic Cooperation
API	American Petroleum Institute
ASEAN	Association of Southeast Asian Nations
BOI	Board of Investment
CAAC	Civil Aviation Administration of China
CAGR	Compound Annual Growth Rate
CKD	Completely Knocked Down
CPC	Central Product Classification
CPU	Central Processing Unit
CSR	Corporate Social Responsibility
DBO	Design-Build-Operate
DENR	Department of Environment and Natural Resources
DOLE	Department of Labor and Employment
DVT	Design Verification Test
EASA	European Aviation Safety Agency

EP	Employment Pass
EPCM	Engineering, Procurement, Construction and Maintenance
EVT	Engineering Verification Test
FAA	Federal Aviation Administration
FDI	Foreign Direct Investment
GDP	Gross Domestic Product
GE	General Electric
GPS	Global Positioning System
GSM	Global System for Mobile Communications
GVCs	Global Value Chains
HDOR	Heavy Duty Off-Road
HP	Hewlett-Packard
IBM	International Business Machines Corporation
IDC	International Data Corporation
IIW	International Institute of Welding
IMMEX	Manufacturing, Maquila and Export Services Industry
IoT	Internet of Things
IP	Intellectual Property
IPO	Initial Public Offering
IPR	Intellectual Property Rights
IQC	International Quality Consultant
ISO	International Organization for Standardization
ISO/TS	ISO Technical Specification
IT	Information Technology
ITA	Information Technology Agreement
JETRO	Japan External Trade Organization
LPI	Logistics Performance Index
LTE	Long Term Evolution
MOM	Ministry of Manpower

MRO	Maintenance, Repair and Overhaul
MTCR	Missile Technology Control Regime
NADCAP	National Aerospace and Defense Contractors Accreditation Program
NAFTA	North American Free Trade Agreement
NEC (n.e.c)	Not Elsewhere Classified
ODM	Original Design Manufacturing
OECD	Organisation for Economic Co-operation and Development
OEM	Original Equipment Manufacturer
OH&S	Occupational Health and Safety
OQA	Outgoing Quality Assurance
PACT	Partnerships for Capability Transformation
PCB	Printed Circuit Boards
PEZA	Philippine Economic Zone Authority
PSU	Policy Support Unit
PVT	Production Process Verification Test
QA/QC	Quality Assurance and Quality Control
RAIDs	Redundant Array of Inexpensive Disks
R&D	Research & Development
RFQ	Request for Quotation
RHQ	Regional Headquarters
SEZ	Special Economic Zone
SMEs	Small and Medium-sized Enterprises
STRI	Services Trade Restrictiveness Index
TESDA	Technical Education and Skills Development Authority
TiVA	Trade in Value-Added
TPP	Trans-Pacific Partnership
UN	United Nations
UNCTAD	United Nations Conference on Trade and Development

USITC	United States International Trade Commission
VAT	Value Added Tax
WIPO	World Intellectual Property Organization
WTO	World Trade Organization

ABOUT THE AUTHORS

Penny Bamber is Senior Research Associate at the Duke University Center on Globalization, Governance, & Competitiveness. Penny's research and consulting focus is on economic development and competitiveness in developing regions. Key industry foci include the agricultural sector, the medical devices industry and the offshore services industry. She has led several projects related to improving national competitiveness in both high-tech and low-tech sectors for developing economies. Recent publications include *Connecting Local Producers to Regional and Global Value Chains* (OECD, 2013), *Global Value Chains, Economic Upgrading and Gender: The Horticulture Industry* (World Bank, 2013), *Costa Rica in the Medical Devices Global Value Chain: Opportunities for Upgrading* (COMEX, 2013) and the *Inclusion of Small and Medium Producers in High Value Agro Chains* (Inter-American Development Bank, 2013). In addition, Penny has led trainings in the use and application of the global value chain (GVC) framework for industry representatives and policy makers in developing economies in Africa, Asia, Latin America and the Caribbean. Prior to joining CGGC, Penny worked in the leadership development field, facilitating programs for business students and executives at the Wharton School of Business, University of Pennsylvania.

Joel Bassig is currently Program Executive at the APEC Secretariat. His portfolio includes providing support for the following sectors: telecommunications and information, electronic commerce, intellectual property, market access, mining and services. Mr Bassig has years of experience in project management as well as trade and international affairs, through his work with the Philippine Chamber of Commerce and Industry (PCCI) in Manila and the Confederation of Asia-Pacific Chambers of Commerce and Industry (CACCI) in Taipei. He has also worked on issues related to Formal and Non-Formal Education while he was with the Asia-Europe Foundation (ASEF) in Singapore. As a facilitator and certified Mediator, he has been exposed to addressing various kinds of conflicts related to business, family and ethnic communities. Prior to being with the APEC Secretariat, he was with the Asian Development Bank (ADB) in Manila as a consultant for the transport sector. Mr Bassig holds a Master of Arts in International Affairs from Ming Chuan University (MCU) in Chinese Taipei and a Bachelor of Arts in Development Studies from the University of the Philippines-Manila (UPM).

Denise Cheung is currently Senior Research Manager at Fung Business Intelligence Centre, the internal research house of Li & Fung, focusing on Asian economies and trade policy. Prior to joining Li & Fung, Denise conducted research on GVCs at the Fung Global Institute (FGI). Prior to that, she worked at the Hong Kong government as an economist, and at the International Finance Corporation of the World Bank Group. Denise holds a Master in Public Administration/International Development from Harvard University and a bachelor's degree from Cornell University.

Deborah Elms is Executive Director of the Asian Trade Centre in Singapore. She is also a senior fellow in the Singapore Ministry of Trade and Industry's Trade Academy. Previously, she was head of the Temasek Foundation Centre for Trade & Negotiations (TFCTN) and senior fellow of international political economy at the S. Rajaratnam School of International Studies at Nanyang

Technological University, Singapore. Her research interests are negotiations and decision making, and her current research involves the Trans-Pacific Partnership (TPP), Regional Comprehensive Economic Partnership (RCEP), ASEAN Economic Community (AEC) negotiations and GVCs. She has provided consulting on a range of trade issues to governments including the United Arab Emirates, Sri Lanka, Cambodia, Chinese Taipei and Singapore. Dr Elms received a PhD in Political Science from the University of Washington, an MA in International Relations from the University of Southern California, and bachelor's degrees from Boston University.

Karina Fernandez-Stark is Senior Research Analyst at the Duke University Center on Globalization, Governance, & Competitiveness. At CGGC she has led several research projects related to economic development and competitiveness in Latin America and other regions of the world. She has consulted for numerous governments and several international organisations: the African Development Bank, Economic Commission for Latin America and the Caribbean, Inter-American Development Bank, OECD, UNCTAD and the World Bank, amongst others. She has published several research reports and articles on industrial upgrading and social and economic development. Publications include *Global Value Chain Analysis: A Primer* (Duke CGGC, 2011), "The Offshore Services Value Chain: Developing Countries and the Crisis", in *Global Value Chains in a Post-crisis World: A Development Perspective* (The World Bank, 2010), "The Offshore Services Value Chain: Upgrading Trajectories in Developing Countries" (*International Journal of Technological Learning, Innovation and Development*, 2011), *Inclusion of Small and Medium Producers in High Value Agro Chains* (Inter-American Development Bank, 2013), and *Connecting Local Producers to Regional and Global Value Chains* (OECD, 2013). In addition, Karina designed and developed a GVCs manual in Spanish for economic development researchers in Latin America and has co-facilitated workshops on GVCs in Africa, Asia and Latin America.

Gary Gereffi is Professor of Sociology and Director of the Center on Globalization, Governance, & Competitiveness at Duke University, where he teaches courses in economic sociology, globalisation and comparative development, and international competitiveness. He received his BA degree from the University of Notre Dame and his MPhil and PhD degrees from Yale University. Gereffi has published numerous books and articles on globalisation, industrial upgrading, and social and economic development in various parts of the world. His books include: *Manufacturing Miracles: Paths of Industrialization in Latin America and East Asia* (Princeton University Press, 1990); *Commodity Chains and Global Capitalism* (Praeger Publishers, 1994); *Free Trade and Uneven Development: The North American Apparel Industry After NAFTA* (Temple University Press, 2002); *The New Offshoring of Jobs and Global Development* (International Institute of Labor Studies, 2006); *Manufacturing Climate Solutions: Carbon-Reducing Technologies and U.S. Jobs* (Center on Globalization, Governance & Competitiveness, Duke University, 2008); and *Global Value Chains in a Postcrisis World: A Development Perspective* (The World Bank, 2010). Gereffi's research interests deal with the competitive strategies of global firms, the governance of GVCs, economic and social upgrading, and the emerging global knowledge economy.

William Haines graduated from the University of Sydney in 2014 with a BA (Hons), majoring in Philosophy and an MA in Economics. He participated in this study while a research analyst for the Fung Global Institute (now the Asia Global Institute) in Hong Kong. Previously, he worked for Mekong Economics, an economic consultancy based in Hanoi and interned with Jardine Matheson, also in Hong Kong. He is currently undertaking further graduate study at the Barcelona Graduate School of Economics.

Arian Hassani has over 14 years of experience as an international development professional. Currently she is based in Hong Kong where she covers J.P. Morgan's philanthropic activities across ASEAN and Australia, focusing on job creation and poverty alleviation. Before joining J.P. Morgan, Arian managed FGI's research on Asian finance, supply chains, governance and sustainability. Prior to moving to Hong Kong in 2013, she was in charge of a cultural development fund at the United Nations Educational, Scientific and Cultural Organization (UNESCO) in Paris, where she oversaw a portfolio of grants across 40 developing economies worldwide. From 2007 to 2009, she headed Hope Street Group's efforts in developing the think tank's Economic Opportunity Index, while advising 2008 US presidential campaigns on education and job creation. From 2004 to 2007, Arian launched and managed UNESCO's Creative Cities Network, a development project that still helps cities across the world integrate cultural industries into their economic growth strategies. Arian started her career in the media and global communications investment banking team at Merrill Lynch in New York. She holds a Master's in Public Administration in International Development from Harvard Kennedy School, and a Bachelor's in Economics and German Studies from Brown University, where she was awarded a Fulbright scholarship.

Patrick Low specialises in research on global supply chains, international trade and global governance. He was the Chief Economist at the World Trade Organization from 1997 to 2013, having previously worked at the GATT Secretariat from 1980 to 1987. After joining the WTO in 1995, he worked on trade in services for two years before his appointment as Chief Economist. From 1999 to 2000, he also served as WTO Director-General Mike Moore's Chief of Staff. Between leaving the GATT and joining the WTO, Dr Low taught economics at El Colegio de México in Mexico City, from 1987 to 1990, and during this time worked as a consultant for a range of governments and inter-governmental institutions. From 1990 to 1994 he worked as a senior economist in the World Bank's research complex (International Trade Division). Dr Low was also Adjunct Professor of International Economics at the Graduate Institute of International and Development Studies, Geneva, from 2004 to 3013. Since 2015, he has been a visiting professor at the

University of Hong Kong and Fellow at the Asia Global Institute. He holds a PhD in Economics and has written widely on a range of trade policy issues.

Gloria O. Pasadilla is Senior Analyst at the APEC Policy Support Unit (PSU) in Singapore and a former research fellow at the ADB Institute in Tokyo. She was the Philippine lead negotiator for services in the ASEAN-Australia-New Zealand free trade agreement negotiation, convenor of the APEC Group on Services from 2007 to 2009, lead economist for services issues at the Philippine Institute for Development Studies, and consultant for several international organisations. She has published journal articles and book chapters, and co-edited books on international trade policy. She has a Master's Degree in Industrial Economics from the University of Asia and the Pacific, Philippines; in International Law and Economics from the World Trade Institute in Bern, Switzerland; and PhD in Economics from New York University.

Emmanuel A. San Andres is currently Analyst at the APEC PSU. He holds a Master of Arts in Economics and a Bachelor's degree in Economics from the University of the Philippines School of Economics, where he is also a PhD candidate in economics. Prior to joining APEC, Mr San Andres was Economics Officer at the Economics and Research Department of the Asian Development Bank (ADB) in Manila mainly working on poverty, inequality, food security and social sector issues such as health, education and employment. He also worked as a consultant for organisations such as the ADB, World Bank and various non-government organisations where he conducted research on migration and remittances, economic development, taxation and public policy.

Ben Shepherd, the Principal of Developing Trade Consultants, is a trade economist and international development consultant. He has worked on a wide range of trade and development issues with organisations such as the World Bank, the OECD, the Asian Development Bank, the Inter-American Development Bank, the

United Nations and the Asia-Pacific Economic Cooperation. He specialises in providing policy-relevant research, as well as capacity-building seminars for researchers working in trade and development. He has published more than 15 articles in peer-reviewed journals, and a similar number of book chapters. Ben's particular areas of expertise include: (i) trade policy, (ii) GVCs, (iii) trade facilitation and logistics and (iv) trade in services.

David Sit is Senior Research Analyst at the Fung Business Intelligence Centre's Business Policy and City Clusters in China team, specialised in economic and business relevant policy in China and the Asia-Pacific region. Prior to joining Fung Business Intelligence Centre, David served as a member in FGI's Global Value Chain project under Dr Patrick Low. He focused on services in GVCs and researched on how government policy in Asia-Pacific region impacts business and value chains. Publications included a series of company case studies which demonstrated the interfaces between government policy and services in GVCs in various Asian economies. David holds a Bachelor of Economics and Finance and a Master of Economics, both from the University of Hong Kong, and a Master of Science in Data Science and Business Statistics from the Chinese University of Hong Kong.

Sherry M. Stephenson is Senior Fellow with the International Centre for Trade and Sustainable Development (ICTSD). Her internationally recognised expertise lies particularly in the area of services trade. Until end of 2012, she was Senior Advisor for Services Trade in the Executive Secretariat for Integral Development of the Organization of American States (OAS) in Washington, DC. While at the OAS, Dr Stephenson held various positions, including Director of the Trade Department, Deputy Trade Advisor and Head of Institutional Relations. Dr Stephenson was an Advisor to the Ministry of Trade in Jakarta for three years where she also trained government officials in the content and implementation of the WTO Agreements. As a member of the Pacific Economic Cooperation Council (PECC), she has provided analytical support to the APEC process. She has done consulting work for many organisations including The World Bank, Asian Development Bank, Commonwealth Secretariat, Forum Island Secretariat, U.S. Agency for International Development and several governments. She has lectured in conferences and workshops, educating government officials and members of the trade and development community on topics including the WTO, regional trade agreements, services trade and GVCs, and has published three books on services and numerous articles.

Katherine Tait is a PhD student of sociology at the University of North Carolina — Chapel Hill. She holds a BA in Sociology and Anthropology from Carleton College and an MA in Economics from the New School for Social Research. Her current work focuses on economic sociology, including GVC analysis and the interconnections between globalisation and development, as well as political sociology and social movements. The chapter included in this volume, co-authored with Professor Gary Gereffi of Duke University, reflects her research interests both in international trade and the strategies employed by global lead firms to maintain competitive advantage under globalisation.

Andre Wirjo is currently Analyst at the APEC PSU where he is engaged in various policy-oriented projects to support the role of APEC in facilitating economic growth, cooperation, trade and investment in the Asia-Pacific region. Some of these projects are: assessing the progress of APEC member economies in facilitating ease of doing business; understanding the linkage between services, manufacturing and productivity in APEC; and examining the relationships between FTA and trade so as to contribute to the ongoing discussions revolving around the establishment of the Free Trade Area of the Asia-Pacific (FTAAP). Prior to joining APEC Secretariat, Mr Wirjo worked as Research Assistant at the Lee Kuan Yew School of Public Policy, National University of Singapore (NUS) where he coordinated the Survey on ASEAN Competitiveness across all ten member states in collaboration with the ASEAN

Business Advisory Council (ASEAN-BAC). He also interned for the UN Economic and Social Commission for Asia and the Pacific (UNESCAP) in Bangkok, Thailand. Mr Wirjo obtained his Master of Science in Economics degree from the London School of Economics and Political Science (LSE).

Yuhua Zhang was previously Analyst at the APEC PSU, providing policy analysis and evaluations of various region-wide initiatives. Prior to APEC, Ms Zhang has worked as a teaching and research assistant at universities in Singapore and China, and as a consultant for the World Bank in the lead up to the inaugural World Cities Summit in 2008. She holds a Master's degree in Public Policy from the NUS and a Bachelor's degree in Economics from Shanghai Jiao Tong University, China.

Huani Zhu is currently Researcher at the APEC PSU, providing research and analysis to support APEC economic agenda. During her time at APEC PSU, Ms Zhu has been working on a broad range of projects such as trade and investment liberalisation, improving business environment, gender equality, energy and environmental issues as well as regional GVCs. Prior to joining APEC, Ms Zhu worked as Economic Analyst in an investment advisory and fund management company in Singapore. Ms Zhu holds a Master of Philosophy (MPhil) degree in Economics from the University of Cambridge. She obtained a Bachelor of Science (Hons) degree majoring in mathematics and Economics from the University of Warwick. Her previous dissertation and research projects have focused on macroeconomics, international economics and development economics.

Chapter 1

MANUFACTURING-RELATED SERVICES

Patrick Low and Gloria O. Pasadilla

1.1. INTRODUCTION

This report contains an analysis of the role of services in manufacturing value chain activities as well as policy issues that affect the supply of these services. It responds to a request by the APEC Committee on Trade and Investment to look at the increasing importance of services in global trade. The approach, for analysing services under this project, was to undertake case studies of the value chains of firms in order to understand how services entered production, trade and consumption, and what functions they performed. The primary reason for this analysis was to identify the policy frameworks and specific policies in various economies within the region that affected both the establishment and operation of manufacturing-related services. The project has compiled 22 case studies from different APEC economies, 14 from Asia and 8 from North America and Chile. Nine out of twenty-two are big multinational firms or their subsidiaries, ten can be considered medium-to-large firms either with growing international operations or as strategic suppliers to multinational firms, and one is a small and medium-sized enterprise (SME) in the technology industry.

The report is organised in three main sections. Section 1.2 discusses the evolving role of services in the global economy. Section 1.3 presents the case studies and draws out some of the ways services are important both in terms of the value they add and also as a vehicle for new opportunities to deepen participation in value chains at the economy level. The final section presents an analysis of how policies have affected the configuration, operation and location of value chains, with particular reference to the contribution of services.

1.2. THE EVOLVING ROLE OF SERVICES IN THE GLOBAL ECONOMY

1.2.1. *Services in a Historical Context*

The role of services has been evolving, driven in particular by technology, globalisation and new business models. Services now make up a dominant share of income in most economies. The global share of gross domestic product (GDP) accounted for by services was 70 per cent in 2012. Unsurprisingly, the prominence of services as a source of income translates into jobs. In the Organisation for Economic Co-operation and Development (OECD) economies, total civilian employment in services in 2011 amounted to 73 per cent of all jobs (OECD, 2014). Shares in developing and emerging economies are lower than in industrial economies for the most part, but rising. In APEC, the average share of employment in services in 2010 is 60 per cent. Notwithstanding variations in the shares of GDP attributable to services, manufacturing, agriculture and mining, in the majority of economies the services share is often greater than that of the other three components of economic activity combined.

Considering what the statistics tell us about the importance of services, it may seem odd that they have received neither greater analytical attention nor policy attention in many cases. One reason for this is the long shadow cast by classical economic thought. Adam Smith famously wrote in *The Wealth of Nations* that

> [T]he labor of a menial servant...adds to the value of nothing... services generally perish in the very instant of their performance, and seldom leave any trace or value behind (Smith, 1776).

The reason services were relegated zero value status turned essentially on their non-storability. Value was equated to the accumulation of capital, which required the production of something physical and storable. This thinking persisted in Ricardian and Marxian thought, and arguably also influenced neo-classical economic analysis. It is true, of course, that services played a less important role in seventeenth- and eighteenth-century economies, and their non-storability would have been a dominant feature.

In the 1960s, a new concern was raised about services, further relegating them to "poor cousin" status in scholarly (Baumol, 1967; Fuchs, 1968) and policy circles. The essential issue was that if services assumed a growing share of production, they would threaten the overall health of an economy. Costs would rise and be unmatched by productivity gains. Services were intrinsically hampered by their incapacity to generate efficiency gains. In other words, they were considered poor generators of productivity growth. Despite rising prices, some services might continue to be supplied because they were considered socially necessary (particularly in the case of social and other services supplied by governments). Wages would match productivity in other sectors and pull up wages in the services industries. This was known as the Baumol Cost Disease.

Recent work has largely dispelled this pessimism. Technology, the internationalisation of production and evolving business models have all contributed to making a significant range of modern-day services, an important source of productivity growth. As services have become a more prominent source of value in modern economies, their contribution to innovation has grown significantly. The innovation may be of a process variety, or it may involve new technology and be bundled with goods.

Several factors account for growing shares of services in global economic activity (Francois and Hoekman, 2010). As incomes have risen, the composition of consumption has shifted in the direction of services. On the supply side, the increased internationalisation of production has intensified the reliance on services. When products can be sourced, made and sold anywhere in the world, services become especially critical. For example, design, research and development (R&D) and prototyping services help decrease the cost of production failure and shorten the product development cycle. For sourcing of intermediate inputs, logistics and transportation services as well as supply chain management services make the geographic dispersion of global value chain (GVC) operations possible (Pasadilla and Wirjo, 2014). Similar structural and compositional shifts have taken place in consumption, both in terms of its internationalisation and changed consumer preferences enabled by technological advances in information and communications technology, including the Internet.

1.2.2. *Contrasts Between Services and Goods*

Some of the neglect of services has arisen from their intangibility, making them analytically and statistically elusive. Systematic efforts to deepen our understanding of the economic role played by services — particularly at the international level — have only started relatively recently. These efforts have intensified with the increased presence of GVCs, where services fulfil a vital and complex role.

Services are typically differentiated from goods in four ways. The first, which troubled the classical economists, is a lack of storability. This means that production and consumption must be simultaneous, the typically cited example of which is a haircut. The second has been the impossibility of transporting them or providing them at a distance, thus requiring that the producer and consumer be in the same place. The haircut comes to mind again. The third is that many services are customised and not commoditised. This variation in the characteristics of units of output makes it difficult to establish reliable value estimates and sometimes product prices, and to settle on an agreed nomenclature for service products. Finally, services are intangible, which makes them hard to see and measure.

The first two of the above characteristics — limits to storability and transportability — have been rendered far less important than they used to be by technological advances in information and communications technology. The third — customisation — is a characteristic that may also apply to some categories of goods, although product differentiation in the goods domain may well be the consequences of the addition of services components into production. Only the final element — intangibility — is unchangeable over time and truly distinguishes goods from services.

The linkages between goods and services in the GVC world, particularly through the bundling of offerings by firms of both goods and services, do raise a question about the wisdom of trying too hard to separate services and goods analytically. The argument for some sort of fusion is even stronger when it comes to the fallout from handling goods and services separately in many international regimes (Low, 2016).

In contrast to goods, services are multi-functional. They are not merely a source of value like any other product. They supply inputs to virtually every economic activity. Producer services, such as transport, telecommunications, financial services, distribution, business and professional services, are essential to the entire operation of an economy. Services are also the 'glue' that enables economic linkages and networks to operate both locally and internationally. Without them, there would be a lot less market integration and a lot more market segmentation.

A growing body of work emphasises how much more we need to understand about the role of services in GVCs. Studies by Sweden's Kommerskollegium (2010a, 2010b, 2012), for example, have introduced the notion of "servicification", a process whereby manufacturing activities have become increasingly service dependent. Other contributions in a similar vein refer to 'servicification' and 'service science' (Low, 2013). The OECD has also contributed to a deeper understanding of how knowledge-based capital is an increasingly important invisible asset that contributes to growth (OECD, 2011, 2012).

1.2.3. *Understanding the Role of Services in Trade*

Traditionally, trade has been measured in gross terms. This contrasts sharply with the way we have always measured domestic production, which is in value-added terms. In other words, GDP is measured — either by returns to factors of production (capital and labour) or by expenditures — through the attribution of value to its source at every stage of the production process. By measuring trade in gross terms, however, the sources of value are unspecified (Ahamed, 2013).

Perhaps, the simplest way of seeing this is to think of an economy's exports. Exports reported in gross terms include foreign sources of value contained in the imports. These are incorrectly counted as value coming from the exporting economy. It amounts to a kind of double counting and overstates the domestic content of exports. Were the import content to be netted out, exports would reflect domestic value-added just as GDP estimates do.

The implications for understanding the nature of trade relationships among economies are far reaching. At least four perceptions can change radically. First, the true nature of bilateral trade balances is obscured. It transpires, for example, that China's much-vaunted current account surplus with the United States was significantly over stated. In 2010, for example, China's trade surplus was estimated at US$176 billion in gross terms but was reduced to US$131 billion in value-added terms—an overstatement of 34 per cent in the bilateral balance (Koopman *et al.*, 2013). Some of that difference was attributable to understated (value added) bilateral trade deficits of Japan and Korea with the United States.

Secondly, if trade flows are recorded in gross terms, it is not difficult to see how the technology content of an economy's exports could be misspecified. Dedrick *et al.* (2010) provide a good illustration of this phenomenon with the feted example of the iPhone putatively made in China. The factory gate price of an iPhone in China was recorded as US$144. But less than 10 per cent of this was Chinese value-added and most of that was attributable to assembly. Taken at face value, however, statistics would suggest that China was an exporter of high-tech smartphones to the United States. In this particular case, the United States was actually re-importing value it had added to the smartphone in the upstream design and engineering segments of the value chain.

Third, in addition to misspecifying the source of technology, gross trade flows also fail to tell us where jobs, and what kinds of jobs, are associated with trade. A final consideration, which follows from the other three, as to why gross trade flows mislead us is that by failing to identify the locus of trade-related economic activity, the data do not identify the true nature on mutual dependency relationships among economies.

The most important insight to emerge from the estimates of trade in value added rather than gross terms is the extent to which services enter trade and generate value. When the services share of trade was being calculated on the basis of balance-of-payments statistics, that share was being regularly reported at less than 25 per cent of world trade. When trade was calculated in value-added terms of an internationally aggregated input–output table, the share of services jumped to 45 per cent (OECD-WTO, 2013). That is a significant difference, and for various technical reasons may still be an under-estimate. The availability of these new data, thanks to the efforts of international agencies, universities, and some government agencies, has raised awareness of an important missing part of our understanding of the role of services in the global economy.

We can no longer think of services as important at home but relatively inconsequential in international trade. Moreover, when it becomes clear that services constitute a greater share of value in trade, it is also apparent that services do more than just enable distance to be bridged. They are an independent source of value in other ways as well. As noted earlier, awareness has grown in recent years of the contribution services make to innovation. In these circumstances, it comes as little surprise that interest is mounting in the ways policy affect services, both in terms of movement of capital (foreign direct investment), movement of people (work permits and visas for foreign eployees) and of services flows in international product markets, regardless of whether these flows are incorporated or bundled in other services or goods, or whether they are supplied in discrete form.

1.2.4. *The Role of Services in Development*

The developmental implications of the emergence of GVCs have reinvigorated the debate about the role of government in economies as they seek to diversify away from agriculture and extractive industries through industrialisation. A lacuna in this discussion is the paucity of debate about where services fit. Given the predominance of services in economic activity and the nature and range of service activities, they offer valuable opportunities for diversification and development in emerging economies.

The importance of the omission is highlighted by the fact that SMEs in developing and emerging economies account for between about one-half (Stein *et al.*, 2010) to two-thirds or more (Berrios and Pilgrim, 2013) of jobs, and a large share of these SMEs appear to be service providers. Normah Mohd (2007) estimates that 90 per cent of Malaysian SMEs are service providers. In Hong Kong, China the number is 94 per cent (HKSAR, 2014).

1.3. A CASE STUDY APPROACH TO UNDERSTANDING MANUFACTURING-RELATED SERVICES

Detailed data on services in economies and international trade are scarce. This is challenging for services because of their intangibility and the resulting difficulties in identifying and measuring the value of services transactions. The case studies have enabled us to dig down to a fine level of detail in order to understand how services and manufacturing interact to produce manufactured output. GVCs that focus on producing services are no less important, and they use goods and services as well in production and trade, but this is not the focus of the present study.

While the advantage of case studies is their attention to detail, they do not lend themselves readily to generalisations or confident extrapolations for explaining the world. This is simply because we are generally looking at a single value chain among many operated by single firms and we are only looking at a small number of firms. The attainment of a statistically significant sample using a detailed case study methodology would simply not be possible within a reasonable time frame and budget. Rather, the case studies give us insights that point a light — not always under the proverbial street lamp — where it may be worthwhile to delve further. Firm surveys or the use of more aggregated secondary data sources are alternative approaches, but they are complementary and less granulated.[1]

In the project, we worked with 22 cases studies from companies around the APEC region. As mentioned, nine are large multinational companies (MNCs) or their subsidiaries, ten are medium-to-large companies with a growing international business and one SME that is purely domestic but is a supplier to GVCs. Eight of the firms are from North America and Chile, while fourteen are located in Asia. These companies have been generous with their time and their insights, enabling us to put together a set of case studies, all of which reveal interesting and different facets of realities on the ground. Table 1.1 lists the economies and products involved in the case studies.

Even though the sample size is small, it covers capital goods, consumption goods, intermediates and natural resource extraction in 12 economies. Two companies that are not strictly manufacturing companies — one is a pure agro product and another a pure service — have been included to provide some contrast (as well as similarities) with manufacturing firms. Before entering into detailed findings from the case studies, the methodological approach is presented.

---

[1]The WTO-OECD work on a Trade in Value Added (TiVA) database has enhanced knowledge about the role of services in manufacturing. However, the TiVA data cannot give further information on more disaggregated service sectors such as business services and its subsectors. Even if complemented by other economic tables like the input–output tables available in the OECD STAN database which are more disaggregated than the TiVA since industries there are divided into 37 sectors instead of the 18 sectors in TiVA, the picture we get remains aggregated. For example, we know that "other business activities" are important for business services, but there is no further information on how "other business activities" are divided into the contribution of the different professional services and other components of this sub-classification (Pasadilla and Wirjo, 2015).

Table 1.1: Sectors and Products Included in the Case Studies.

Sector	Economy where firm is located
Aircraft control systems	Philippines
Automotive components	Japan
Brake hose end fittings	Mexico
Car antenna	Mexico
Car assembly	Philippines
Chassis parts	Mexico
Computer servers	Chinese Taipei
Construction machinery	United States
Consumer electronic appliances	Indonesia
Industrial welding	Thailand
Mining and construction equipment	Japan
Oil and gas equipment	Singapore
Power generation equipment	Japan
Precision die and machine parts	Thailand
Printed circuit boards	Canada
Refrigerators	Japan
Telecommunications equipment	China
Wastewater treatment	Thailand
Watch	Hong Kong, China
Wine	Chile
Fresh cherry	Chile
Integrated logistics	Mexico

Source: Authors.

1.3.1. *The Methodological Approach to the Case Studies*

1.3.1.1. *Defining the parameters of the value chain in the case study*

The first task is to identify the particular value chains that will be followed. In many instances, this entails choosing from among alternative operations within the firm offering the case study. Once the particular product and production sequence has been identified, a decision is required on where the chain should be deemed to begin and end. This is essential because of the networked nature of all production activity within an economy. For ease of analysis, it is more straightforward to keep the parameters of the analysis within a single ownership structure — in other words, to work with a single lead firm. Where a switch of ownership in a value chain occurs, changing the identity of the lead firm, it becomes more complicated to map the value chain with continuity.

The second cut-off point defining the value chain to be analysed relates to how the arms-length purchase of inputs is to be managed. For analytical ease, only first-tier outsourced inputs into the lead firm's value chain are included. The reason for this is that each seller at the level of the first-tier outsourced supplier is at the final downstream point of another value chain. It is the production of another lead firm. It would, of course, be possible to trace the production of the outsourced supplier back down the value chain to build a more comprehensive picture of the networked environment in which our initial lead firm is operating. Such an exercise would yield rich insights into the interdependency intrinsic to networks, and possibly the role of SMEs in value chains. But the work would be very time-consuming. Moreover, eventually it would be necessary to establish analytical cut-off points.

Supply chain managers seeking to assess the robustness of the value chains and potential sources of risk are most likely to trace multiple tiers of outsourced suppliers in order to identify potential weak points in their chains. As part of their risk management strategies, they will often envision potential alternative arrangements in case an unforeseen disruption occurs. Other managers prefer to diversify sources of supply at the outset as a risk mitigation strategy, but this depends on the scale of the lead firm's operation and will also incur additional transaction costs.

For our purposes, however, we simply cut off our definition of the value chain to be analysed at the level of first-tier outsourced suppliers. This approach clearly reduces the completeness of our understanding of all the inputs into the value chain, but as noted above, there will always be somewhere in out networked world where a cut-off decision is required for analytical purposes. The alternative would be the impossible task of tracing the web of every production process back to its primal inputs of labour and capital.

The implication of truncating the domain within which the analysis occurs is important from the perspective of comparing different case studies in the set of 22. They simply become less comparable on account of ownership structures defining the start and finish of the value chain, and on account of the relative frequency of outsourcing, or the 'thickness' or frequency of the in-house-supplied value chain.

1.3.1.2. *Mapping Services Inputs*

Having defined the dimensions of the value chain to be analysed, the next step is to enumerate all services entering the value chain. The source reference for services entering production is the United Nations (UN) Central Product Classification (CPC), Version 2, Sections 5–9 (United Nations, 2013). The exercise at this point is straightforward. It is merely a matter of identifying each of the services entering production on the value chain whose dimensions we have already delineated. The reason for using the UN CPC is that it allows us to keep tabs on the degree of disaggregation used in defining individual service inputs. This helps for comparison purposes across case studies.

Some of the identified services will be upstream from the manufacturing phase of production, some will be involved in the manufacturing phase, and yet others will be downstream, usually linked to conveying the product to the final consumer (it may be a product for consumption or an intermediate or capital good that subsequently finds its way into another production process). In mapping the services inputs, a distinction is required between in-house and outsourced supply. This distinction is crucial for later analysis.

Companies who agreed to be interviewed were not expected to reveal cost or price data in this exercise. Nevertheless, where possible we attempted, not always successfully, to obtain an empirical sense of the major service-based sources of value in the lead firm's chain, along with an estimate of what the combined value was of services in the total cost of the production. In most cases these numbers were estimates, either where services inputs embodying value were not explicitly priced or where inputs of certain services were supplied across many value chains within the same organisational structure. Examples of the latter include research and development expenditures, personnel services, security, telecommunications and utilities within shared premises. Where a firm is part of a conglomerate, the picture can be further complicated as a result of diffuse decision-making and a lack of knowledge on certain matters at the individual firm level.

In order to facilitate the analysis and comparisons among case studies, value chains were generally divided into six separate categories, not all of which were necessary relevant in all cases. Table 1.2 lists the categories and provides some brief illustrative

Table 1.2: An example of How the Stages in the Value Chain can be Divided.

Production stage	Examples of services inputs
Establishment	Government negotiations and license applications; land clearing and preparation; construction
Pre-manufacture	Design; R&D, procurement and transport of inputs; customs services; licensing requirements for health, safety and environmental compliance of inputs; storage of inputs
Manufacture	Transport and handling of raw materials; testing; maintenance and repair of equipment; cleaning; utilities; compliance inspections for environment, health, safety and working conditions
Post-manufacture	Packaging; transport; installation; advertising; marketing; branding; retail; quality control; standards assessment
Post-sales services	Repair and maintenance of machinery and other facilities; inventory and warehouse services for parts
Back office services	Back office accounting; legal services; personnel; insurance

Source: Authors.

examples of the numerous services falling within each of the categories.

A notable difference between the approach in this study and the UN's CPC is that we have taken government interventions, mostly in the form of public policy, as part of the value-added in services story. The reason for taking this approach is twofold. First, from an analytical perspective, government services are either directly or indirectly intended to provide social value in areas where the private sector or individuals acting alone or in communities cannot necessarily supply essential social goods, such as shared infrastructure, health, safety, environmental quality and social well-being more generally. These are sources of added value that it is incumbent upon governments to supply because these products embody spillovers, or what are sometimes referred to as public goods or externalities. For our purposes they may be seen as additive to the private sources of value generated by services along value chains that we are seeking to identify.

The second reason for identifying and treating policy interventions in this manner is that it provides the basis upon which to conduct the policy analysis that constitutes the final main section of this report. Moreover, it allows us to observe that although public policy in principle adds value, it may fail to do so where government services are provided inefficiently or at excessive cost (both in terms of money or time). This can occur as a consequence of inadequate resourcing of government services in terms of infrastructure, personnel or training; a lack of commitment and dedication to good service among personnel; and to poor governance that gives rise to perverse incentives for officials to seek rewards that impose private and social costs on society at large. Improved performance in any of these areas can generate policy innovation and become a source of productivity gains.

1.3.1.3. *Additional information sought: outsourcing of services*

As already noted, after defining the value chain under study the procedure requires identification of all services entering production (including upstream and downstream activities) and a determination of which services are outsourced to third parties and which ones are supplied in-house. The source of supply information provides the raw material for an analysis of the reasons and the degree to which firms outsource in our selected value chains — why they do so or refrain from doing so, and if they do, to whom. Other information sought in interviews was more difficult to acquire, mostly because interviewees did not have it to hand, and obtaining it would have been more onerous than it was reasonable to expect from managers who were already giving generously of their time.

We sought information on bundling — the creation of a composite offering in the marketplace. A bundled offering is one that includes more than a single product. Strictly speaking, any arms-length transaction between independent parties comprises a bundle, which may be both services and goods combined, or an exclusive combination of one or the other. The reason that all third-party transactions can be considered bundles is that products offered for sale are made up of a combination of other inputs. The reason for being concerned about bundling is that it can be used strategically. Moreover, the possibility of bundling renders all services inputs tradable in principle. These issues will be discussed in relation to the case studies.[2]

---

[2] In the initial interviews, questions were also asked about innovation, employment and policy impact. The last of these will be addressed in the section on policy. As previously noted, traditional thinking used to have it that manufacturing, mining and agriculture provided the most significant contributions to productivity growth and technological advancement, while services lagged behind. More recently, recognition has grown of the scope for advances in technology and innovation to foster productivity growth in services, as a standalone sector or as an integral part of others. The aspiration of the project had been to examine services inputs along value chains in terms of their technological content and their potential to contribute to innovation. It soon became apparent that it was infeasible to address innovation as an add-on to the core issues examined in the interviews, so the issue did not form part of the analysis. A separate project would be required to research this question.

Finally an initial attempt was made to collect data on jobs attributable to services and to classify such jobs in terms of skill levels. The intention was to assess

1.3.2. *Patterns Relating to Services Emerging from the Case Studies*

For reasons already discussed in relation to the determination of the bounds of individual value chains, comparisons among the 22 case studies should be treated with some caution. Nevertheless, it is clear that the number and range of services entering the value chain as sources of value is considerable. Figure 1.1 depicts the number of services entering the value chains of each of the case studies. They range from 37 in the case of automotive components to 74 in the case of power plant equipment.

It proved difficult to obtain systematic information on the share of total value (cost) attributable to services, but numbers obtained ranged from anywhere between 30 per cent and 90 per cent.

Figure 1.2 looks at the distribution of service inputs across the different stages of the case study value chains. Once again, allowing for differences among the case study value chains that limit their comparability, back office services — which are mostly support services of one sort or another (see Table 1.2 for an illustration) — account for one-third of service inputs. The manufacturing stage accounts for a further one-fourth of the total, followed by pre-manufacturing, post-manufacturing and sales, after-sales support, and establishment-related services in descending order of importance. The count of numbers of services involved in each production stage says little about their respective value contributions. In the case of after-sales services, those companies producing capital equipment with lengthy lifespans could generate

---

the level of technological sophistication associated with particular services inputs. Several problems arose from this. First, it was not easy in the space of time-limited interviews to secure an accurate attribution of jobs to specific service inputs. Secondly, the classification of skill levels was also problematic in that we lacked a standard metric to use across cases studies. Thirdly, it was not obvious what we would be able to use the information in the absence of any trend analysis, which was impossible to obtain within this framework. The effort was abandoned with the realisation that the data would be difficult to obtain, an adequate analytical framework was lacking, and a different methodological approach was indicated.

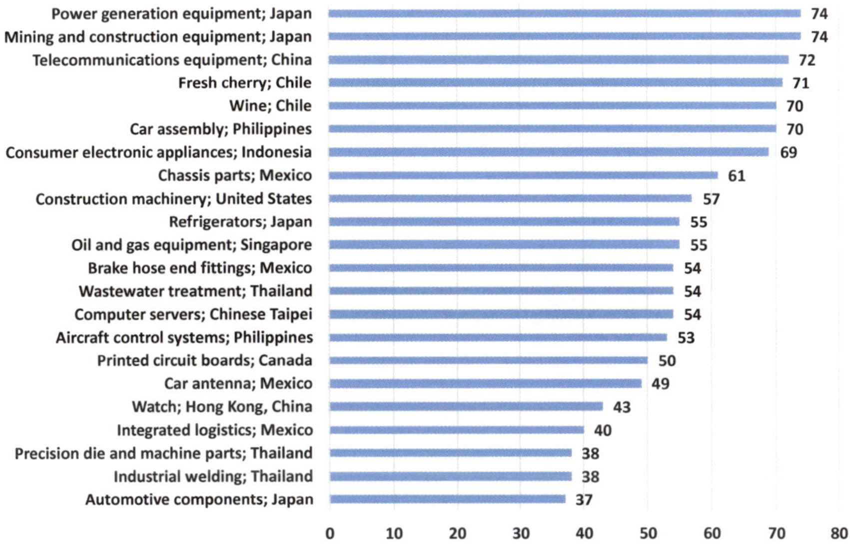

Figure 1.1: Number of Services Entering the Case Study Value Chains.

Note: The count in this figure and in Figure 1.2 slightly differs from the count in the individual chapters because authors of each chapter had free rein to count the more disaggregated services classifications or the more aggregated one. The method in this summary report is to recount services based on the more aggregate grouping, in particular, the "services" column in the tables of services that enter the value chain from each case study. For example, one case study counted CPC 8313 (IT consulting and support services), CPC 8314 (IT design and development services) and CPC 8316 (IT infrastructure and network management services) separately, while this summary report counts this as only one (i.e., IT services). The main findings that manufacturing makes intensive use of different services, however, remain the same regardless of the slight differences in the counts.
Source: Authors.

more value from looking after the manufactures it has produced over 20–30 years than from any other single source.

1.3.3. *Outsourcing*

For the case studies where data were available, a considerable amount of outsourcing of services occurred. The incidence of outsourcing was greater than we anticipated. Figure 1.3 highlights

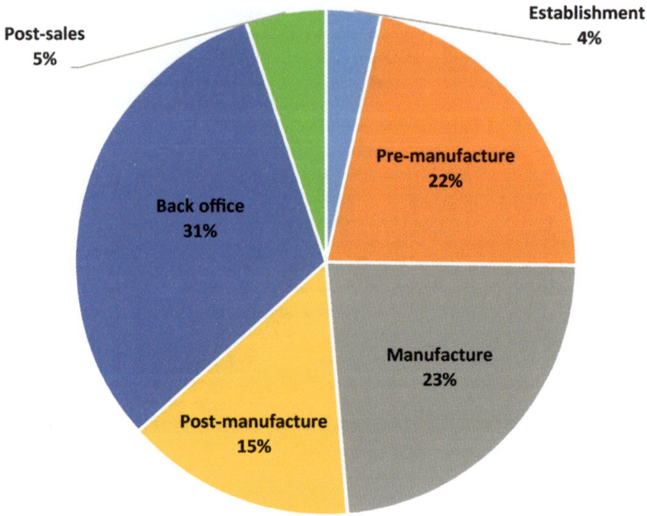

Figure 1.2: Incidence of Service Inputs at Different Stages in the Case Study Value Chains.

Note: This figure reclassifies some services in the different case studies into one of the six stages: establishment, pre-manufacturing, manufacturing, post-manufacture, back office or post-sales, because some authors used different terminologies for the stages in the value chain.
Source: Authors.

the results. Of the 22 case studies, 21 provided complete information on the share of services entering the value chain were supplied in-house, the share fully outsourced and the share partially outsourced. No information on outsourcing shares was available for one of the case studies. Figure 1.3 reveals that for the case studies where information was available, there are only four cases where the share of outsourcing is less than 50 per cent. In many instances the share exceeded 50 per cent by a considerable margin, rising to 90 per cent in one instance.

Systematic information on the identity of these third parties was not available. In some cases, the outsourced supplier was a government entity or an entity specifically designated by the government to carry out particular regulatory tasks. In other

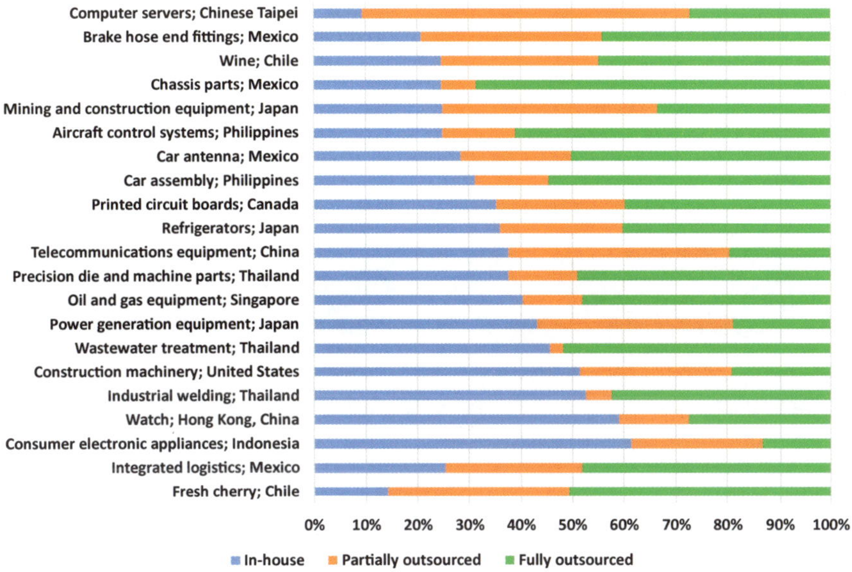

Figure 1.3: Shares of Services Supplied In-house or Outsourced (%).

Source: Authors.

instances, outsourced suppliers were part of a conglomerate structure to which the case study firm belonged. For the rest, outsourced suppliers were most frequently local enterprises, many of them small in size.

Both from a policy and a business perspective, it is instructive to consider why firms may or may not favour outsourcing when it comes to services inputs on value chains. In interviews with the case study firms, a range of reasons were offered, often expressed in different ways, but covered by the categorisation contained in Fig. 1.4.

As already mentioned, some outsourcing is the result of mandatory requirements. Regulations will typically specify the competent authorities to undertake functions such as auditing and compliance with certain health, safety, and environmental services.

A core consideration when it comes to firms' outsourcing decisions is cost. In effect, if we were to take cost as an overriding consideration we could subsume a number of the other factors

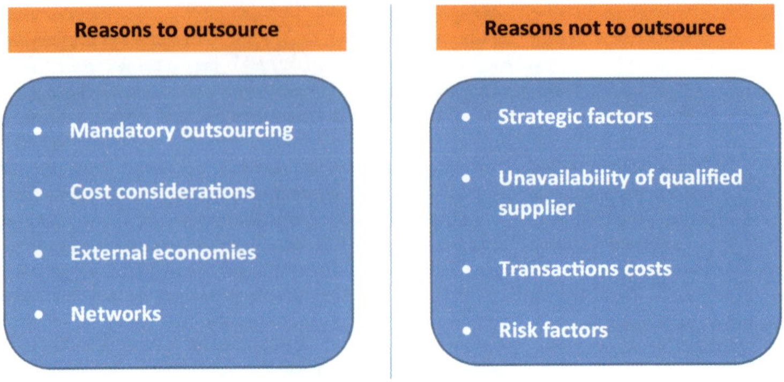

Figure 1.4: Considerations Affecting Outsourcing Decisions in GVCs.
Source: Authors.

listed in Fig. 1.4 under the cost rubric. But this would obscure both the real motivation for outsourcing decisions as well as clues as to what actions governments might take to secure more value chain-related activity in their own economies. For the present purposes, the starting point is that firms have finite resources and must decide how best to deploy them. Even if a firm is absolutely more efficient at competing in two activities, for example, it may pursue its comparative advantage and outsource the one that it is relatively less efficient at producing compared to third-party suppliers. In other cases, an absolute cost advantage may exist for an outsourced supplier, and so other things equal, it would make sense for the firm to outsource.

Finally, a related argument concerns internal economies of scale. Production sometimes requires high fixed costs in the form of plant and equipment, for example, and production is zero without large initial investment. If the firm's requirement is for a small amount of the product, then high fixed costs will have to be allocated to a low level of production, raising the unit cost of output. In this case, it is better for a firm to outsource to a supplier who can attain economies of scale.

Another set of reasons for outsourcing is to do with the existence of external economies. Certain enterprises may specialise in particular activities that allow them to reap the benefits of focused attention on these activities along with others who do the same. These are interactive benefits from clustering that are beyond the control of any single firm and they render those involved in the particular activity more cost efficient and competitive. Typical examples of this in the services sector might involve aspects of information technology and other knowledge-intensive activities.

A related reason for outsourcing has to do with the advantages of networks in supply. A readily apparent example of this is security services. A single guard on a premises employed internally is likely to be less effective in dealing with security-related eventualities than a single guard with access to a network upon which to call as the need arises. This is why virtually all the firms in our sample outsourced security services.

Turning to reasons that cause firms to pause before outsourcing, or indicate that there are some services they would never outsource, strategic considerations loom large. Many interviewees identified key operations or activities in the value chains that were either too important or too sensitive to outsource. The important activities tended to be those considered defining features of the operation. The sensitive parts of the operation often involved proprietary information. For the most part, it is safe to assume that these are activities that are unlikely ever to be outsourced to third parties.

The availability of qualified supplier is another reason for eschewing outsourcing. Firms may want to outsource because it would allow them to focus more closely on key elements of their operations, but be unable to do so because they cannot find suppliers who are sufficiently low cost in relation to specified quality requirements.

Other cost-related factors that may inhibit outsourcing concern expenses incurred by the firm in dealing with arm's-length suppliers that would not arise with in-house supply. These can be thought of as transaction costs. The two most obvious are search costs and supervisory costs. Search costs are to do with the identification of suitable third-party suppliers, which can be time-consuming and costly. Supervisory costs are ongoing and require that the firm spends resources on making sure that outsourced suppliers comply with their contractual obligations.

Finally, a series of risk factors associated with ceding some control to outsiders will weigh on the firm and influence outsourcing decisions. The lead firm has to be confident that outsourced suppliers will be reliable and consistently so in matters such as quality and timeliness. A weak link in the chain can carry severe costs for the parent firm. More generally, firms worry about their reputation, which takes time to build and can be destroyed in an instant. They have to be assured that third parties are no more likely to prejudice reputation than in-house actors. Finally, an additional and separate source of risk is failure to comply with government regulations. If the matter is in the hands of a third party, the firm wants to know that the likelihood of a problem is minimal.

For governments and businesses in whose interest it is to reap the benefits of fuller participation in GVCs, one channel for doing so is through participation via outsourcing. This is not a topic for the present study, but a range of new participation and upgrading opportunities are likely to present themselves in appropriate policy environments. For governments in particular, actions aimed at ensuring an operating environment conducive to local competitiveness can make a significant difference. This is an area that may be of interest to APEC in the context of its work on SMEs.

1.3.4. *Bundling*

As previously argued, any arms-length transaction in the marketplace will involve products made up of composites of other products. The point here, however, is that bundling can involve an element of strategic choice. Four considerations are the most pertinent in this regard. The most obvious, perhaps, is bundling as a means of minimising cost, where, for example, joint production occurs and there is demand for more than one output. Second, bundling may serve as an instrument simply of adding more value by rendering the offering more complex. Third, bundling can provide a means of gaining temporary market advantage through product differentiation. Finally, bundling can be a means of fostering innovation, and if the innovation is proprietary, the product differentiation advantage will be more long-lasting. As noted earlier, since most of the observed bundling in the case studies occurred with arm's-length inputs provided by others, it was not possible to systematically trace incidences of strategic bundling.

1.4. POLICY IMPLICATIONS

Services, as discussed above, are embedded in manufacturing value chains, from the pre-production phase to the delivery of goods. The case studies show that in some manufacturing value chains, services extend all the way up to the disposal of the product (or re-manufacturing in others), with maintenance and repairs service contracts sometimes lasting up to its end-life. Such intertwining implies that manufacturing and services policies affect each other, that is, policies restricting one will have an effect on the competitiveness of the other and on the efficiency of the value chain.[3] The studies in this volume therefore asked the firms about the policy issues they face both in host economies and in markets where they sell their products. In addition to policy constraints, attention was also paid to noteworthy instances of cooperation between industry and government. The results from discussions with firms are summarised in this section.

The analysis of policy that follows focuses primarily on the perspectives of firms — particularly but not exclusively foreign firms — operating in various APEC economies. This chapter does not attempt to analyse in any systematic way the public policy objectives underlying the interventions of governments. It is

[3] For example, some case studies find substantial correlations between investments in services and growth of manufacturing labour productivities (see Arnold, et al., 2011; Fernandes and Paunov, 2008).

understood that such objectives are diverse. No judgement is intended as to the legitimacy of the objectives behind these interventions. Rather, the focus is more on how policies are designed and whether they embrace the most efficient, cost-minimising manner of achieving their putative goals. Policies can be self-defeating if they are poorly conceived or excessively heavy-handed in application. Another key policy issue that the study touches upon is policy coherence among multiple jurisdictions. Diverse policy approaches and duplicative requirements can add significantly to costs and undermine competitiveness. In sum, the approach adopted in this chapter is about efficiency and effectiveness, not legitimacy.

A general sentiment encountered among firms was the wish that services sectors were more open for foreign investment so manufacturing activities could grow better.[4] Labour mobility facilitation was another constant refrain because firms find that where they need to obtain visas or work permits for their employees, delays and uncertainty in obtaining them in a timely manner unnecessarily raise costs. The cost arises not only in terms of time and money paid for the application and processing, but also more importantly, in terms of missed business opportunities or inadequate service to clients. Local content requirements are sometime unclear or are difficult to comply with when global firms cannot find local partners that meet their stringent technical standards. Skills development and matching were also a concern and is an area fertile for joint private and public sector collaboration. Effective enforcement of intellectual property protection and infrastructure improvements are other elements of firms' desirables. Finally, SME participation in GVCs is often stymied not only by inadequate access to finance, but also by the difficulty of obtaining accreditation for sector-specific international standards.

[4] An APEC Business Advisory Council (ABAC) survey of business executives in the region corroborates this sentiment. For details, see ABAC (2014).

1.4.1. *Investment Policy Incoherence: Manufacturing, Yes; Services, No*

Equity restrictions in services are more prevalent and stringent than restrictions in manufacturing because many economies have preferential investment policies towards manufacturing investments. According to the OECD foreign direct investment (FDI) restrictiveness index, the overall equity restriction index in the services sectors is higher than the overall FDI index (Fig. 1.5) and this is a trait common to both developed and developing economies. The thinking behind this duality in investment policies might be that manufacturing FDI is deemed to be employment generating, while services FDI can eliminate jobs and compete more directly with domestic service providers. Think, for example, of professional services or retail distribution services, where most domestic participants are micro-, small- and medium enterprises that tend to benefit from the protective hand of governments. Whatever the reasons may be for the different treatment of manufacturing and services FDI, the GVC literature that has emerged is casting doubt on the usefulness of such policies, particularly because services are

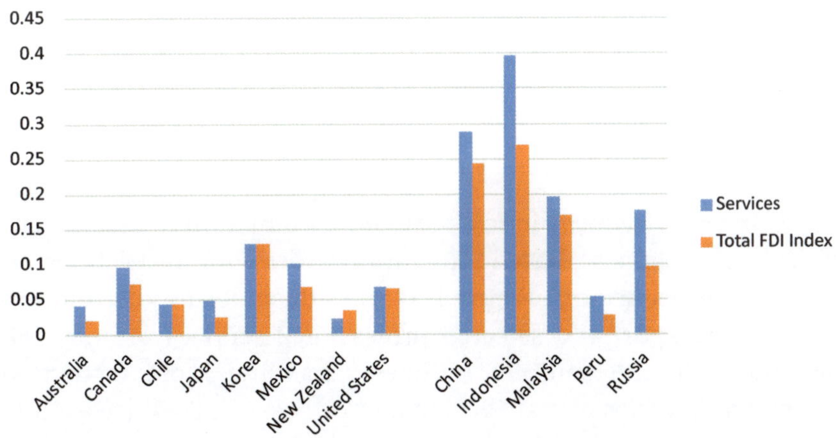

Figure 1.5: Equity Restrictiveness Index.

Source: Authors' computation based on OECD FDI restrictiveness index.

important for the competitiveness of manufacturing (see, e.g., Arnold *et al.*, 2006, for empirical evidence).

1.4.1.1. *Forms of investment restrictions*

Besides equity limitations, other forms of FDI restrictions are screening of foreign investments, restrictions on key personnel, and limitations on firms' operations. Screening of foreign investments can include the stipulation of minimum capitalisation requirements, the need for cabinet-level approval or the imposition of an economic needs test. Related to these are restrictions on key personnel (Section 1.4.2), restrictions on land ownership, and other sector-specific limitations. For example, in construction services, even if there is no statutory limit on equity ownership, if the conditions for granting a contractor license or an engineering license are too restrictive for foreigners, the sector becomes effectively closed to foreign investment. Tables A.1–A.3 list restrictions in selected APEC economies in some important manufacturing-related services, such as logistics, and wholesale and retail distribution. The tables show that in domestic transportation services and customs clearance, the majority of the economies allow only partial foreign ownership, while warehousing is mostly open for foreign investment. More APEC economies allow 100 per cent foreign equity in distribution services, but with conditions in some economies

1.4.1.2. *FDI restrictiveness index*

The FDI restrictiveness index provides a comparative assessment of lack of openness based on various forms of restrictions in different service sectors. Figure 1.6 shows the index for distribution services, transport and selected professional services. Among the economies in the sample, Indonesia has the most restrictive investment policies in distribution and professional services, while Mexico and the United States are restrictive in land and maritime transport, respectively.

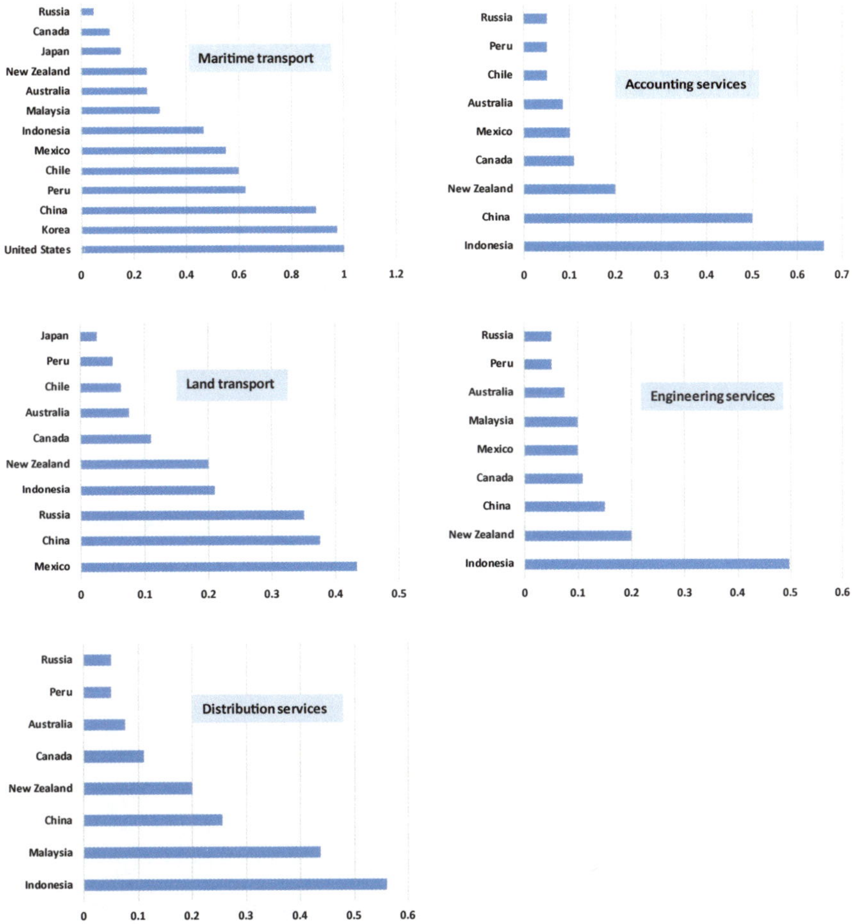

Figure 1.6: FDI Restrictiveness Index: Distribution, Transport and Professional Services.

Source: Authors' computations based on OECD FDI restrictiveness index.

Of the different elements of the index, equity restrictions are the biggest contributor to the high FDI index in economies like Indonesia or Mexico (for land transport) or the United States (for maritime). Among the sectors with the highest barriers to FDI are distribution, construction, media, communications, financial services, business services and real-estate investment (Appendix C).

Except for media services, all these service sectors figure considerably as manufacturing-related services in the case studies in this volume and, as shown in Appendix B, have significant shares of value added in the industry sectors to which the case study firms belong.

1.4.1.3. *Effect of policy incoherence*

How does investment policy incoherence affect the operations of regional and GVCs? The example of Japanese GVCs can illustrate the problem. Tier 1 suppliers of Japanese MNCs usually try to follow the GVC lead firm wherever it invests. Thus, for example, major component suppliers for automobile manufacturing follow a major car manufacturer to where it locates, thus replicating the supply chain efficiencies in the home economy. However, service suppliers of the lead firm may be unable to follow because of FDI restrictions. While partnering with domestic service providers also benefits foreign companies because they provide local knowledge and a network of domestic contacts, for many lead firms, the challenge is to find the right local business partner.

Differences in FDI restrictions between manufacturing and services and among service sectors prevent foreign companies from operating what they deem the most efficient business model. Figure 1.7 provides an example of how a company that can have an integrated operation — from the manufacture of elevators to installation, maintenance and repair (right panel) — may be prevented from doing so because of equity limitations in installation, or maintenance and repair services.

It is, however, also possible that some companies prefer not to establish an integrated operation but rather outsource some services to third parties for reasons cited in Fig. 1.4. For example, construction equipment manufacturers,[5] even as they take upon themselves the obligation of supplying machine parts and components needed for repair, nevertheless prefer to rely on a

[5] See Pasadilla (2010) and Tait and Gereffi (Chapter 19, this volume).

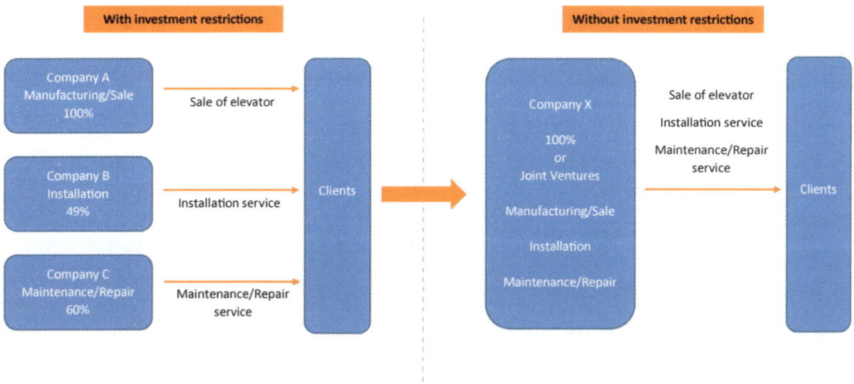

Figure 1.7: A Forced Business Model?

Source: Adapted from Shiino (2015).

network of trusted distributors in different economies to take charge of equipment maintenance and repairs. In this case, the firm invests heavily in the network relationships as well as in providing the technical training of its distributors in order to provide their customers with the same quality service they would receive if the lead firm itself were to carry out the service.

Box 1.1 provides other examples of how FDI policies affect foreign operations and how changes in investment policies can upset business plans. These experiences, which are taken from the case studies in this volume, illustrate how sudden changes in investment policy can freeze expansion plans, or how they increase costs by forcing companies to do things they would not have otherwise done based on efficiency considerations.

1.4.2 *Labour-Related Restrictions*

The need for non-resident service suppliers to enter foreign markets, whether for short- or long-term periods, has become more important with the internationalisation of production. Global companies have organised themselves into specialised hubs and draw skills from these different hubs when they develop new projects, undertake research, or supply customer services including

Box 1.1: Business Effects of FDI Restrictions: Examples from Case Studies.

- *Sharing of back office facilities by affiliated companies prohibited.* On efficiency grounds, the case study firm wanted its newly established subsidiary to share some back office functions with the group's existing company. Local regulations forced the plan to be scuttled because the law requires any registered company to be completely equipped with all back office functions and a stand-alone office address. If two companies within the same group were to share certain functions, formal transactions would need to take place, meaning that formal contracts would be required. From the company's perspective, this regulation simply augments costs and lowers efficiency (Cheung, Chapter 7, this volume).

- *Requirement to have a local partner.* An oil and gas equipment manufacturer said that to be able to bid for a contract or to respond to a request for quotation, it is sometimes legally required to have a local partner. Depending on the market involved, this requirement can lead to an increase in the firm's costs and higher price bids, thus potentially making it less competitive. Even if it won the contract, the cost would have been higher for consumers. However, in some cases, the firm itself prefers to work with a local agent because the latter has specific knowledge of the customer and local market conditions. The challenge in many places is finding suitable local partners who can add value to what the firm offers (Wirjo and Pasadilla, Chapter 8, this volume).

- *Unpredictable policy changes.* A newly enacted government policy in one ASEAN economy restricts foreign equity ownership for wholesale, distribution and after-sales services to 33 per cent. This change affected the planned recapitalisation by this case study firm that manufactures "white goods". The firm had established a local distribution subsidiary in the economy years before the new rule was promulgated. The firm's existing equity is 67.5 per cent while its local partner has 32.5 per cent. While grandfathered from the application of a one-third maximum foreign equity holding as per the new policy, any change to the ownership proportions would

(*Continued*)

Box 1.1: (*Continued*)

eliminate grandfather rights. The firm's plan to recapitalise could not be carried out because the local partner was unwilling to inject more capital into the enterprise (Sit, Chapter 12, this volume).

- *License restrictions and limited availability of proper services.* The law in one ASEAN economy restricts foreign licenses for services companies, for example, wastewater treatment services. A die manufacturer located in the economy, however, tells of the limited number of licensed industrial waste disposal companies which can properly treat the type of waste that the firm produces (Cheung and Wirjo, Chapter 11, this volume). In some cases, firms have to build their own wastewater disposal system within the factory, with the help of their long-standing environmental service provider from their home economies, to remove any reputational risk of not being environmentally friendly.

maintenance and repairs. This means that large flows of people crossing frontiers have become a normal feature of modern global business and are necessary for the flow of goods, services and knowledge.

1.4.2.1. *Forms of labour restrictions*

Restrictions on labour mobility take various forms. They can appear as labour quotas, economic needs tests or the lack of recognition of qualifications. At times, procedures for obtaining visas and permits can be costly and complex, requiring copious documentation, and involving long delays. Opacity in the decision-making process and uncertainty arising from discretionary decision-making procedures for granting or rejecting visa applications are often impediments in themselves. Table 1.3 provides some information on labour restrictions from the case studies.

1.4.2.2. *Inflexible labour policies*

Though not related to labour entry in itself, some firms in the case studies also mention inflexible labour policies in certain

Table 1.3: Samples of Labour Mobility Restrictions.

Type of labour policy	Description
Quota	• Singapore imposes a maximum to the number of foreign workers that a firm can hire, depending on the sector.[a] For the manufacturing sector, the quota for foreign workers is set at 60 per cent of employment. Once a firm has reached its quota for foreign workers, it has to employ more local workers to be able to hire additional foreign ones. The shortage of personnel sometimes leads to delays in project completion or to poor service delivery • Mexico requires that 90 per cent of employees are Mexicans
Economic needs test/ conditions for granting work permit	• Under Thailand's Alien Working Act, B.E. 255 (2008), which governs the employment of foreign labour, the government stipulates that for every THB 2 million paid-up capital, a company can employ one additional foreign employee up to a maximum of 10. If a company wishes to employ additional foreign workers, it has to employ more than 100 local personnel, or pay at least THB 3 million in corporate income tax, or export at least THB 30 million worth of goods, or have brought in more than 5,000 foreign tourists during the previous year • Firms in Singapore have to consider nationals for professional, management and executive positions first. The job needs to be advertised in a jobs data bank administered by the Singapore Workforce Development Agency for at least 14 calendar days before applying for an employment pass application • Indonesia's parliament is considering a language test requirement for managers and directors, which, if approved, effectively puts restrictions on foreign directors, putting unnecessary burden on them in carrying out their responsibilities

(*Continued*)

Table 1.3: (*Continued*)

Type of labour policy	Description
Complex requirements for visa and permits, and discretionary decisions	• One case study firm said that in their host economy, applications for both work permits and non-immigrant visa for foreign workers require the submission of 11 supporting documents, are processed slowly and are ultimately awarded on a discretionary basis
Recognition of qualifications	• ASEAN has mutual recognition agreements for some professional services, but none for technicians like welders who have to pass a professional examination in Thailand based on standards laid out by the International Institute of Welding (IIW), and must be trained in occupational health and safety (OH&S) practices. Though neighbouring economies also use IIW standards, they do not recognise each other's government licenses and OH&S processes

[a]More information about the quota for foreign workers can be obtained at: http://www.mom.gov.sg/Documents/services-forms/passes/Guidelines_on_Levy_Bill_Computation.pdf
Source: Compiled by authors.

economies, especially with respect to the hiring of contractual or temporary workers. Contractual hires can be useful during peak periods of production. However, in the Philippines, the current labour law prohibits the hiring of contractual workers if they do not receive the same benefits accorded to permanent employees. Likewise, only a maximum of 20 per cent of workers are allowed to be on contractual terms. Worker lay-offs are not possible without a 30-day notice to the Department of Labour and without consulting the labour union (Stephenson, Chapter 9, this volume). In Indonesia, rigid labour laws on compensation reward seniority over performance. For example, Article 156 of the Manpower Act[6] imposes strict rules on severance compensation pegged to years of service rather than performance, making it difficult for firms to incentivise productivity among workers. The case study firm thinks that more flexibility

[6]Act of the Republic of Indonesia Number 13 Year 2003 Concerning Manpower.

on compensation issues could help to promote productivity among its workers (San Andres, Chapter 20, this volume).

1.4.2.3. *Effects of labour restrictions*

Having different visa restrictions for different economies makes the playing field uneven for business competitors in the region. Consider Fig. 1.8, which shows that nationals of some economies can practically travel anywhere in APEC because they face very few visa restrictions. Citizens of Australia, Brunei Darussalam, Canada, Japan, Korea, Malaysia, Singapore, and the United States encounter only one or at most four territories where they are required to obtain an entry visa. On the other side of the spectrum are citizens of China, Indonesia, Mexico, Papua New Guinea, the Philippines, Russia and Viet Nam who face visa restrictions in almost all other APEC economies. This difference in access to territories results in an uneven level playing field for firms located in these economies. For example, if a company wants to submit a bid in response to a tender where the presence of an all-star team

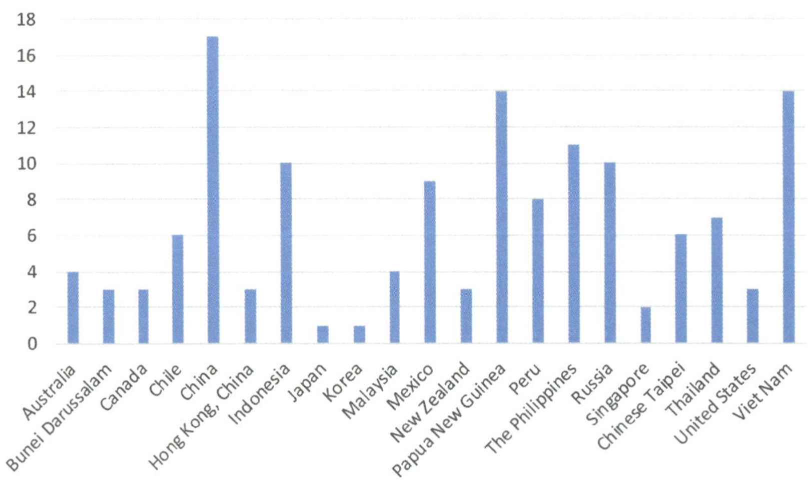

Figure 1.8: Number of Economies that Require Visa for APEC Nationals.

Source: Authors.

of experts during the tendering process can boost the chance of winning a contract, the inability to secure a visa for everyone in its technical team can lead to the loss of a potential contract (see Box 1.2, for examples, of effects of labour mobility restrictions).

Box 1.2: Anecdotal Evidence of Labour-related Pains.

- *Missed opportunities.* A telecommunications equipment manufacturing firm based in China sometimes participates in a call for tender in other economies. But visa applications affect the firm's ability and effectiveness to participate in the overseas tendering processes. When a project is sizeable, a technical team with personnel from various divisions is normally required to be present to assist in the proposal submission and to demonstrate what the firm is capable of doing. But Chinese nationals need a visa to go almost anywhere. The cost and uncertainty associated with obtaining visas in time has made it difficult for the case study firm to deploy the necessary number of persons for tendering and has hampered its ability to exhibit the firm's full capabilities (Zhu and Pasadilla, Chapter 15, this volume).

- A multinational subsidiary based in Thailand engaged in the manufacture of welding alloys and cladding services finds regional expansion within ASEAN difficult because of a lack of recognition of qualifications of welders. In Thailand, welders must pass a professional examination based on industrial standards laid out by the International Institute of Welding (IIW). On top of this, welders must be trained in Thai OH&S practices. These requirements consume a significant number of person-hours plus idle equipment because these have to be used for the audit instead of being used in the factory. For the firm to operate in neighbouring economies, it has to send its welders and equipment away to repeat the licensing and OH&S processes again, even though places like Malaysia follow the same IIW standards. Internationally consistent industrial standards, along with mutual recognition of licenses and safety training based on these standards, are regarded by the firm as a policy reform that would most facilitate additional regional business through trade (Haines, Chapter 3, this volume).

(*Continued*)

Box 1.2: (*Continued*)

- *"Sub-optimal" service.* A thermal equipment manufacturing firm based in Japan sometimes needs to dispatch an engineer within 24 hours to fix malfunctioning equipment and to avoid a power plant shutdown. But in a few economies where Japanese nationals need a visa, it takes time to send the necessary personnel. Though the cost and the application procedures are not considered burdensome, the time the process takes creates a problem even for big MNCs. The firm thinks that minimising the visa-processing period would greatly help, particularly for short-term and temporary stays of intra-corporate transferees (Pasadilla, Chapter 10, this volume).

- A firm that manufactures equipment for oil and gas extraction based in Singapore needs its staff to travel seamlessly from one location to another to provide installation and commissioning services, do repairs and maintenance, or undertake training services. All these after-sales services make up a significant share of the firm's business. However, it often encounters challenges in obtaining visas, resulting in situations where the most qualified staff are unable to provide the requisite services (Wirjo and Pasadilla, Chapter 8, this volume).

Box 1.2 provides anecdotal evidence of how visa delays and restrictions or a lack of mutual recognition of qualifications hamper companies' ability to provide the best service they could give. In some cases, this has resulted in the loss of a good business opportunity, like winning a contract or the possibility of regional expansion. Almost all firms in the case studies have indicated problems associated with moving their technicians abroad because of visa issues. In one car component manufacturing firm located in Mexico, for example, its Japanese technicians had to exit several times to the United States to obtain re-entry visas in order to remain in Mexico for the period required by the company (Wirjo, Pasadilla and Bassig, Chapter 14, this volume).

1.4.3. *Localisation and Human Capital Constraints*

Many governments provide incentives and subsidies to attract FDI. At times, however, protectionist policies force firms to locate and increase operations domestically through the imposition of local content requirements. Such regulations add to operational costs especially when local suppliers are not necessarily the most efficient or the ones who provide the most competitive services. However, given government regulations on local content, foreign firms are sometimes forced to accept an uncompetitive local suppliers' bid, knowing that its production costs will increase in the process, and that these costs will either have to be absorbed through lower profits or passed on to customers.

An example of a specific local content requirement serves to further illustrate the costs that such policies can produce. In Indonesia, the government passed a regulation that requires importers of 4G LTE smartphones, starting in 2017, to use 30 per cent of local content in production. For the telecommunication equipment manufacturing firm in the case study, complying with such a regulation would mean that its sales subsidiary in Indonesia will undertake not only sales and distribution but also the assembly of components and parts imported from its parent firm. Since assembly costs are lower in its Chinese factories, this arrangement will result in higher production costs and make the smartphone uncompetitive.

For other firms, satisfying local content requirements means involving more local companies in the value chain. While most interviewed firms indicated that, where possible, they prefer to work more with local companies, they cannot always find companies in the host economy that satisfy their stringent technical requirements. For example, the manufacturing subsidiary of a "white goods" producer disclosed that local firms could only provide basic manufacturing services. At present, the firm needs to ship plastic powder to China or Malaysia where it will be transformed into, say, air-conditioning casings, before importing the more elaborated product back to be assembled in the ASEAN economy concerned. Simply put, local capability is missing.

Part of the reason for this is the small pool of quality engineers in developing economics and experienced managers relative to those in developed economies. In turn, this is connected to the quality of education in the respective host markets. For example, a subsidiary of a Japanese manufacturing company in Mexico said that it has to pay significantly higher costs to hire skilled engineers because of their scarcity value. Another firm indicated that it is not easy to persuade Japanese staff to relocate even with strong incentives, so it hopes to find competent local managers to eventually take over from the current Japanese staff (Wirjo, Pasadilla and Bassig, Chapter 14, this volume). In Indonesia, a pending local content requirement that would apply to software, R&D and design development will augment costs because of the difficulties of finding enough qualified researchers locally.

Human capital development and skills matching are important to support manufacturing and service facilities. There are examples in the case studies of fruitful collaboration between government and private firms in improving the availability of skilled persons to hire in the economy (Box 1.3). Some firms independently provide in-house training to upgrade the skills of their workers. Besides sending Japanese engineers to conduct training and give technical guidance, the firm concerned has set up local training centres to

Box 1.3: Public–Private Partnership in Human Capital Development.

- *Collaboration with vocational schools.* Skilled and semi-skilled labour are important for the Philippine-based firm engaged in manufacturing aircraft components for Boeing, Airbus and other aircraft manufacturers. Being close to the educational center in the northern part of the Philippines, the firm has access to a potentially skilled supply of labour. For additional training, the firm sends its employees, many of them with a minimum 3 years of engineering education, for 6 months of schooling at the government-run Technical Education and Skills Development Authority (TESDA) schools to become machinists. The firm's wage structure is competitive but, occasionally, some staff move to other firms in the Philippines or abroad. To ensure the supply of machinists, the firm collaborates with the TESDA schools by providing inputs

(*Continued*)

Box 1.3: (*Continued*)

into its curriculum, thereby strengthening TESDA's vocational training program and by taking in some TESDA students for paid internships. In the process, TESDA graduates become a reserve pool of labour. The firm provides possibilities for promotion and career growth to keep good workers, especially because it takes years to build the necessary expertise in their workforce (Wirjo and Pasadilla, Chapter 2, this volume).

- *Skills certification system.* Chile has established a program providing a framework for the recognition of competencies. Both public and private stakeholders from 15 different industries are involved in the program. Since 1998, more than 500 occupational standards have been established and 10,000 workers have been certified annually. In the wine industry where the case study firm belongs, the system has identified the skill profiles required for the majority of jobs performed in the industry. Since 2010, the firm has awarded more than 4,500 workers with certified skills in the wine sector. Skills certification in the wine industry or in any industry facilitates the portability of skills, decreases uncertainty in the hiring process, confers credibility in relation to workers' abilities and most importantly generates pride among certified workers in a culture of lifelong learning and development (Fernandez-Stark and Bamber, Chapter 17, this volume).

nurture local engineers as well as sales and support staff. The company has also promoted locals to top management, including in China, Indonesia, and Thailand. Foreign companies often point to the high cost of hiring expatriates as opposed to locals, which gives them the incentive to hire and train local staff wherever possible.

1.4.4. *SMEs in GVCs and Standards Conformity*

Standards, both public and private, have an important role to play not only in promoting public policy objectives (e.g., safety and sanitation) but also in addressing information asymmetries and often in signalling quality. For example, in the oil and gas industry, firms must adhere to the standards and requirements of the American Petroleum Institute (API), which is a private trade association with more than 500 standards and recommended practices. API's Monogram Program verifies that manufacturers of equipment are operating in compliance with these standards and its Witnessing Program provides witnesses to observe critical material and equipment testing and verification. Many manufacturers like to have the API monogram mark on their products because it serves as a form of quality assurance to their customers, some of whom even require this mark.

However, standards can impose additional unnecessary costs. Sometimes, a lack of clarity in standards requirements adds uncertainty in testing and certification requirements and procedures. At other times, duplicative testing and accreditation in various jurisdictions multiply costs both in terms of time and money without actually yielding additional information or public policy advantage. An example is the number of certifications and accreditations that the aircraft component manufacturer in the case study sample has to undergo with different civil aviation authorities to provide after-sales service to airline companies. Complying with these requirements means that the firm has to be audited by approximately six to seven different regulatory agencies in the region annually. To obviate the need for repeat accreditation processes, some economies could enter into mutual recognition agreements. For example, the Department of Civil Aviation Malaysia and the Civil Aviation Authority of Singapore do not need to audit the case study firm's repair station in the United States (US) because: (i) the US repair station has certifications from the Federal Aviation Administration (FAA), the European Aviation Safety Agency (EASA) and the Civil Aviation Administration of China (CAAC) and (ii) importantly Malaysia and Singapore each has a bilateral agreement with the US for recognition of these certifications. In the case of the same US company's repair station in the Philippines, despite also having certifications from FAA,

EASA and CAAC, Malaysian and Singaporean authorities carry out the regulatory audits because the three economies do not have a mutual recognition agreement. The case study firm holds that the issue is no longer about capacity or standards because this had been proven through the FAA, EASA and CAAC certifications, but is merely "regulatory bureaucracy".

If big MNCs find certification processes burdensome in some cases, the situation can only be worse for SMEs with far less resource capacity to meet the cost of repeated audits, particularly if the auditors have to be imported. Standards and certification procedures can represent a hurdle sufficiently high to frustrate the hopes of SMEs for a chance of doing business with large global firms. For example, SMEs wishing to supply basic manufacturing business to the aircraft component manufacturing company in the Philippines need to pass the certification of the National Aerospace and Defense Contractors Accreditation Program and demonstrate compliance with different category-specific (e.g., conventional machining and welding) checklists. In turn, each category-specific checklist involves detailed and complicated requirements. Additionally, the audit is not one-off but repeated every few years, depending on the requirements of the accrediting organisations.

Going through certification for GVC participation also sometimes implies acquiring costly equipment that the limited size of the local market cannot justify. Thus, a standstill situation results, where MNCs find a lack of capability among local companies and local companies (especially SMEs) find doing business with GVCs beyond their reach.

Domestic and institutional issues also saddle SMEs far more than foreign firms. While foreign firms have the capacity to comply with regulations and enjoy preferential benefits from being located in special economic zones, and have access to better infrastructure, SMEs often have to contend with the shortcomings of the local institutional context, such as poorly functioning financial markets and poor infrastructure, among other factors (Bamber *et al.*, 2015).

1.4.5. *Intellectual Property*

For developmental purposes, some economies want local transfer of technology (and intellectual property) (IP) whether for imported components or for manufacturing. The compliance requirements resulting from regulations associated with these objectives have either spurred creativity in contracting agreements between foreign and local partners or have barred the host economy from obtaining advanced technology.

A Japanese company manufacturing thermal equipment resolved the potential difficulty of technology transfer by forming a joint venture company in China. The local company is then licensed for the use of its proprietary technology with restricted conditions, that is, only for the purpose of manufacturing for a specific market (exclusive market agreement). As long as the local partner is reliable and qualified both in the technical and business sense, and the regulatory framework and government enforcement of Intellectual Property Rights (IPR) is effective, a simple licensing agreement is sufficient to protect intellectual property (with or without a joint venture agreement). In the case referred to above, the joint venture structure added an additional layer of protection for the company. This company is aware, however, that other enterprises have experienced leakage in the use of their IP by their local partners. These latter companies have found machines being sold in other Asian markets that made use of their licensed technology when it was supposedly restricted to goods produced for China.

Another example of how technology transfer policy has subverted the attainment of its avowed objective is the case of one economy that implemented new standards on 13 categories of Information Technology (IT)-related products in 2008. The new standards regime required that certain information relating to source codes be submitted for the (consumer) products that use these IT components. The guidelines that are not clear on the extent source codes shall be disclosed. But in order to minimise risk and protect its most up-to-date technology from source codes disclosure, one case study firm decided not to sell its most modern product models in the economy concerned. The effect was to deprive the market of the technology and knowledge transfer it wanted.

Intellectual property rights enforcement capability is another issue that several case study firms brought out. Counterfeit products are commonplace in Asia and have become a threat to one case study firm's car accessory business. Other firms complain about trademark infringement because local competitors (possibly deliberately) use logos, industrial design, or packaging that can be confused with their own products. Lengthy patent application processes is another issue that begs resolution.

1.4.6. *Infrastructure Bottlenecks*

There is a widely recognised urgent need to upgrade port and road infrastructure so that port services and transport services can be more efficient in coping with current and future needs. Bottlenecks from congested port and transport infrastructure add significantly to cost and time delays. Cheap and constant electricity supply is also indispensable for manufacturing operations, but in some economies outages disrupt manufacturing activities.

Many APEC economies have already come a long way in improving their infrastructure. Mexico, relative to the situation 20 years ago, now has several large ports and allows foreign investors to manage its ports. Indeed, Mexican ports now serve as good substitutes for saturated US ports such as those in Los Angeles, California or Houston, Texas. During the recent labour turmoil affecting the Californian port, one case study firm found it cost efficient to bring its products from Japan to the Mexican port and then send the products from Mexico to the United States by land.

Chile gives another good example of how infrastructure-related reforms have shortened its distance from the Asian markets (Box 1.4).

1.4.7. *Government Services and Trade Policy Affecting Goods Trade*

The above discussions have alluded to the role of government policies and services in either hampering or helping growth. All case study firms brought out issues related to goods trade and

Box 1.4: Chile's Infrastructure Reform.

Despite considerable distance from its major markets, logistics costs are lower in Chile than in many competing economies. High import and export volumes and a liberalised transportation sector have fostered competition amongst logistics providers, which has helped reduce costs. The modernisation of the Chilean ports following their privatisation, which began in the late 1990s, has significantly contributed to the economy's capacity to rapidly export large quantities of products. There are numerous ports along the economy's coast, reducing costs and distances for land transportation.

Online processing with the implementation of the single-window system has improved the efficiency of customs clearance, reducing unnecessary delays at the ports. This has helped to circumvent issues such as strikes of customs staff at the ports. The firm is able to complete all customs formalities online and deliver the wine it produces to the international transport company. This helps to prevent disruptions in the on-time delivery of wine to foreign distributors.

Source: Fernandez-Stark and Bamber (Chapter 17, this volume).

these are summarised below. Such issues also touch upon regulations that affect services because, as discussed, services are embedded in goods and it is no surprise that even though questions addressed to firms were on policies affecting manufacturing-related services, some of their answers applied largely to goods.

1.4.7.1. *Transparency, multiple layers of authority and predictability of regulations*

Foreign companies have to spend a lot on information, particularly information on regulations, because government policies are either not transparent or the requisite regulations applicable to their business involve multiple government agencies. For example, the energy generation equipment manufacturing firm in the case studies needs to deal with environmental issues, land ownership, various permits, and local tax laws, to cite a few. In some economies, it is not straightforward where, how, or from whom to obtain the relevant information. Similarly, a car assembly company based in the Philippines has to deal with no fewer than 10 different government departments involved in various aspects of vehicle assembly and subsequent sale, ranging from issues on incentives, the granting of various permits, implementing environmental and technical standards, determining and enforcing fuel standards, labour laws and vehicle registration, and other issues (Stephenson, Chapter 9, this volume).

While some governments are more efficient than others, inefficiency and delays in government services can disrupt company plans and targets, and impose significant costs. An example is the delay in the construction of a power plant caused by the inability of a government to evict or transfer dwellers from the construction site. If the power plant equipment manufacturer were left unaware of the project delay, it would be saddled with unplanned inventory storage and maintenance costs of bespoke equipment, not to mention transportation costs if the equipment had been already shipped. If the manufacturer were informed of the postponement, the costs to the company would cascade down to all their suppliers and also affect the schedules of its global subsidiaries involved in the manufacture of components for the bespoke equipment. Most of the factors causing delays are already difficult for foreign firms to observe or anticipate, and it is worse if the government is not transparent.

Another typical issue with government services is frequent changes in laws and regulations. For example, a subsidiary based in Mexico cited new tax laws and regulations promulgated in December 2013. The firm had been trying to comply with these revisions but indicated that constant revisions had made it challenging for its staff to understand and comply. There were instances when the company thought that it had already complied with everything only to find that the laws and regulations had been revised yet again. Another example is the recent requirement to maintain electronic accounting records and make a monthly submission of a general ledger to the tax authorities as well as submitting digital invoices of transactions involving all taxpayers. In the process, the government revealed it was unprepared to receive the resulting quantity of information. Consequently, the regulation is again being revised.

The enforcement of the regulation, or the lack of it, is another source of uncertainty. In the wastewater treatment services case study (Hassani and Wirjo, Chapter 6, this volume), the lack of enforcement of environmental laws and regulations, for example, regarding maximum allowable limit for treated wastewater discharge in public domains, results in a poor take up of environmentally friendly technologies by domestic firms. Because government regulators rarely check the quality of industrial discharge into public domains, only the firms that care about the reputational risk — usually from their own home economies — from not using proper wastewater treatment processes, or have corporate social responsibility, end up following the legal limits.

1.4.7.2. *Customs issues: procedures, single-window, re-manufacturing goods, standards*

Customs issues such as disagreements over valuation — for example, whether royalties should or should not be part of the dutiable value of the product, or a lack of effective implementation of single-window projects after millions has been spent on the project — are familiar issues in trade policy. Complex customs procedures along with the discretionary aspects of some customs decisions may necessitate the employment of customs brokers who know how to work the imperfect system. Thus, despite automation and the adoption of single-window systems, costs of human intervention in customs transactions persist — the very costs that modernisation was designed to eliminate.

The absence of agreed international standards on the definition of goods, especially re-manufactured goods, frustrates one case study firm's re-manufacturing business worldwide operation and obliges it to sell remanufactured parts and equipment in fewer economies (Tait and Gereffi, Chapter 19, this volume). Trade in re-manufactured goods is banned in some economies. The avowed intent is to protect customers from low-quality products and prevent the importation of electronic waste under the guise of used products. Clear standards and definitions for re-manufactured goods would go a long way to promote the resource and environmental advantages of re-using materials and machines that still possess a useful life.

In the case of the IT industry, a company has to meet numerous standards to be a qualified supplier, manufacturer or distributor. There are both public and private standards, though private standards are more dominant in this industry. Public standards, under rules and regulations, mainly concern public health, environment, consumer safety, and safety at the work place. The difficulty in complying with public standards lies in cross-border trade, when different economies have different public standards that companies are required to follow. Substantial resources are needed to ensure compliance with these public standards.

1.4.7.3. *Security-related restrictions, cyber security, a secured business environment*

Export restrictions of highly advanced equipment that has a dual purpose for both industrial as well as missile production complicate the importation of such equipment for use in manufacturing in some economies that are not signatories to the Missile Technology Control Regime. The restriction, however, is not insurmountable under certain conditions such as providing reports if the equipment is moved from one location to another, or promising to ship back the equipment if it is no longer needed for industrial production. Presumably, the fact that the subsidiary that will use the equipment is fully owned by a company headquartered in the source economy also facilitates the easing of export restrictions.

Similarly, the need to trace the origin of raw materials to ensure that none of them comes from economies under sanctions imposes additional administrative burdens. Firms need to have advanced supply chain management systems to keep track of each and every component to ensure compliance and to avoid any potential penalties.

These restrictions are relatively benign compared to outright bans on telecommunications equipment because of an alleged potential cyber security threat. By invoking security exemptions, some governments prohibit the purchase of equipment for use in government-funded broadband networks from one of the case study firms based in China.

Though unrelated to terrorism, another form of "security" threat is criminal activities that lead to the loss of cargo during transport, for example. In Mexico, delays in logistics are attributed by more than one-third of firms to the experience of criminal activities and solicitation of informal payments. Such security concerns add to the cost of logistics in the form of extra insurance and additional measures such as live tracking of trucks or the use of armed escorts (Wirjo and Pasadilla, Chapter 18, this volume).

1.5. SUMMARY AND CONCLUSION

This chapter seeks to understand the importance of manufacturing-related services in different firms. Through face-to-face interviews, the study collected various experiences relating to services in different value chain configurations as well as some policy restrictions that affect these services and the companies' businesses in general. It finds that value chains use a wide range of services, numbering from as few as 37 and as many as 74. Moreover, 38 per cent to 90 per cent of services are outsourced to third parties, usually depending on cost considerations and on whether they form part of the core activity of the firm.

Policy issues discussed with firms included restrictions on investment and labour mobility, policies relating to skills development, a range of other regulatory issues, customs facilitation, and security-related threats. For the most part, the policy issues were not merely about services but also highly relevant to manufacturing operations. This attests to the need to think about policies in terms of their overall impact on economies in a world where manufacturing and the supply of services are increasingly co-dependent. The chapter discusses the nature of restrictions in different economies, the costs they impose on firms and customers, and their general economic effects.

The policy discussions are also relevant to institutional and infrastructure development. Examples of positive experiences of economies that have liberalised port management and modernised ports and transport facilities show how these have minimised the distance disadvantage of some economies from their markets and fostered competitiveness. The chapter has also brought out the need for effective intellectual property rights protection. Importantly, it highlights the effects of the policy incoherence inherent in opening up manufacturing for investments but maintaining restrictions on services.

Although perhaps a distant objective at the moment, providing visa waivers for more economies, facilitating labour mobility, and relaxing foreign labour quotas can lead to large benefits for the receiving economy (Kommerskollegium, 2015). Such policies are a strong complement to foreign investment openness and can support the transfer of technology and technical know-how.

Helping SMEs with international standards accreditation would boost their ability to participate in GVCs. Standards and high conformity assessment costs, along with other institutional disadvantages, such as a lack of access to finance, are among the other hurdles confronting SMEs as they seek to reap the benefits of being part of GVCs.

In order to improve government services and to avoid creating a negative impact on trade, it is also imperative to build and maintain constructive stakeholder relationships. This may include providing opportunities for public consultation in the regulatory process and constantly reviewing the regulations with stakeholder involvement. Such measures can contribute to enhancing transparency, predictability and consistency in the business environment and minimise the costs described in the various case studies.

Finally, the case studies illustrate how policies carrying unnecessary costs merit careful consideration in each economy's political and economic calculus as it crafts its vision and plans for generating growth, jobs and development.

APPENDIX A

Table A.1: Restrictions on Foreign Ownership for Logistics Sector.

	Domestic transport service			
	Full foreign equity ownership[a]		Partial foreign equity ownership[b]	
	Without conditions	With conditions[c]	Without conditions	With conditions[c]
China	✓		✓	
Chile		✓		
Indonesia			✓	
Malaysia		✓	✓	✓
Mexico			✓	✓
Peru			✓	✓
The Philippines			✓	
Russia	✓			
Thailand		✓	✓	
Viet Nam			✓	
	Warehouse			
China	✓		✓	
Chile		✓		
Indonesia			✓	
Malaysia	✓	✓		✓
Mexico	✓	✓		
Peru		✓		
The Philippines		✓	✓	
Thailand		✓	✓	
Viet Nam	✓			

[a]Full foreign equity ownerships in domestic transport service are allowed for: all sub-sectors (Chile and Russia); freight transport by road (China); some business operations (Thailand); transportation of goods held by a company (Malaysia).

[b]Partial foreign equity ownerships in domestic transport service are allowed for: all sub-sectors (Thailand); freight transport by railway (China); commercial vehicles licensing of freight transport and container transport (with conditions) and domestic shipping licensing (Malaysia); general cargo transport, domestic shipping business and freight forwarding business (Indonesia); companies that are considered as operating and managing public utilities

	Customs clearance service			
	Full foreign equity ownership[g]		Partial foreign equity ownership[h]	
	Without conditions	With conditions[i]	Without conditions	With conditions[j]
China	✓			
Chile		✓		
Indonesia			✓	
Malaysia				✓
Peru		✓		
The Philippines			✓	
Thailand			✓	
Viet Nam			✓	

Source: Authors' compilation based on Japan External Trade Organization (JETRO) data.
(The Philippines); freight transport by land and sea (Viet Nam); domestic commercial air and marine transport (with conditions) (Peru); domestic air transport, air-taxi transport, and special air transport, port transport services such as towing, mooring and chartering, shipping companies engaged in marine operations of ocean transportation and provision of public railway services (with conditions) (Mexico).

[c]Conditions vary between economies and can include obtaining approval from the Cabinet or relevant Ministry, meeting certain minimum capital/investment requirements, having its nationality or residents in the management, captain and/or crew.

[d]Full foreign equity ownerships in warehouse service are allowed for: all sub-sectors (Chile; Mexico (but there are conditions to be met if foreign share is more than 49 per cent and investment exceeds US$271 million); Peru; the Philippines; and Viet Nam); warehousing of international marine transport and freight transport by road (China); logistics centre (Thailand); non-bonded warehouse and private bonded warehouse operators (with conditions) (Malaysia).

[e]Partial foreign equity ownerships in warehouse service are allowed for: all sub-sectors (Indonesia and Thailand); warehousing of freight transport by air (China); ordinary warehouse operators (Malaysia); companies that are considered as operating and managing public utilities (The Philippines).

[f]Conditions vary between economies and can include obtaining approval from the relevant commission or committee, meeting certain minimum capital/investment requirements, and/or utilisation of latest technologies.

[g]Full foreign equity ownerships in customs clearance service are allowed for: all sub-sectors (Chile and Peru); agents of international freight transport (China).

[h]Partial foreign equity ownerships in customs clearance service are allowed for: all sub-sectors (Indonesia; Malaysia; and Viet Nam); harbour cargo handling companies that are considered as operating and managing public utilities (the Philippines); customs service and freight forwarding business (Thailand).

[i]Conditions vary between economies but generally include meeting certain minimum capital/investment requirements.

Table A.2: Restrictions on Foreign Ownership for Retail Services.

	Retail service			
	Full foreign equity ownership[a]		Partial foreign equity ownership[b]	
	Without conditions	With conditions[c]	Without conditions	With conditions[c]
China	✓			
Chile		✓		
Indonesia		✓		
Malaysia				✓
Mexico	✓	✓		
Peru	✓			
The Philippines		✓		
Russia	✓			
Thailand		✓	✓	✓
Viet Nam		✓		

Source: Authors' compilation based on JETRO data.

[a]Full foreign equity ownerships in retail service are allowed for: all sub-sectors (China; Chile; Mexico (but there are conditions to be met if foreign share is more than 49 per cent and investment exceeds US$271 million); Peru; The Philippines; Russia; Thailand; and Viet Nam); minimarkets, supermarkets and department stores (Indonesia).

[b]Partial foreign equity ownerships in retail service are allowed for: all sub-sectors but there are conditions to be met if more than 50 per cent foreign equity are desired (Thailand); hypermarkets and supermarkets (Malaysia).

[c]Conditions vary between economies and can include obtaining approval from the relevant committee or commission, meeting certain minimum capital/investment requirements, having certain minimum floor spaces, handling certain luxury goods, undertaking economic needs test for opening additional stores, and/or being members of certain preferential agreement/FTA.

Table A.3: Restrictions on Foreign Ownership for Retail Services.

	Wholesale service			
	Full foreign equity ownership[a]		Partial foreign equity ownership[b]	
	Without conditions	With conditions[c]	Without conditions	With conditions[c]
China	✓			
Chile		✓		
Indonesia			✓	
Malaysia	✓	✓		
Mexico	✓	✓		
Peru	✓			
The Philippines	✓	✓	✓	
Russia	✓			
Thailand		✓	✓	✓
Viet Nam	✓			

Source: Authors' compilation based on JETRO data.

[a]Full foreign equity ownerships in wholesale service are allowed for: all sub-sectors (China; Chile; Malaysia (but require permission for importation and selling new completely assembled vehicles); Mexico (but there are conditions to be met if foreign share is more than 49 per cent and investment exceeds US$271 million); Peru; Russia; and Thailand); import-export businesses and domestic wholesale services (with conditions) (the Philippines); all except if materials handled are cigarettes, books, newspapers, magazines, video recording devices, precious metals, medicines and sugar (Viet Nam).

[b]Partial foreign equity ownerships in wholesale service are allowed for: all sub-sectors but there are conditions to be met if more than 50 per cent foreign equity are desired (Thailand); distributor, warehouse and cold storage businesses (Indonesia); domestic wholesale services (the Philippines)

[c]Conditions vary between economies and can include obtaining approval from the relevant Ministry or committee, meeting certain minimum capital requirements, and/or being members of certain preferential agreement/FTA.

APPENDIX B

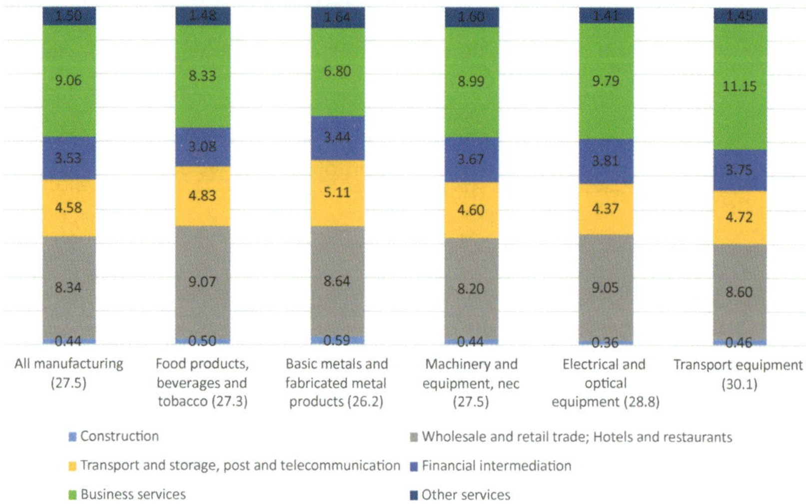

Figure B.1: Share of Services in Manufacturing Value Added.

Note: Numbers in parentheses are the shares of services in exports of respective industries.

Source: Authors' computation based on TiVA data.

APPENDIX C

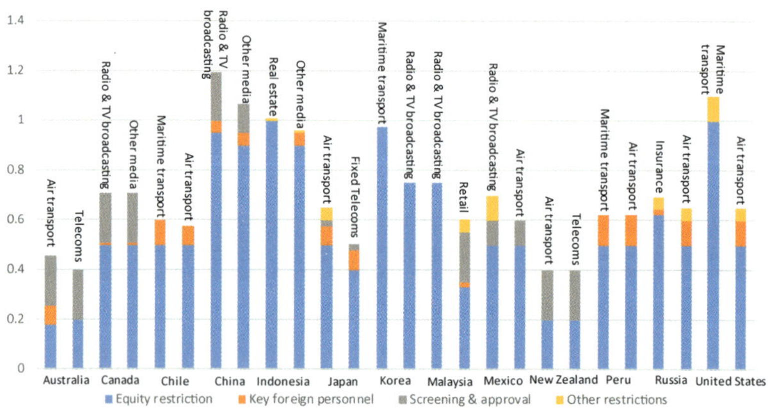

Figure C.1: FDI Regulatory Restrictiveness Index (2014).

Note: Two most restrictive sectors per economy.

Source: Authors' computations based on OECD FDI regulatory restrictiveness database.

REFERENCES

Ahamed, N (2013). Estimating trade in value-added: Why and how. In *Global Value Chains in a Changing World*, DK Elms and P Low (eds.). Geneva: World Trade Organization.

APEC Business Advisory Council (2014). *Accelerating Investment in Services in APEC: Growing Businesses Globally for the Benefit of Economies.* University of Southern California Marshall School of Business. November.

Arnold, J, BS Javorcik and A Mattoo (2011). Does services liberalization benefit manufacturing firms? Evidence from the Czech Republic. *Journal of International Economics*, 85(1).

Bamber, P, K Fernandez-Stark, G Gereffi and A Guinn (2015). Connecting local producers in developing countries to regional and global value chains. OECD Trade Policy Paper No. 160. Paris: OECD. December 3.

Baumol, WJ (1967). Macroeconomics of unbalanced growth. *American Economic Review*, 57, 415–426.

Berrios, M and M Pilgrim (2013). *Is Small Still Beautiful? Analysis*. Geneva: International Labour Organization, Small Enterprises Unit.

Cheung, D (2016). Manufacturing of automotive components in the ASEAN Region. In *Services in Global Value Chains: Manufacturing-Related Services*. Singapore: APEC Secretariat.

Dedrick, J, KL Kraemer and G Linden (2010). Who profits from innovation in 91 global value chains? A study of the iPod and notebook PCs. *Industrial and Corporate Change*, 19(1), 81–116.

Fernandes, A and C Paunov (2008). Services FDI and manufacturing productivity growth: There is a link. World Bank Working Paper. April.

Fernandez-Stark, K and P Bamber (2016). Wine industry in Chile. In *Services in Global Value Chains: Manufacturing-Related Services*. Singapore: APEC Secretariat.

Francois, JF and B Hoekman (2010). Services trade and policy. *Economic Literature*, 48(3), 642–692.

Fuchs, VR (1968). *The Service Economy*. New York: Columbia University Press.

Haines, W (2016). Industrial welding services in Thailand. In *Services in Global Value Chains: Manufacturing-Related Services*. Singapore: APEC Secretariat.

Hassani, A and A Wirjo (2016). Wastewater Treatment Services in GVCs. In *Services in Global Value Chains: Manufacturing-Related Services*. Singapore: APEC Secretariat.

HKSAR (Hong Kong Special Administrative Region) (2014). Trade and Industry Department. Support and Consultation Centre for SMEs. http://www.success.tid.gov.hk/english/aboutus/aboutus.html.

Kommerskollegium (2010a). *Servicification of Swedish Manufacturing*. Stockholm: National Board of Trade.

____ (2010b). *At Your Service: The Importance of Services for Manufacturing Companies and Possible Trade Policy Implications*. Stockholm: National Board of Trade.

____ (2012). *Everybody is in Services — The Impact of Servicification in Manufacturing on Trade and Trade Policy*. Stockholm: National Board of Trade.

____ (2015). *Trade Costs of Visas and Work Permits: A Trade Facilitation Perspective on Movement of Persons*. Stockholm: National Board of Trade.

Koopman, RB, M Tsigas, D Riker and W Powers (2013). The implications of using value-added trade data for applied trade policy analysis. In *Global Value Chains in a Changing World*, DK Elms and P Low (eds.). Geneva: World Trade Organization.

Low, P (2013). The role of services in global value chains. In *Global Value Chains in a Changing World*, DK Elms and P Low (eds.), Geneva: World Trade Organization.

____ (2016). Rethinking services in a changing world. E15 Expert Group on Services. International Centre for Trade and Sustainable Development and World Economic Forum. http://e15initiative .org/publications/rethinking-services-changing-world/ [January 2016].

Normah Mohd, A (2007). *SMEs: Building Blocks for Economic Growth*. Kuala Lumpur: Department of Statistics. www.statistics.gov.my/portal/images/stories/files/journal/smes.pdf.

Organisation for Economic Co-operation and Development (2011). New sources of growth: Intangible assets. www.oecd.org/sti/inno/46349020.pdf.

____ (2012). New Sources of Growth: Knowledge-Based Capital Driving Investment and Productivity in the 21st Century. Interim Project Findings. http://www.oecd.org/sti/50498841.pdf.

____ (2014). Factbook 2014. www.oecd-ilibrary.org/docserver/download/3013081e.pdf?expires=1409450861&id=id&accname=guest&checksum=BFEB69CB655F79C4F6AD3DF952D9526B.

OECD and WTO (2013). Measuring trade in value added: An OECD-WTO joint initiative. http://www.oecd.org/sti/ind/measuringtradein-value-addedanoecd-wtojointinitiative.htm.

Pasadilla, G (2016). Manufacturing of thermal power generation equipment. In *Services in Global Value Chains: Manufacturing-Related Services*. Singapore: APEC Secretariat.

Pasadilla, G and A Wirjo (2014). Services and manufacturing: Patterns of linkages. APEC Policy Support Unit Policy Brief No. 10. 31 July.

____ (2015). Services, manufacturing and productivity. Issues Paper No. 9. APEC Policy Support Unit. January.

San Andres, EA (2016). Manufacturing of consumer electronic appliances in Indonesia. In *Services in Global Value Chains: Manufacturing-Related Services*. Singapore: APEC Secretariat.

Shiino, K (2016). How manufacturing-related services contribute to trade in goods? Presentation at the APEC Public-Private Dialogue on Services. Boracay, Philippines. May 17.

Sit, D (2016). Manufacturing of refrigerators. In *Services in Global Value Chains: Manufacturing-Related Services*. Singapore: APEC Secretariat.

Smith, A (1776). An inquiry into the nature and causes of the wealth of nations. In *Library of Economics and Liberty*. Book II.3.1 of the *Accumulation of Capital*, or of *Productive and Unproductive Labour*, E. Cannan (ed.). www.econlib.org/library/Smith/smWN8.html.

Stein, P, T Goland and R Schiff (2010). *Two Trillion and Counting: Assessing the Credit Gap for Micro, Small, and Medium-size Enterprises in the Developing World*. International Finance Corporation and McKinsey & Company.

Stephenson, S (2016). Manufacture of car in the Philippines. In *Services in Global Value Chains: Manufacturing-Related Services*. Singapore: APEC Secretariat.

Tait, K and G Gereffi (2016). Remanufacturing services in the construction machinery value chain. In *Services in Global Value Chains: Manufacturing-Related Services*. Singapore: APEC Secretariat.

United Nations (2013). United Nations Central Product Classification, Version 2, Sections 5–9. https://unstats.un.org/unsd/cr/registry/docs/CPCv2_structure.pdf.

Wirjo, A and G Pasadilla (2016a). Manufacturing of aircraft control systems in the Philippines, In *Services in Global Value Chains: Manufacturing-Related Services*. Singapore: APEC Secretariat.

____ (2016b). Manufacturing of oil and gas industry equipment in Singapore. In *Services in Global Value Chains: Manufacturing-Related Services*. Singapore: APEC Secretariat.

____ (2016c). Integrated logistics solutions provider in Mexico. In *Services in Global Value Chains: Manufacturing-Related Services*. Singapore: APEC Secretariat.

Wirjo, A, G Pasadilla and J Bassig (2016). Manufacturing of automotive components in Mexico: Perspectives from three firms. In *Services in Global Value Chains: Manufacturing-Related Services*. Singapore: APEC Secretariat.

Zhu, H and G Pasadilla (2016). Manufacturing of telecommunications equipment. In *Services in Global Value Chains: Manufacturing-Related Services*, Singapore: APEC Secretariat.

Chapter 2

MANUFACTURING OF AIRCRAFT CONTROL SYSTEMS IN THE PHILIPPINES

Andre Wirjo and Gloria O. Pasadilla

2.1. INDUSTRY OVERVIEW

The manufacturing process of an airplane is a complex undertaking involving myriad global partners. It is a real example of a global value chain (GVC) in action. The Boeing B787 illustrated in Fig. 2.1 shows how many different parts and components originating from different economies make up the aircraft. For instance, the wings are manufactured in Japan, the movable trailing edges are produced in Australia, while the passenger entry doors are from France, and so forth. These tier 1 suppliers, in turn, have their own GVCs to be able to produce the wings or the door or the landing gear. It is no surprise, therefore, that Boeing has relationships with more than 5000 supplier factories and sub-tier suppliers around the globe.[1] One such company, not depicted in Fig. 2.1, is the case study firm — a supplier of components that are crucial for how planes fly.

[1]http://www.boeingblogs.com/randy/archives/2013/02/supply_chain.html.

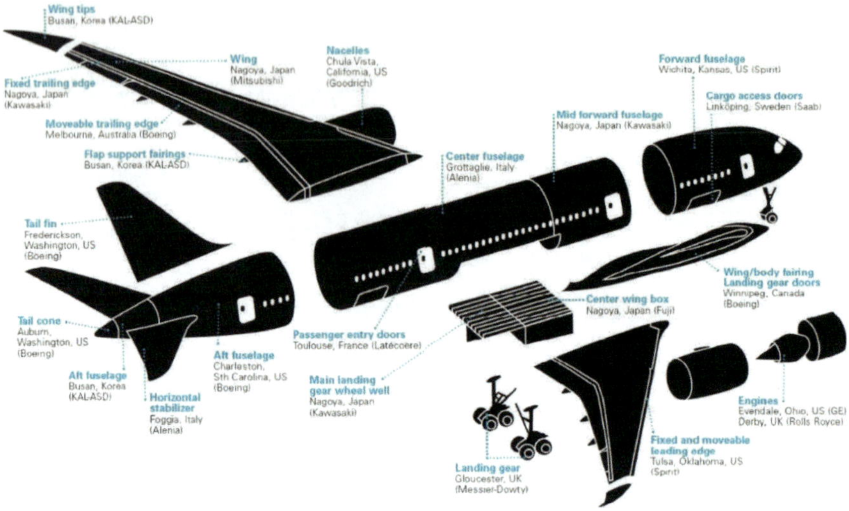

Figure 2.1: Simplified Supply Chain of a Boeing B787.

Source: Courtesy of Boeing.

2.2. BACKGROUND INFORMATION ON THE FIRM[2]

The firm is a designer, manufacturer and integrator of high performance precision motion control products and systems. From an original application for aircraft, it has since expanded its products and applications to satellites and space vehicles, launch vehicles, missiles, industrial machinery, wind energy, marine applications and medical equipment. Its business segments are grouped into five, namely aircraft controls, space and defence controls, industrial controls, components, surveillance camera systems and medical devices. The firm is headquartered in the United States and has international operations (wholly owned foreign subsidiaries) that are predominantly located in Europe and Asia Pacific.

Among the products that the firm manufactures and for which the firm was founded are primary and secondary flight controls

[2]All information about the firm in this case study is from its corporate website as well as from the interview.

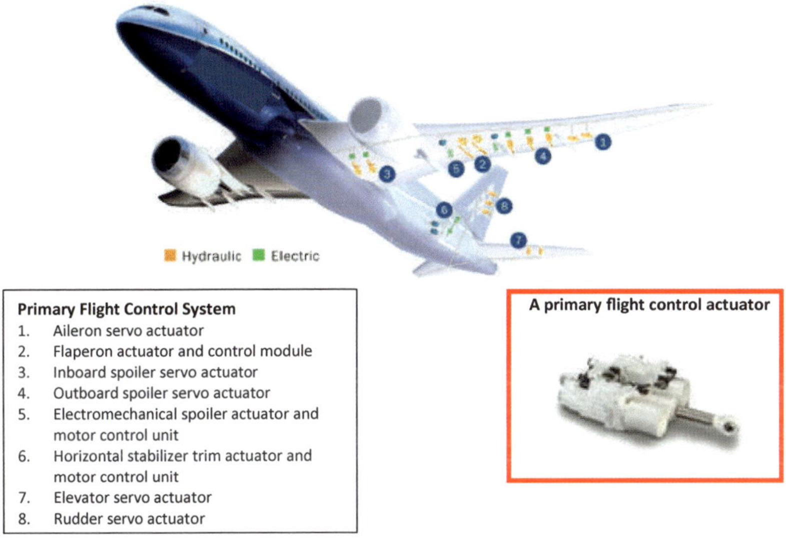

Figure 2.2: The Location of Servo Actuators in a Boeing B787.

Source: Courtesy of the firm.

systems and components for commercial aircraft such as servo valves and servo actuators. This is what the Philippines operation does. A servo valve is a device which controls hydraulic pressure for fine control of actuators. A servo actuator is a mechanism to induce or control motion in mechanical systems. They are devices that transform an input signal (usually electrical) into motion. Today, these products have various applications but initially, they were supplied to the airplane manufacturers such as Gulfstream, Airbus and Boeing, which install them in various models such as Gulfstream G280, Airbus A350 and Boeing B787 (see Fig. 2.2). From the perspective of these airplane manufacturers, the firm is a first-tier supplier.

The firm has expanded the range of products that it manufactures over the years. It started with the manufacture of servo valves which went into another company's servo actuators. It then moved to the manufacture of its own servo actuators, followed by a family of servo valves and servo actuators. Before long, it started supplying fully integrated flight control systems. The firm is involved in the development and production of flight control systems of almost all major airplane models and is typically part of the early stage team of any major airplane model development because servo valves, servo actuators and flight control systems are critical parts of airplanes.

In addition to selling manufactured products, the firm has a segment focusing on the provision of after-sales services. Its five customer support centres spread across the world have been certified by leading civil aviation authorities such as the US Federal Aviation Administration (FAA), the European Aviation Safety Agency (EASA) and the Civil Aviation Administration of China (CAAC) to provide services such as maintenance, repair and overhaul (MRO) services of servo valves and servo actuators as well as inspection, modification and testing of components for auto flight, flight controls, hydraulic power and landing gear systems. Sometimes the company group partners up with existing plane maintenance and repair services providers for the servicing of their own servo valves and actuators. It is reckoned that its after-sales business could provide revenue for the firm for 20–30 years, providing a strong anchor for revenues in a highly cyclical aircraft manufacturing industry.

2.3. DESCRIPTION OF THE VALUE CHAIN

This case study will focus on the firm's Philippine subsidiary, which is the largest manufacturing site of the firm for aircraft controls. The product of focus for this study is a set of servo actuators for Boeing B787. For a single Boeing B787, it supplies over 200 separate line replaceable units classified into two groups. Primary actuators are typically hydraulic or electro-hydraulic actuators while secondary actuators usually include hi-lift and gear-type actuators. They come in various sizes and the largest one can weigh up to 200 pounds.

For the purpose of this case study, the value chain begins when the design is provided to the case study firm by its UK- or

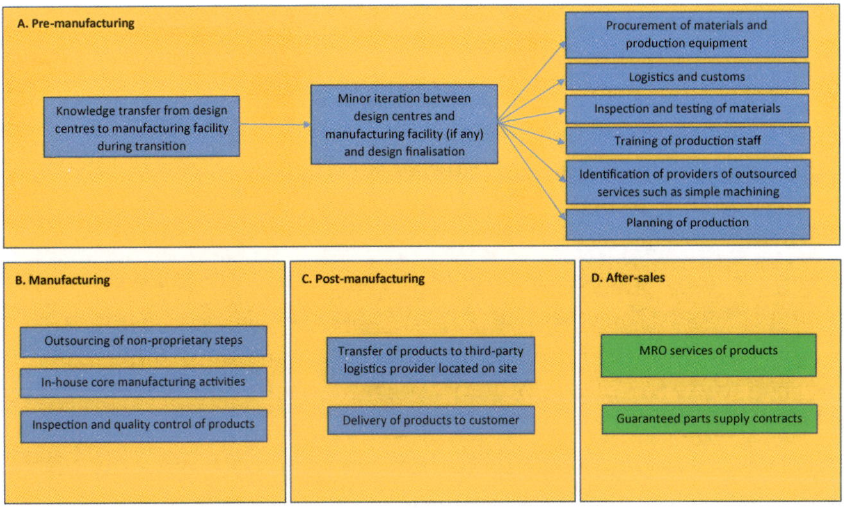

Figure 2.3: Dimension of the Value Chain Covered by the Case Study.

Note: Optional activities in the value chain are indicated by green boxes.

Source: Authors.

US-based design centres/subsidiaries within the company. The chain then moves on to the importation of raw materials from its suppliers up to the provision of after-sales services to the customer (see Fig. 2.3).

2.3.1. *Pre-manufacturing: Design, Materials Sourcing and Logistics*

The case study firm's job starts with the sourcing and importation of raw materials into the Philippines. However, before describing this part of the value chain in greater detail, it may be worthwhile to take a few steps back to better understand the complex processes that occur prior to this stage.

The actual starting point for any airplane model is when the airplane manufacturer provides the detailed specifications and asks its first-tier suppliers to produce parts and components to meet these specifications. Once chosen by the plane manufacturer, the company group's design centres are typically part of the early stages of the design and development process of the plane.

Everything from design, initial development and prototype production of parts and components for the Boeing B787 is carried out by one of the company group's four design centres. Before then, to even qualify to become suppliers, the process is highly involved requiring myriad certifications for the firm. Once the design is approved and finalised, it moves to large-scale manufacturing which is where the case study firm comes in.

Significantly, though not formally part of the design process, the case study firm contributes important inputs into the finalisation of the design of servo valves or servo actuators. Because of its experience in actual manufacturing, the firm can have a say on the manufacturability of the designed product and provides such feedback to the design group. During the transition from the design centres to the large-scale manufacturing facility, engineers from both sites work closely together to ensure a smooth handover.

Moving on to the sourcing and importation of material inputs, the firm's local supply chain staff handle the operational and tactical aspects of procurement, such as issuing purchase orders, and maintaining and managing supplier relationships. The company group has its own supply chain management team that vets the qualification of its global suppliers. Once selected, this group[3] works with selected suppliers and regularly visits them to ensure that the quality of material inputs is always up to the standard, tracking them to their source to ensure compliance with international and domestic laws. The main reason for centralised supply chain management is direct and indirect cost savings made possible by resource sharing, but centralization also facilitates compliance with aviation standards for safety and security. The traceability requirement, for example, means that all parts or materials meet standards and that they should have been sourced from approved suppliers. At times, change in suppliers (and hence the source of material inputs) may, depending on its nature, have

[3]Or sometimes also the local supply chain staff.

to be treated as modifications to the existing process, and therefore may require re-qualification and re-certification. Many changes, however, do not require re-qualification and re-certification.

The logistics of importation, including offloading and storage in the firm's stockroom, is mostly outsourced to third-party logistics providers. Once the materials reach the Philippines, these providers will facilitate the customs clearance process through their local agents. The raw materials are then trucked to the firm's facility where they are offloaded, inventoried, and sent for storage in the stockroom. The outsourced supplier of logistics may also be engaged within the factory, delivering materials from the stockroom to the fabrication, assembly and testing areas.

2.3.2. *Manufacturing Process*

Products are manufactured in-house although certain non-proprietary steps such as basic machining may be outsourced to third parties for cost reasons. These outsourced activities, however, are very few. Being at the low technological end of the spectrum when it comes to aircraft manufacture, they are an almost insignificant source of value. The outsourced suppliers will typically be part of the local metal fabrication industry.[4] The firm prides itself on its engineering expertise, where each product and each part is built to stringent tolerances and precision, inspected and controlled for quality at every step of the manufacturing process.

---

[4]These third-party firms are usually members of Metalworking Industries Association of the Philippines, Inc. (MIAP). The MIAP is a trade association whose members comprise firms engaged in metalworking and related activities. Among its activities are the establishment of product standards with accredited technical/professional societies and appropriate government agencies as well as provision of training to industry staff in collaboration with government agencies such as the Technical Education and Skills Development Authority (TESDA). More information about MIAP can be obtained at: http://hdm.miapnational.com/.

The in-house staffs are organised around product line/business units, each with its own organisation chart. Each unit's management team oversees the segment of manufacturing or manufacturing-related activity that has been assigned to it such as materials planning, fabrication, assembly and testing of specific servo actuators. Each unit also has staff that oversee shops, planning, engineering and quality matters.

Once the manufacturing process is completed, the products are inspected before being transferred to the logistics provider for post-manufacturing handling. Concurrently, the sales/marketing team is notified of the completion of the manufacturing process so that they can inform the customers.

2.3.3. *Post-manufacturing*

In this case study, post-manufacturing refers to certification, packaging, handling and transport/logistics services to customers. This is mostly handled by the third-party logistics provider which has an office on-site at the firm's manufacturing facilities. Its agreement with the logistics firm is for both internal and external logistics, which means that it includes the movement of materials on-site. The logistics company designs or procures the packaging materials and readies all the products for loading onto the truck, and thereafter for transport to the destination. Because the case study firm's Philippine manufacturing facility has certifications from a wide range of international regulations and standards, it undertakes certifications in-house, making the products ready for shipment direct to the end-user.

Because of the highly regulated natured of its business, the firm's logistics provider also has to follow closely the firm's processes which have been certified by US FAA and EASA.

2.3.4. *After-sales Services*

From the company group's perspective, after-sales services are among the core segments of its business. The firm can potentially extend the revenue stream from a single aircraft model by 20–30 years. The services offered include MRO, as well as parts supply. The case study firm's servo actuators are "built to last" but there may be a need for maintenance to fix damaged parts due to corrosion and use. The mean time between failures of actuators ranges between 50,000 hours (for primary actuators) and millions of hours of flying time (for hi-lift actuators).

The customers for the firm's after-sales activities are airline companies, not the aircraft makers who are the clients for the firm's manufactured output. The service offered is therefore different and undertaken by a different business entity within the firm's structure. Certification by civil aviation authorities is a pre-requisite for providing after-sales services to airlines. In this market, response cycles are usually timed in days rather than months.

The way the firm provides after-sales services has changed over time in order to maintain competitiveness. While airline companies used to purchase spare parts upfront and store them in their own warehouses to ensure the supply of critical parts, many are beginning to choose a different arrangement. Airline companies are buying access to a pool of spare parts within a guaranteed period instead of buying them upfront, essentially outsourcing the stock management to the parts supplier.

2.4. SERVICES ALONG THE VALUE CHAIN

2.4.1. *Services Identification and Value Contribution*

A total of 53 services, which have been further disaggregated into at least 92 separate services (see Fig. 2.4) are identified in this value chain and they are categorised according to the various stages within the chain: (i) pre-manufacturing services, (ii) services during manufacturing, (iii) post-manufacturing services, (iv) after-sales services and (v) back office services.

Although information on which are the most important services inputs in terms of added value is not available, it may be inferred that engineering services are perhaps the most critical

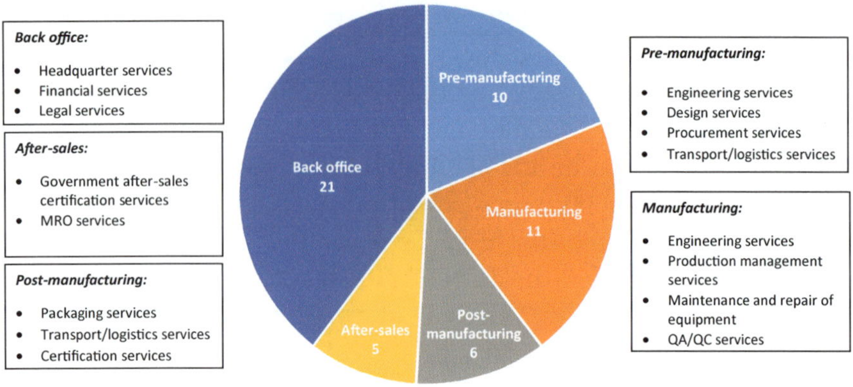

Figure 2.4: Breakdown of Services by Stage and Examples of Key Services.

Source: Authors.

services input. With the exception of post-manufacturing, engineering services feature in all stages covered in this case study and in fact, play a very central role even when an airplane model is still in its conception stage. Indeed, the firm considers itself first and foremost an engineering entity. The large number of services that enter the manufacturing value chain of servo actuators conforms to a pattern of pervasive reliance on services observed across all productive sectors subject to value chain case studies in this project.

The importance of services can also be observed in the share of employees devoted to service activities. The case study firm employs approximately 1500 people[5] on-site, of which 400 are contractual (i.e., non-permanent) workers. Of the 1100 full-time employees, 750 are manufacturing-specific staff such as machinists, while the remaining 350 are involved in the provision of services such as management and other support services, planning, supply chain services, facilities maintenance, and security. Service-related employees are therefore approximately one-third of the firm's full-time workforce. This segment of employees is likely to account for more than one-third of added value since it includes

[5]Globally, the firm employs around 10,000 employees.

most of the company's executives. Moreover, some of the personnel categorized as manufacturing personnel are also supplying services such as testing, inspection and other quality control services.

The share of services in the firm's value chain is expected to remain significant because the firm believes that in addition to price competitiveness, the principal determinants of success in the industry are factors such as product quality and reliability, design and engineering capabilities, product development, conformity to customer specifications, timeliness of delivery, effectiveness of the distribution organisation, and the quality of after-sales support. Many of these are essentially services in character, or at least heavily reliant on services.

2.4.2. *Outsourcing, Bundling and other Aspects of Services Supply*

Of the 92 services identified in the value chain, 23 services are supplied in-house, 13 are partially outsourced while 56 are fully outsourced (see Appendix A).[6]

Services provided entirely in-house are generally those that can be considered core services activities such as engineering services, production management, technical testing services as well as MRO services.[7] These services are conducted in-house because

---

[6]Some of the reasons for firm to outsource services include: (i) government services, such as certification for provision of after-sales services, visa and immigration services and inspections pertaining to environmental, health and safety (EHS); (ii) required by laws and regulations, such as external auditing services by third party providers; (iii) lack of expertise or specialisation in-house to provide certain services, such as consulting services, medical services and legal services; (iv) lack of feasibility to supply service in-house, such as freight insurance and utilities services; (v) economies of scale, such as transport/logistics services; (vi) need for strong relationships with government agencies, such as customs clearance services; and (vii) network economies, such as recruitment services.

[7]Although firm has also began to form partnership with third-party service providers in the aviation industry to provide MRO services.

they are necessary to ensure the quality and/or involve proprietary technology.

The firm outsources about 20–30 per cent of its fabrication activities to global and regional suppliers as well as local firms. However, it is noted that the share captured by local firms is relatively low because of the certification hurdles. Besides, the main activity of the case study firm is "precision machining", which requires skilled labour as well as very specific and specialised equipment. The local fabrication industry (includes many small and medium enterprises (SMEs)) does not possess the skills nor the resources to purchase the necessary technology that would open the way to participating in large segments of the firm's supply chain.

In addition, the firm makes use of outsourced bundled services. For example, the logistics provider not only supplies transportation services but also customs clearance and even loading and unloading of raw materials as well as inventory services. In fact, as mentioned above, the agreement between the firm and the third-party provider also covers movement and handling of materials/products within firm's premises. In addition, the same logistics provider designs and procures custom-made packaging for the firm's products.

2.5. POLICIES AFFECTING THE VALUE CHAIN

An essential part of this study is the analysis of how policies, both government and private, influence the value chain (see Appendix B). The firm identified a number of policy issues in the areas of trade and labour, which will be further elaborated upon below. However, it must be noted that the firm generally faces few challenges. One probable reason is that the firm has the necessary preparedness and resources to address problems as they arise. As a large multinational firm that makes a significant contribution to output and employment in a high technology, high profile sector with high profile international customers, the firm has the ear of government when problems arise. The firm receives benefits from being located in a government-designated special economic zone.

The firm is also recognised for its contribution to the local economy, including the provision of technical training of the local workforce.

2.5.1. *Multiple Certifications and Need for Mutual Recognition*

Provision of after-sales services is one of the businesses the firm wants to grow in Asia. Globally, aftermarket sales accounted for 16 per cent of its total sales in 2014, and specifically for the aircraft controls segment, these activities represented about one-third of sales in the last three years (i.e., 2012–2014).[8] In order to serve this market, the firm requires certification from the regulatory authorities of each individual economy before its repair station in the Philippines is allowed to provide after-sales services to the carriers or airlines established in a specific economy. For example, in order to provide services to carriers established in Singapore and Malaysia, the firm has to be certified by the Civil Aviation Authority of Singapore (CAAS) and the Department of Civil Aviation of Malaysia (DCA Malaysia) respectively. Complying with these requirements means that the firm has to be audited by approximately six to seven different regulatory agencies annually.

These audits are necessary to ensure that the services provided are of the highest standards. Nevertheless, the certification process is a cost to the firm because time and effort are needed to ensure that every audit proceeds smoothly. In addition to the on-site visit which usually takes between two and three days, a typical audit also takes days or weeks to prepare. The challenge is to keep up with the high number of audits and to minimise the potential disruptions that such audits may cause in its operations.

One possibility is to reduce the frequency of each audit. The firm indicated that its repair station has to be re-certified at any

[8]The firm defines aftermarket sales for the aircraft controls segment as consisting of maintenance, repair, overhaul and parts supply to both military and commercial aircraft. Although this case study focuses on commercial aircraft, data specific to that operation are not publicly available.

time between one and five years, although the most common re-certification takes place every two years. If the regulatory authorities were able to issue certificates with a longer time validity, the cost to firm could be significantly reduced.

Another alternative is to reduce the number of regulatory authorities conducting the audit by mutually recognising each other's certifications. This has in fact been practised in different economies. For example, the firm noted that DCA Malaysia and CAAS do not need to audit the firm's repair station in the United States because: (i) the repair station has certifications from the FAA, the EASA and the CAAC; and (ii) Malaysia and Singapore each has a bilateral agreement with the United States for recognition of these certifications. In the case of its repair station in the Philippines, the firm also has certifications from FAA, EASA and CAAC but there is no mutual recognition agreement between the Philippines, Malaysia and Singapore. This is certainly an issue that could be resolved considering that the repair station has been named one of the best in the world by the FAA and EASA. The issue in this case is not so much of capacity nor of standards, but more of perception and regulatory bureaucracy which can be minimised.

2.5.2. *Security-related Export Restrictions*

On the manufacturing side, some of firm's imported equipment are highly advanced and can also be used for missile production — so-called dual-use equipment. Importation of such equipment into the Philippines is complicated by many security-related export restrictions in source economies, especially because the Philippines is not a signatory to the Missile Technology Control Regime (MTCR).[9]

The MTCR is an association of economies which seeks to coordinate national export licensing policies so as to prevent the proliferation of unmanned delivery systems capable of carrying

[9]Philippines is a signatory of the UN Treaty on the Non-Proliferation of Nuclear Weapons (http://disarmament.un.org/treaties/t/npt) and Comprehensive Test-Ban Treaty (http://ctbto.org/map/#status).

weapons of mass destruction.[10] Its 38 partner economies, which exclude the Philippines, adhere to common export policy guidelines applied to controlled items listed in the MTCR Equipment, Software and Technology Annex. Export transactions are considered on a case by case basis. The list includes a broad range of military and dual-use equipment and technology relevant to missile development, production, and operation.[11]

To address security concerns, the equipment manufacturer is required to install a mechanism in the equipment for it to automatically shut down if the company ever changes or increases measurements beyond a certain threshold. More recently though, the source economy has relaxed some export restrictions on condition that the firm informs them if the equipment is moved to another location or has its function changed. The equipment also has to be shipped back to the source economy if it is no longer needed for industrial production.

The same problem is faced by the firm in the importation of some materials. Export restrictions to the Philippines are in place in several source economies for materials such as steel beyond a certain threshold quality because they can be used to make missiles.

Related to the above issues is the need for firm to also trace the origin of raw materials to ensure that it does not source materials from economies that are under sanctions. However, once the system is in place, these policies are not considered burdensome by the firm because it benefits from the company group's supply chain management system.

2.5.3. *Availability of Skilled Labour*

Skilled and semi-skilled labours are very important for the firm because these employees are the backbone of the production process.

---

[10]More information about the Missile Technology Control Regime (MTCR) can be obtained at: http://www.mtcr.info/english/index.html.

[11]The MTCR Equipment, Software and Technology Annex can be accessed at: http://www.mtcr.info/english/annex.html.

Being close to the educational centre in the northern part of the Philippines, the firm does not encounter any significant problem with a potentially highly skilled supply of labour. However, the firm still needs to send its employees, many of them with a minimum three years of engineering education, for six months of vocational training at TESDA schools to become machinists. The firm's wage structure is competitive but from time to time some staff move to other firms in the Philippines or abroad. To minimize the disruption from employment attrition, the firm provides inputs in TESDA's vocational training program so that TESDA graduates, especially those that have been selected for paid internships with the firm, become a reserve pool of labour. Once hired at an entry-level position, it still takes three to five years of learning-by-doing to reach an expert level for machinists. The firm provides possibilities for promotion and career growth in order to keep good workers, especially because it takes years to build the necessary expertise in their workforce.

2.5.4. *Labour, Environment, Health and Safety Standards Compliance*

A requirement for operating manufacturing facilities in many economies is certification and licensing by government agencies in areas such as EHS standards and also labour standards. In the Philippines, environmental issues are overseen by the Department of Environment and Natural Resources (DENR),[12] while labour issues (including the health and safety of workers) are overseen by the Department of Labour and Employment (DOLE).[13] In addition to inspections conducted by these two departments, the firm is also inspected by its estate manager, the Philippine Economic Zone Authority (PEZA). The firm has no major issues with regard to these regulations because it follows the high environmental and

[12]Details of laws and policies overseen by the DENR can be obtained at: http://www.denr.gov.ph/laws-and-policies.html.

[13]More details about the Labor Codes of the Philippines overseen by the DOLE can be obtained at: http://www.dole.gov.ph/labor_codes.

labour standards of its parent company. Nevertheless, in the case of environmental policies, for example, it noted shortcomings in government regulation whereby companies observe strict environmental standards even as "jeepneys"[14] ply the road belching pollution seemingly without sanction.

2.5.5. *Electricity Supply and Cost*

Constant electricity supply is indispensable for the firm's manufacturing operations but the firm has had three outages in the last seven months and had to rely on its standby generator during the period. While the firm does not consider the cost of electricity a core issue, greater reliability and lower costs would be helpful. Electricity costs in the Philippines are among the highest in ASEAN.

2.5.6. *Limited Capacity of SMEs*

Private standards are rarely an issue for established industries, especially for multinational companies which in fact participate in crafting industry standards. But for the aviation industry, regulations abound. For example, for some aerospace customers, firms in its value chain must adhere to the many standards and requirements set by the National Aerospace and Defense Contractors Accreditation Program (Nadcap) at different stages of the manufacturing process. Nadcap is a private-led approach to establish requirements for accreditation, accredit suppliers and define operational program requirements. It provides independent audits for various activities, which it has grouped into 17 categories, among which is conventional machining and welding.[15]

Except for the time and resources spent on preparing for and attending to regular and multiple standards- or accreditation-related audits, the firm does not encounter any significant issues in

[14]Typical public transportation vehicle in the Philippines.

[15]Complete list of auditable activities by Nadcap can be obtained at: http://p-r-i.org/nadcap/accreditation/.

meeting the requirements of the certification programs. However, this is not the case for its suppliers, in particular SMEs, which also need to be certified if they want to be part of the firm's manufacturing value chain. The scope covered by each of the 17 categories of Nadcap standards is broad and requires the audited firm to demonstrate compliance with different category-specific checklists. Indeed, several of the firm's suppliers have found the process too complex and costly and have withdrawn, or not ventured to attempt to supply the value chain in the first place.

The firm's product requirements likewise sometimes call on suppliers to acquire special kinds of very costly equipment. Considering the relatively small market in the Philippines for products that need the same equipment, local metal fabricators cannot justify such big capital expenditures. In short, local companies have a hard time inserting themselves into this value chain because of the costs involved.

2.5.7. *The Way Forward*

This case study has endeavoured to present an aviation component manufacturing firm's perspectives on the importance of services in its operations. Policies clearly impact the firm's access to and use of services as well as its ability to provide them. Policymakers can therefore play a significant role in supporting businesses by providing the right enabling environment.

A case in point, which can serve as a model, is the firm's positive experience in its dealings with the government, including the economic zone management authorities. The firm has open communication channels with the relevant government agencies, including the Department of Trade and Industry, the Board of Investments, the PEZA and the TESDA, ensuring that issues brought to their attention are generally resolved. The problem-solving mechanisms in place to address the case study firm's concerns should be the norm for the productive sector as a whole, regardless of firm size.

However, there are also areas where the policy environment could be improved. Concerns over security issues, for example, lead to export restrictions for equipment and raw materials with dual functions. Though not completely insurmountable, restrictions nevertheless add to costs and cause delays in companies' operations. By being part of international agreements, host economies can facilitate the importation of security-related products and equipment needed for manufacturing.

Likewise, the noodle soup of certification requirements and regulatory bureaucracy sometimes results in deadweight losses for the firm and the economy, particularly if these are repeated several times for different accrediting institutions and for different economies. Mutual recognition of many of these accreditations could minimise such costs. Host government participation in international agreements, whether multilateral, bilateral, or among a few economies, lessen the need for numerous international audits and help improve the business environment, for both foreign and domestic companies.

If MNCs find some certification processes burdensome, the situation can only be worse for SMEs with far less resource capacity to meet the costs of repeated audits. More importantly, to be certified and participate in GVCs often requires that SMEs upgrade their capacity and purchase costly equipment or technologies. Lack of capital and access to finance remain major hurdles for their insertion in GVCs.

APPENDIX A

Table A.1: Pre-manufacturing Stage Including Sourcing and Importation of Raw Materials.

Service	Corresponding CPC ver. 2 code	Supplied in-house	Outsourced to affiliated companies and reasons	Outsourced to third-party suppliers/ government and reasons	Bundled
Pre-manufacturing stage					
1 Product research and development to facilitate transition of production from design centres to manufacturing facility	81129 — Research and experimental development services in other engineering and technology	Yes	No	No	
	81400 — Research and development originals	Yes	No	No	
2 Product design services during transition from design centres to manufacturing facility	83920 — Design originals	Yes	No	No	
	83912 — Industrial design services	Yes	No	No	

3	Engineering services during transition from design centres to manufacturing facility	83310 — Engineering advisory services	Yes	No	No	
		8332 — Engineering services for specific projects	Yes	No	No	
4	Procurement services	83116 — Supply chain and other management consulting services	Yes	No	No	
5	Customs clearance services and logistics of raw materials	67110 — Container handling services	No	No	Yes, efficiency; strong relationship with government agencies; government services	Bundled as part of the logistics agreement
		85999 — Other support services n.e.c.	No	No	Yes, efficiency; strong relationship with government agencies; government services	Bundled as part of the logistics agreement

(*Continued*)

Table A.1: (*Continued*)

Service	Corresponding CPC ver. 2 code	Supplied in-house	Outsourced to affiliated companies and reasons	Outsourced to third-party suppliers/ government and reasons	Bundled
6 Technical testing of raw materials	83441 — Composition and purity testing and analysis services	Yes	No	No	
7 Transport services of raw materials	651 — Land transport services of freight	No	No	Yes, efficiency; economies of scale	Bundled as part of the logistics agreement
	652 — Water transport services of freight	No	No	Yes, efficiency; economies of scale	Bundled as part of the logistics agreement
	6531 — Air transport services of freight	No	No	Yes, efficiency; economies of scale	Bundled as part of the logistics agreement
	67910 — Freight transport agency services and other freight transport services	No	No	Yes, efficiency; economies of scale	Bundled as part of the logistics agreement

8	Freight insurance of raw materials	71333 — Freight insurance services	No	No	Yes, not possible to supply in-house	
9	Storage and warehousing services of raw materials	67220 — Bulk liquid or gas storage services	No	No	Yes, efficiency	Bundled as part of the logistics agreement
		67290 — Other storage and warehousing services	Yes	No	Yes, efficiency	Bundled as part of the logistics agreement
10	Training services for staffs	92919 — Other education and training services, n.e.c.	Yes, learning by doing	No	Yes, in collaboration with TESDA	

Source: Authors.

Table A.2: Manufacturing Stage

	Service	Corresponding CPC ver. 2 code	Supplied in-house	Outsourced to affiliated companies and reasons	Outsourced to third-party suppliers/ government and reasons	Bundled
		Manufacturing stage				
11	Production administration — Production management	83115 — Operations management consulting services	Yes	No	No	
12	Maintenance and repair of factory equipment	87156 — Maintenance and repair services of commercial and industrial machinery	Yes	No	Yes to equipment suppliers	
13	Utilities (electricity, gas and water supply)	691 — Electricity and gas distribution (on own account)	No	No	Yes, not possible to supply in-house	
		692 — Water distribution (on own account)	No	No	Yes, not possible to supply in-house	
14	Manufacturing services provided in-house and by suppliers of activities such as basic machining	886 — Basic metal manufacturing services	Yes	No	Yes, cost consideration	

		887 — Fabricated metal product, machinery and equipment manufacturing services	Yes	No	Yes, cost consideration
15	Engineering services during manufacturing	83310 — Engineering advisory services	Yes	No	No
		8332 — Engineering services for specific projects	Yes	No	No
16	Warehousing services for intermediate goods	67220 — Bulk liquid or gas storage services	No	No	Yes, efficiency
		67290 — Other storage and warehousing services	Yes	No	Yes, efficiency
17	Quality control and assurance as well as compliance with ISO	8344 — Technical testing and analysis services	Yes	No	No
18	Product testing to obtain certification at export market	8344 — Technical testing and analysis services	No	No	Yes, required by laws and regulations
19	Sewage water treatment services	94110 — Sewerage and sewage treatment services	No	No	Yes, efficiency; not possible to supply in-house

(*Continued*)

Table A.2: (*Continued*)

Service		Corresponding CPC ver. 2 code	Supplied in-house	Outsourced to affiliated companies and reasons	Outsourced to third-party suppliers/ government and reasons	Bundled
20	Specialised cleaning services for machines and equipment	85340 — Specialised cleaning services	Yes	No	No	
21	Waste collection and recycling services	942 — Waste collection services	No	No	Yes, efficiency; not possible to supply in-house	

Source: Authors.

Table A.3: Post-manufacturing Stage.

Service	Corresponding CPC ver. 2 code	Supplied in-house	Outsourced to affiliated companies and reasons	Outsourced to third-party suppliers/government and reasons	Bundled
Post-manufacturing stage					
22 Certification and commissioning services of equipment	8344 — Technical testing and analysis services	Yes	No	No	
23 Packaging services	83919 — Other specialty design services	No	No	Yes, efficiency	Bundled as part of the logistics agreement
	85400 — Packaging services	No	No	Yes, efficiency	Bundled as part of the logistics agreement
24 Warehousing services for products	67220 — Bulk liquid or gas storage services	No	No	Yes, efficiency	Bundled as part of the logistics agreement

(Continued)

Table A.3: *(Continued)*

Service	Corresponding CPC ver. 2 code	Supplied in-house	Outsourced to affiliated companies and reasons	Outsourced to third-party suppliers/ government and reasons	Bundled
	67290 — Other storage and warehousing services	Yes	No	Yes, efficiency	Bundled as part of the logistics agreement
25 Customs clearance services and logistics of products	67110 — Container handling services	No	No	Yes, efficiency; strong relationship with government agencies; government services	Bundled as part of the logistics agreement
	85999 — Other support services n.e.c.	No	No	Yes, efficiency; strong relationship with government agencies; government services	Bundled as part of the logistics agreement

26	Transport services of products	651 — Land transport services of freight	No	No	Yes, efficiency; economies of scale	Bundled as part of the logistics agreement
		652 — Water transport services of freight	No	No	Yes, efficiency; economies of scale	Bundled as part of the logistics agreement
		6531 — Air transport services of freight	No	No	Yes, efficiency; economies of scale	Bundled as part of the logistics agreement
		67910 — Freight transport agency services and other freight transport services	No	No	Yes, efficiency; economies of scale	Bundled as part of the logistics agreement
27	Freight insurance of products	71333 — Freight insurance services	No	No	Yes, not possible to supply in-house	Bundled as part of the logistics agreement

Source: Authors.

Table A.4: After-sales Services.

Service	Corresponding CPC ver. 2 code	Supplied in-house	Outsourced to affiliated companies and reasons	Outsourced to third-party suppliers/government and reasons	Bundled
After-sales services					
28 Government certification and licensing to provide after-sales services	91134 — Public administrative services related to transport and communications	No	No	Yes, government services	
29 Telephone-based technical support services	85931 — Telephone call centre services	Yes	No	Yes, efficiency	
30 Warranty, MRO services	8714 — Maintenance and repair of transport machinery and equipment	Yes	No	Yes, partnership with service providers for aviation industry	
	Options contract for parts supply	Yes	No	No	
31 Travel services for engineers and other staffs pertaining to after-sales services	8551 — Reservation services for transportation	No	No	Yes, efficiency	

		85521 — Reservation services for accommodation	No	No	Yes, efficiency
32	Visa and immigration services for staffs	91290 — Public administrative services related to other public order and safety affairs	No	No	Yes, government services

Source: Authors.

Table A.5: Business Processes (Back Office Support).

Service	Corresponding CPC ver. 2 code	Supplied in-house	Outsourced to affiliated companies and reasons	Outsourced to third-party suppliers and reasons	Bundled
Business Processes (Back Office support)					
33 Company registration and licensing services (obtaining permit to operate)	91138 — Public administrative services related to general economic, commercial and labour affairs	No	No	Yes, government services	
34 Government licensing and inspections on fire prevention, health hazards, environmental protection and other aspects	91133 — Public administrative services related to mining and mineral resources, manufacturing and construction	No	No	Yes, government services	
	91290 — Public administrative services related to other public order and safety affairs	No	No	Yes, government services	
35 Headquarter services	83118 — Head office services	No	Yes, economies of scale	No	

36	Management services	83111 — Strategic management consulting services	Yes	Yes, economies of scale	No
37	Accounting, auditing and bookkeeping services	82210 — Financial auditing services	No	No	Yes, required by laws and regulations
		8222 — Accounting and bookkeeping services	Yes	No	No
38	Financial services	71121 — Deposit services to corporate and institutional depositors	No	No	Yes, not possible to supply in-house
		71313 — Group pension services	No	No	Yes, not possible to supply in-house
39	Legal services	82120 — Legal advisory and representation services concerning other fields of law	No	No	Yes, lack of expertise
		82130 — Legal documentation and certification services	No	No	Yes, lack of expertise

(*Continued*)

Table A.5: (*Continued*)

Service	Corresponding CPC ver. 2 code	Supplied in-house	Outsourced to affiliated companies and reasons	Outsourced to third-party suppliers and reasons	Bundled
40 Insurance services (commercial life and accident/health insurance, property insurance for the factory compound, product quality insurance, management liability insurance)	7131 — Life insurance and pension services	No	No	Yes, not possible to supply in-house	
	7132 — Accident and health insurance services	No	No	Yes, not possible to supply in-house	
	71334 — Other property insurance services	No	No	Yes, not possible to supply in-house	
	71335 — General liability insurance services	No	No	Yes, not possible to supply in-house	

41	Human resources services	91320 — Administrative services related to government employee pension schemes; old-age disability or survivors' benefit schemes, other than for government employees	Yes	No	Yes, government services
		91330 — Administrative services related to unemployment compensation benefit schemes	Yes	No	Yes, government services
		8511 — Personnel search and referral services	Yes	No	No
		8512 — Labour supply services	No	No	Yes, network economies
		83113 — Human resources management consulting services	Yes	No	Yes, lack of expertise
42	Business and management consultancy services	8311 — Management consulting and management services	No	No	Yes, lack of expertise

(Continued)

Table A.5: (*Continued*)

Service	Corresponding CPC ver. 2 code	Supplied in-house	Outsourced to affiliated companies and reasons	Outsourced to third-party suppliers and reasons	Bundled
43 Corporate communications and public relationship	83114 — Marketing management consulting services	No	No	Yes, economies of scale	
	83121 — Public relations services	Yes	No	No	
44 Courier, postal and local delivery services	681 — Postal and courier services	No	No	Yes, not possible to supply in-house	
45 Information technology services	8313 — Information technology (IT) consulting and support services	Yes	No	No	
	8314 — Information technology (IT) design and development services	Yes	No	No	
	83151 — Website hosting services	No	No	Yes, lack of expertise	
	8316—IT infrastructure and network management services	Yes	No	No	

46	Telecommunication services	841 — Telephony and other telecommunications services	No	No	Yes, not possible to supply in-house
		84210 — Internet backbone services	No	No	Yes, not possible to supply in-house
		8422 — Internet access services	No	No	Yes, not possible to supply in-house
47	Uniform design, alteration and laundry	83919 — Other specialty design services	No	No	Yes, efficiency
		97130 — Other textile cleaning services	No	No	Yes, efficiency
48	Transport services for staffs	641 — Local transport and sightseeing transportation services of passengers	No	No	Yes, efficiency; not possible to supply in-house
49	Property management services	72212 — Non-residential property management services on a fee or contract basis	No	No	Yes, efficiency; lack of expertise
50	Medical services	93121 — General medical services	No	No	Yes, lack of expertise

Table A.5: (*Continued*)

Service	Corresponding CPC ver. 2 code	Supplied in-house	Outsourced to affiliated companies and reasons	Outsourced to third-party suppliers and reasons	Bundled
51 Catering services	63393 — Other contract food services	No	No	Yes, lack of expertise	
52 Security services	85230 — Security systems services	Yes	No	Yes, lack of expertise	
	85250 — Guard services	No	No	Yes, lack of expertise	
53 Cleaning services	853 — Cleaning services	No	No	Yes, lack of expertise	

Source: Authors.

APPENDIX B

Table B.1: Policies Affecting Services in the Value Chain.

Government and private policies/services	Authority(ies) in charge	Details	How the policy affects services in the value chain
Certification by regulatory agencies of individual economies for provision of after-sales services	Various jurisdictions	Different certification is needed to provide after-market services to carriers/airlines based in different jurisdictions and most need regular re-certification which ranges between 1 and 5 years	The policy restricts the coverage and extent of after-sales services that the firm can provide to its customers
Export restrictions for equipment and raw materials with dual function in source economy	Signatories of Missile Technology Control Regime (MTCR)	Export restrictions are applied by the 34 partner economies on controlled items listed in the MTCR Equipment, Software and Technology Annex. Partner economies consider each transfer on a case by case basis and may require additional steps to be taken by the importer before granting the transfer	Where the policy is not streamlined to the extent possible, it may unnecessarily restrict the firm's access to critical equipment and raw materials for its operations. It may also need to explore alternative options to access these items, hence increasing costs to the firm

(*Continued*)

Table B.1: (*Continued*)

Government and private policies/services	Authority(ies) in charge	Details	How the policy affects services in the value chain
Provision of labour training	Philippines TESDA	Firm collaborates with TESDA to select potential candidates and sends them to centres run by TESDA for 6-month vocational training	Best practice examples of how collaboration helps to minimise labour issues faced by the firm such as insufficient experience of new hires and replacement of ex-staff.
EHS and labour compliance inspections	Philippines DENR Philippines DOLE PEZA	Regular inspections are carried out by the different authorities on areas affecting environment, health as well as safety of the workers	Although the firm benefits from these inspections, it also incurs cost from having to prepare for the inspections. In addition, regulations are sometimes non-transparent and the firm is found not to have adhered to them although it intends to follow strictly all regulations
Electricity supply and cost	Philippines Department of Energy (DOE)	Firm has had three run-outs in the last 7 months and had to rely on its standby generator during the period. Electricity cost is competitive but does not adjust in response to firm's needs	Firm's manufacturing operations are affected by the run-outs and may have to adjust its delivery time if run-outs happen regularly

Audits for different categories of manufacturing activities are required	National Aerospace and Defense Contractors Accreditation Program (Nadcap)	The highly regulated nature of the aviation industry makes it necessary that the industry players and their suppliers be audited for various activities that they undertake during the manufacturing process.	The audit is a challenge for firm's suppliers, particularly SMEs, who may opt not to do business with the firm. This affects firm's outsourcing strategy and limit the pool of suppliers that firm has access to. Perhaps ways could be found of assisting SMEs to develop the necessary capacity.

Source: Authors.

REFERENCES

About Nadcap. Performance Review Institute. http://p-r-i.org/nadcap/about-nadcap/[26 March 2015].

Metalworking Industries Association of the Philippines, Inc. http://hdm.miapnational.com/[26 March 2015].

Missile Technology Control Regime. http://www.mtcr.info/english/[26 March 2015].

"Nadcap Audit and Accreditation Process." Performance Review Institute. http://p-r-i.org/nadcap/accreditation/[26 March 2015].

Philippines Department of Environment and Natural Resources. http://www.denr.gov.ph/[26 March 2015].

Philippines Department of Labor and Employment. http://www.dole.gov.ph/[26 March 2015].

Philippines Technical Education and Skills Development Authority. http://www.tesda.gov.ph/[26 March 2015].

Tinseth, R (2013). "Our supply chain," Randy's Journal (blog). http://www.boeingblogs.com/randy/archives/2013/02/supply_chain.html[26 March 2015].

Chapter 3

INDUSTRIAL WELDING SERVICES IN THAILAND

William Haines

3.1. FIRM BACKGROUND

The parent company of the firm under study is a global leader in the production of welding consumables; the design and manufacture of sophisticated welding equipment; and the provision of welding services. Due to a history of innovation, their products are market leaders and are protected by a significant portfolio of intellectual property rights. The parent company operates in over 150 economies and has majority shares in 30 subsidiaries.

The group's dual role as both manufacturer and service provider, and moreover as a service provider that relies on its manufacturing activities to gain competitive advantage, has impacted the geographic distribution of its subsidiaries. Traditionally, manufacturers invest in production facilities in locations that allow them to take advantage of factor prices, ease of trade, low transportation costs and possibly the local supply of certain inputs. Production is then concentrated in these areas to take advantage of scale and network economies. The goods may then be exported to meet demand globally. The movement of labour however is generally far more restricted than the movement of goods, and thus an *in situ* service provider must generally establish multiple local subsidiaries in order to compete in multiple markets, each of which must operate at a smaller scale.

Regionally, the parent company has worked with these conflicting constraints and incentives by creating a network of Asian service centres that also act as: (i) distributors of both the high-value welding equipment and consumables the group produces in Europe and the low-value welding consumables it produces in Asia and (ii) a specialised producer of one of the firms low-value products such as low-alloy cored wires, electrodes or wear plates to take some advantage of scale economies in production. Intra-group trade then supplies each subsidiary with the inputs they require for their service operations.

The subsidiary under study, located in Thailand, thus participates in three distinct but related value chains. Like most group members, it acts as: (i) a welding service provider, (ii) a distributor of welding wires and machines and (iii) a manufacturer of a range of the group's trademarked wear plates, which are exported globally. This specialisation was selected due to Thailand's mature steel industry, which are core inputs in the production of the wear plates. Thailand also has a mature cement industry, which is a big user of the firm's products.

The firm is a small, relatively capital-intensive company. Roughly half of its employees work in production and engineering, and half in administration and sales. The small number of staff and the range of the firm's activities mean that the three value chains are managed by the same individuals and make use of many of the same resources. Though the three deliverables are non-bundled and share only a few physical inputs, this resource sharing synergy minimises cost and allows the firm to focus its present activity on whichever market displays robust demand.

3.2. BASIC OPERATION OF THE VALUE CHAIN

This case study focuses on the value chain associated with the *in situ* provision of welding services, which accounts for approximately 70 per cent of the firm's revenue, and determined its classification as a Tier Two Service Provider upon its registration with the Thai Board of Investment (BOI). The *in situ* welding services most commonly supplied by the firm are cladding and hard facing. These are surface welding operations designed to increase

Figure 3.1: Cladding Under Way.

Source: Courtesy of the firm.

industrial components' resistance to abrasion and corrosion through the application of alloy metals. There are two basic physical inputs for an *in situ* cladding or hard facing operation (welding equipment and welding consumables) and two basic service inputs (engineers and technicians, and logistics and project management) (Fig. 3.1).

Other corporate members provide the physical inputs. All tasks associated with the production of the welding equipment necessary to the firm's operations are outsourced to group members in Europe and imported by the firm. This includes stock equipment for which the firm keeps an inventory, and bespoke equipment required for a specific contract. The key welding consumable is a product called flux cored wire. It represents the method for delivering the alloy to be fused to the surface under operation into the welding device. The production of the wires is outsourced to a group member in Malaysia and imported by the firm; again, this includes both inventories of common varieties and the production of novel alloy wires for unusual contracts. The production bases in

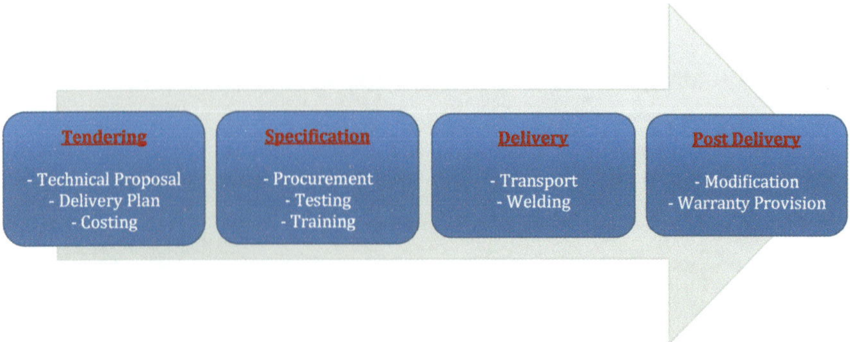

Figure 3.2: An *In Situ* Welding Service Value Chain.

Source: Author.

Europe and Malaysia not only produce customised versions of their products at the firm's request, but also design and test them based on the client's requirements as communicated by the firm. The firm is not involved in the research and design process.

The service inputs, as the firm's core business competencies, are provided in-house. The engineers and welding technicians who complete the *in situ* welding itself are all full-time employees of the firm. This is partly because the firm uses its own equipment and thus welders must be trained in its use, and partly as a quality-control measure. The firm's engineers complete the technical aspects of project management themselves, and the administrative staff handle the logistical elements. The ownership and driving of the trucks used to transport the firm's equipment and technicians however is outsourced.

We will define the value chain as beginning with the firm's decision to tender for a specific project, and as ending with any repairs the firm is obliged to provide under their warranty. For ease of analysis, the value chain may then be divided into the following stages (Fig. 3.2).

3.2.1. *Tendering*

While clients may simply invite the firm directly for small routine jobs, large contracts involve a tender, open to pre-qualified companies. This usually takes the form of a one-time blind auction, where contractors submit delivery plans including technical specifications, timelines and prices and one is then awarded the job. The firms key corporate partners in Europe and Malaysia may need to come on board at this stage if it is clear that customised machinery or unusual cored wires will be required for the job. If so their advice will be needed in the writing of the technical proposal.

In some cases the tender takes the form of a bidding contest, where contractors are aware of the prices their competitors are offering and are given the opportunity to undercut them. Bidding contests are more common in technologically simple jobs where price is the client's key determinant in awarding the contract. The firm is not competitive in this market, due to its capital intensity and highly trained personnel. Rather, it tenders for complex jobs that require a degree of equipment and process customisation and a high level of technological sophistication. In these cases, the quality of the technical plan is the key determinant and so the tendering mechanism is designed to produce detailed technical plans rather than fierce price competition.

3.2.2. *Specification*

Upon being awarded the contract, the firm's plan undergoes a series of tests, the results of which are discussed with the client. Before these tests can be completed other group members must construct any bespoke equipment or alloy wire and local staff must be trained in its use. Mock-up test results are then examined by the firm's in-house metallurgic laboratory, and process inspections carried out by the client. Throughout this process, results are shared with the client and the price continuously negotiated as the client requests process and/or input adjustments based on the test results.

3.2.3. *Delivery*

Once the client is satisfied with the product and a price has been agreed, the welding operation itself begins. Because the firm uses its own machinery and its own engineers and technicians, this often involves a substantial logistics operation. The equipment, personnel and welding consumables must be transported to the site, housed for the duration of the project and returned to the firm's factory on completion.

3.2.4. *Post-delivery*

Upon the delivery of the service, the firm will adjust its process and its machinery's software in consultation with the client. Though these are usually minor adjustments the firm believes they are important in developing its reputation and network of clients, with whom repeated business is crucial for growth. The firm also provides a warranty for its welds. Inspection is the responsibility of the client, but upon the identification of an issue the firm redeploys a team to address it.

3.3. SERVICES ALONG THE VALUE CHAIN

From the preceding discussion of the value chain, it is clear that there are a substantial number of service inputs required in the fulfilment of a welding contract. If we extend the analysis one stage further upstream then the number expands dramatically for two reasons.

The first is that automated welding equipment and cored wires are highly knowledge-intensive goods. Each requires substantial amounts of scientific and engineering research, both into the product itself and into the method of fabrication, as well as legal support in obtaining intellectual property rights. As they are often customised for a specific job, further research and testing are required on an ongoing basis. This means that their value is largely constituted by services entering into their production.

The second is that the welding technicians themselves are the product of a number of service inputs. Some of these, such as basic training, are necessitated by the technical nature of their work; however, heavy government regulation of services and industry standards add to the number of service inputs required in the creation of a ready-to-work welding technician. These might include the design and administration of professional examinations, licensing procedures and health and safety inspections. Even getting onto their client's approved vendor list, which allows the firms to tender for a job, requires technical inspections and audits.

Thus a value chain consisting fundamentally of two physical and two service inputs is revealed to involve large numbers of distinct services when the production of these four inputs is taken into account. A total of 38 services were identified as being involved in the provision of an *in situ* welding application, as documented in Table A.1 of Appendix.

3.3.1. *Outsourcing and Bundling*

Figure 3.3 shows that of the 38 services entering the value chain 25 are at least partially outsourced, 7 to other group members and 18 to independent contractors.

The elements outsourced to independent contractors fall into categories that are widely acknowledged as profitable outsourcing opportunities: back office functions, professional services, low value services and the ownership of certain physical assets. The

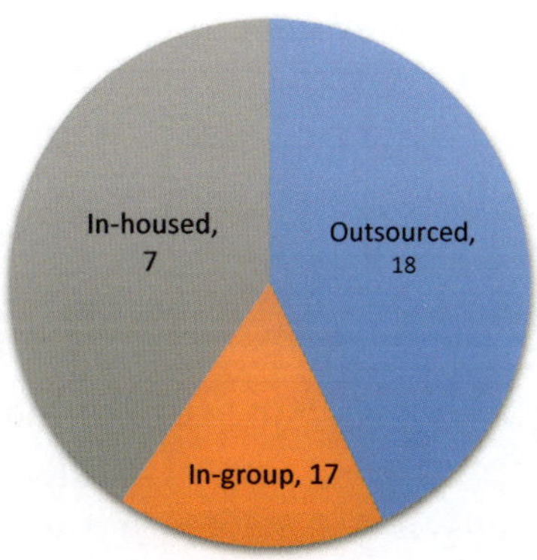

Figure 3.3: Modes of Supply by Number of Services.

Source: Author.

key motivations for outsourcing these elements were cost minimisation and legal obligation.

The cost minimisation achievable by outsourcing is primarily generated in one of three ways: the spreading of investment costs; the increased efficiency gained by specialisation and experience; and the efficiency associated with buying a bundle. One classic example is that by outsourcing the ownership and operation of the trucks used in the firm's transport operations the firm avoids incurring an investment cost for underutilised assets. A specialist trucking firm will be able to use its trucks more efficiently and cover its investment costs more quickly than one which owns the trucks for occasional use. This is an economy of scale effect brought about by spreading the fixed costs of production over a larger quantity of output.

On the other hand, the outsourcing of customs-related services to forwarding agents is made profitable by the time savings implicit in their network of relationships and detailed knowledge of procedures. As the agent increases output these relationships are cemented further, the agent becomes more efficient in navigating the regulatory landscape and transactions costs may also fall. This is also an economy of scale effect but is brought about through the agent's production function displaying greater dynamic efficiency increases than the firms. It is also worth noting that as the practice of outsourcing customs-related tasks is widespread, this may be considered a form of an external economy of scale.

An example of a service being outsourced due to the attractiveness of a bundle is the firm's contracting of an independent logistics firm to handle its international freight requirements. Importing the cored alloy wires from Malaysia requires the coordination of multiple transport stages with customs procedures and related delays. Rather than dealing with a trucking firm in Malaysia, a Malaysian customs agent, a shipping company, a Thai customs agent and then a Thai trucking company, it is much more convenient to buy a bundle from a contractor who can simply guarantee delivery.

Bundling is much more fundamental to the firm's business model than this sort of "outsourcing for convenience", however.

Though a service provider, the firm's competitive advantage is gained primarily through its use of sophisticated welding equipment and bespoke cored alloy wires, neither of which it produces nor is capable of producing. As mentioned previously, the firm is not price competitive when it comes to run-of-the-mill welding contracts, but rather competes for complex jobs that require technological sophistication and often a degree of product customisation. The firm is only able to supply these thanks to its ability to outsource the design, testing and production of bespoke manufactured goods to its corporate partners. This is an example of a service firm that gains much of its value through its use of a specific set of manufactured goods, which it imports bundled with technical advice and R&D.

It would not do the situation justice to invert the analysis and think of the parent firm as outsourcing the use of its equipment to the subsidiary in the hope of simplifying matters; the relationship is too deep for that. The firm maintains a well-trained staff of technicians, some of whose training is provided by the parent firm specifically in the use of its machinery. Simply outsourcing the equipment to independent welding firms would not provide clients the same level of value as the firm is currently offering them. Furthermore, if local independent contractors were more comfortable using competitors' equipment, the parent might lose market share. The solution to this is an intimate relationship with a local service provider; its subsidiary.

This analysis reveals the heavily bundled nature of the final deliverable, a sophisticated welding service. Strategic bundles can give firms competitive niches where the impersonal outsourcing of input manufacturing or post-production services could not. This is a testament to the parent firm's business model, which creates synergistic outsourcing relationships between group members, but with sufficiently complete contracts, there is no reason this model could not be emulated outside corporate structures.

More generally, it reveals that the disaggregation of a final deliverable's value into the contributions made by goods and services components may not tell the whole story, as the disaggregated value contributions may not sum to the value of the final deliverable. The residual is in part a measure of the synergy created by the bundling of the goods and services.

3.4. POLICIES AFFECTING SERVICES ALONG THE SUPPLY CHAIN

The firm is registered with the BOI as a tier two services firm. This status gave them a corporate tax holiday for seven years, which has now ended, and allowed their parent firm to take a majority position in the firm's ownership. This would not have been legal for a services firm in Thailand otherwise. Their BOI status also exempts them from import duties on the goods they import from corporate partners in Europe, or on those they import from the group member in Malaysia (though due to the ASEAN Economic Community these would not be taxed regardless of the firm's BOI status).

Though no doubt partly due to these exemptions, the firm believes that import duties would not substantially alter their supply chains were they obliged to pay these charges. What minimal frustrations the firm did express regarding trade in goods regulation was process based. For example, the firm explained that in general shipping the cored wires they import from Malaysia is slower but significantly cheaper than trucking them. Nevertheless, there are times when trucking is favourable, as the wires might be needed urgently. The firm does this less often than it would on a purely cost-determined basis; however, because the two-day delay in Singapore as trucks bringing the wire from the Malaysian firm are unloaded and reloaded onto ships correlates perfectly with the time taken for the necessary paperwork to be approved by customs. When trucking the whole way the cargo is simply held up by this two-day administrative delay, and thus the extra cost incurred for the sake of speed is partially wasted.

Of far greater concern to the firm is the regulation of labour movement, which is quite severe. A foreign worker must obtain both a work permit and a non-immigrant visa; these require the submission of 11 supporting documents, are processed slowly and are ultimately awarded on a discretionary basis. The conditions relate the number of foreign workers a firm may hire to the firm's registered paid up capital, tax payments and number of local workers employed; a 10:1 ratio being required. As welding is a protected industry, even if the formal requirements are met, the permits may not be granted due to the discretionary nature of the process.

Overall the foreign worker permits make it difficult to the point of impossible to bring in trained technicians from other group members in order to meet spikes in demand. The firm cannot outsource such work to local technicians, as they are not trained in the use of its equipment. As the demand is for sophisticated services, which local firms may not be able to deliver, this generally results in the demand going unmet rather than in local welders being employed to meet it. Even ASEAN regulations, which allow 30-day tourist visitations, do not make short-term work permits more easily obtainable. In a recent government initiated policy consultation the firm highlighted this issue as its primary obstacle to local expansion.

Regional expansion is made even more difficult by the fact that neighbouring economies do not recognise Thai government-issued licenses or safety inspections. In Thailand, welders must pass a professional exam based on the industrial standards laid out by the International Institute of Welding (IIW). On top of this, welders must be trained in Thai OH&S (occupational health and safety) practices, which require several days worth of man-hours and an equipment audit.

Overall, these processes represent a significant opportunity cost to the firm. Though some neighbouring economies also use IIW standards, they do not recognise each other's government licenses, OH&S training or equipment audits. The firm cannot afford to send its welders and equipment away to repeat the licensing and OH&S processes and thus its capacity to operate in neighbouring economies is severely limited. Internationally consistent industrial standards, along with mutual recognition of licenses and safety training based on these standards, is regarded by the firm as the policy reform that would most facilitate additional business through trade.

Interestingly, this reveals that the greatest obstacle to trade in welding services is not trade regulations as such, but the lack of harmonisation or mutual recognition in domestic industrial regulations. Even if worker permits were easily obtainable, licensing barriers would remain. It is also significant to note that registration with the BOI, which has all but eliminated tax-related barriers, does nothing to alleviate these difficulties, which have not been the focus of policymakers.

A tabulation of these policy interfaces can be found in Table A.2 of Appendix.

APPENDIX

Table A.1: Services Entering the Value Chain and their Outsourcing Status.

Service	CPC Code	Mode of Supply			Nature of Contractor	Reason for Mode of Supply
		In-house	In-group	Outsourced		
		Establishment stage				
Government liaison services	Division 9113 — Public administrative services related to the more efficient operation of business	✓		✓	BOI	The BOI acts as a facilitator of the firm's interactions with various government bodies
Company registration and licensing services	91138 — Public administrative services related to general economic, commercial and labour affairs	✓				
Visa and immigration services	91290 — Public administrative services related to other public order and safety affairs	✓				

(*Continued*)

Table A.1: (*Continued*)

Service	CPC Code	Mode of Supply			Nature of Contractor	Reason for Mode of Supply
		In-house	In-group	Outsourced		
Professional training/exams	92919 — Other education and training services	✓		✓	Industrial standards organising body	Necessitated by law
Safety standards and inspection	91290 — Public administrative services related to other public order and safety affairs			✓	Government agency	Necessitated by law
Personnel search and referral services	85112 — Permanent placement services, other than executive search services	✓				
Pre-production stage — Services relating to physical inputs						
Procurement agent	85999 — Other support services n.e.c.	✓				
Customs-related services	85999 — Other support services n.e.c.			✓	Customs agent	Time and cost savings brought about by specialis-ation

Quality assurance (of raw materials)	83441 — Composition and purity testing and analysis services		✓		Group members	Bundled with the supply of physical inputs
Freight transportation	Group of 65 — Freight transport services			✓	International logistics firm	Time and cost savings brought about by specialisation
Storage of raw materials	67290 — Other storage and warehousing services	✓				
Design of welding equipment	81129 — Research and experimental development		✓		Group members	Technical capacity
Design of welding consumables	81129 — Research and experimental development services in other engineering and technology		✓		Group members	Technical capacity

(*Continued*)

Table A.1: (*Continued*)

Service	CPC Code	Mode of Supply			Nature of Contractor	Reason for Mode of Supply
		In-house	In-group	Outsourced		
Production stage — Services used directly in production						
Welding process design	83920 — Design originals	✓	✓		Group members	Certain process design is started by other subsidiaries and later shared with the rest
Intellectual property acquisition	83960 — Trademarks and franchises n/a		✓		Group members	Bundled with the supply of physical inputs
Quality assurance	83441 — Composition and purity testing and analysis services	✓	✓		Group members	The firm provides quality assurance for its own welds, but the quality of the alloy produced by group members is assured by the producer

Compliance management	83190 — Other management services, except construction project management services	✓				
Testing and trialling including laboratory testing	83449 — Other technical testing and analysis services	✓	✓		Group members	The firm does metallurgic testing as part of its testing regime, the research process of creating a bespoke alloy wire is carried out by the group member producing these wires
Cleaning services (factory and warehouse)	85330 — General cleaning services			✓	Independent cleaning contractor	Cost savings

(Continued)

Table A.1: (*Continued*)

Service	CPC Code	Mode of Supply			Nature of Contractor	Reason for Mode of Supply
		In-house	In-group	Outsourced		
Security services	85250 — Guard services			✓	Independent security contractors	Cost savings
Waste treatment	Group of 942 — Waste collection services			✓	Independent waste treatment firm	Cost savings and technical expertise
Repair and maintenance services of machines and equipment	871 — Maintenance and repair services of fabricated metal products, machinery and equipment	✓				
Logistics	85519 — Other transportation arrangement and reservation services n.e.c.	✓				
Truck hire	660 — Rental services of transport vehicles with operators			✓	Independent trucking company	Cost savings
Welding services	8332 — Engineering services for specific projects	✓				

Utilities	Group of 863 — Support services to electricity, gas and water distribution		✓	Local utilities providers	Cost savings and technical expertise
Work gear design	882 — Textile, wearing apparel and leather manufacturing services		✓	Specialised work gear manu-facturers	Cost savings and technical expertise
Laundry services	97130 — Other textile cleaning services		✓	Independent laundry contractors	Cost savings
Accommodation services	632 — Other accommodation services for visitors and others	✓		Independent hotels	Cost savings
Back Office					
Auditing services	Group of 822 — Accounting, auditing and bookkeeping services	✓		Accounting firm	Necessitated by law
Insurance services	Group of 713 — Insurance and pension services (excluding reinsurance services)	✓		Insurance firm	Cost savings and risk assurance

(Continued)

Table A.1: (*Continued*)

Service	CPC Code	Mode of Supply			Nature of Contractor	Reason for Mode of Supply
		In-house	In-group	Outsourced		
Accounting services	8222 — Accounting and bookkeeping services	✓				
Banking services	71121 — Deposit services to corporate and institutional depositors			✓	Bank	Cost savings and risk assurance
Legal services	82120 — Legal advisory and representation services concerning other fields of law			✓	Legal partnership	Cost savings and technical expertise
General management	8311 — Management consulting and management services	✓				
Communications and marketing	82130 — Legal documentation and certification services		✓		Group members	

Estate management	72112 — Rental or leasing services involving own or leased non-residential property			✓	Owners of the industrial estate on which the firm's factory is located	Cost savings and legal benefits of locating on a designated industrial estate
IT and information system management	8316 — IT infrastructure and network management services	✓				

Source: Author.

Table A.2: Government Policies and Services Affecting the Supply Chain.

Government Policy/ Service	Authority(ies) In-charge	Details	How the Policy Affects Services in the Value Chain
Worker safety — Occupational health and safety licenses and inspections	Provincial Labour Office under the guidance of the Department of Labour Protection and Welfare	Inspections are carried out on the basis of an undisclosed annual planning system or on receipt of a complaint. They consist of two stages, a general conditions inspection and then OH&S inspection. SMEs may self-report by filling in a 50 question survey. Inspections are only carried out if less than 70% of the answers indicate compliance	The process is costly enough that the need to do it repeatedly for different regulatory regimes acts as a barrier to international expansion
Foreign worker permits	Ministry of Labour	Foreign workers must apply for "non-immigrant visa B" and foreign worker permit. This requires the provision of proof of education, blood type and medical fitness as well as official applications forms. Also requires exiting Thailand to activate on re-entry	As there is no exemption despite BOI registration, even for workers employed by another corporate partner, the firm finds it very difficult to meet spikes in demand as it cannot outsource work to local welders due to specialised training requirements

Professional Standards	The IIW has an Authorised National Body in Thailand, which is responsible for monitoring Approved Training Bodies, such as the Thai Institute of Welders	The IIW-International Authorisation Board (IAB) Qualification System is recognised by the ISO under 14731 — Welding coordination. The oversight of industry standard compliance is the responsibility of the Thai Industrial Standards Institute. The system involves the training of all personnel, the auditing of equipment and ongoing safety procedure compliance	Meeting these standards is considered very costly by the firm as it requires not only travel expenses and man hours but represents an opportunity cost as machinery must be out of action while being audited. The process is too costly to carry out multiple times and thus prevents international expansion
Customs Clearance Process	Thai, Malaysian and Singaporean Customs Authorities	Standard border customs processing procedures	At-the-border measures had a minimal impact on the firms operations. However, there were cases of transport methods being chosen on the basis of correlation with administrative inefficiencies which caused delays in the supply of imported inputs

(Continued)

Table A.2: (*Continued*)

Government Policy/ Service	Authority(ies) In-charge	Details	How the Policy Affects Services in the Value Chain
BOI Exemptions	BOI	Thailand limits foreign investment in the services sectors through the Foreign Business Act of 1999 (FBA) and other sector-specific legislation. Currently, foreign investment in most services sectors is subject to a 49% equity cap (with only a case-by-case provisional exemption). The BOI provides exemption from these laws however on a discretional basis	The crucial impact of BOI support was that it allowed the formation of the company in the first place
Industrial estate	Industrial Estate Authority of Thailand	The firm is located on a registered Industrial Estate managed by a private firm	The Industrial Estates allow the firm to benefit from network economies and bundles. The clustering of industry into estates ensures that utilities are efficiently and reliably supplies and that good road access can be ensured. For their fee the firm received not only the right to locate themselves on the estate but received waste removal services, security services, electricity and water supply bundled into the cost

Source: Author.

Chapter 4

MANUFACTURING OF MINING AND CONSTRUCTION EQUIPMENT

David Sit and Patrick Low

4.1. BACKGROUND OF THE FIRM

The Japan-based company under study is a key global manufacturer of mining, construction and utility equipment. These products account for some 90 per cent of the firm's production, with the rest comprising industrial machinery and other equipment. The company operates in over 150 economies, with around 50 manufacturing/assembly facilities and 200 distributors located around the world. The company has 179 subsidiaries involved in all manner of goods and services activities. Some of the group companies are fully owned subsidiaries and some are part of the conglomerate by virtue of equity holdings.

The company has global sales of more than USD 17 billion, in a global market estimated to be worth more than USD 170 billion (Businessvibes, 2013). In descending order of importance, Japan, North America, Latin America and Asia (other than China) represent over 60 per cent of the firm's market. China accounts for an additional 9 per cent.

According to industry estimates, the world market for the firm's main products has been growing at about 6.5 per cent on average for a number of years and is set to continue in a range around this rate. Growth is expected to be strong in Asia in the coming years.

The company engages in activities at all stages of the value chain, which include design, manufacturing, assembly, distribution, remanufacturing, and after-sales support services to customers. This case study will follow the value chain for manufacturing heavy machinery from production to sales and post-sales service (Fig. 4.1).

The rest of the study is structured as follows. Section 4.2 will briefly describe the value chain for manufacturing, selling and post-sales services for heavy machinery. Section 4.3 will discuss a potential list of services which the company may have recourse to in its value chain and examine the attributes of services in relation to outsourcing, bundling and innovation. Section 4.4 will discuss polices that affect the firms value chains, specifically in relation to foreign investment constraints, labour restrictions, customs regimes and re-manufacturing restrictions. Information and data are collected both in interviews with the executives of the company as well as from publicly available information.

4.2. DESCRIPTION OF THE VALUE CHAIN

4.2.1. *Contours of the Pre-production and Manufacturing in the Value Chain*

Before the value chain studied here begins, the company must set up assembly plants, distribution centres and sales centres in different economies. In many cases, the parent company's subsidiaries that set up these facilities do so in partnership with local businesses in the markets they intend to serve. A range of professional services, such as legal advices and business consultancy are outsourced to specialists at this stage of the operation.

Where the company's value chain begins with the procurement of raw materials (Fig. 4.2), key components — such as engines, hydraulic valves and motors — will have been manufactured in one of the company facilities from metal, plastic, chemicals and other necessary inputs. These parts are usually technologically advanced components which define the cutting edge of machine capabilities. As a good deal of proprietary technology is involved, the company often prefers to manufacture in Japan. There is also the question of precision engineering and the need to guarantee the highest quality,

Figure 4.1: Earth Moving, Construction and Moving Equipment.

(*Upper* left: Bulldozer; Upper right: Wheel Loader; Lower left: Harvester; Lower middle: Mining Equipment; Lower right: Hybrid eccavator

Source: Courtesy of the firm.

which also militates in favour of in-group fabrication. It is estimated that plants in Southeast Asia import 30 per cent of their inputs from Japan (excluding key components). The other parts may be sourced from third-party suppliers as well. All inputs are transported by logistics services providers to assembly plants and stored in nearby warehouses.

Assembly then takes place in manufacturing plants that put together the parts and build the machines. Their fundamental policy is to assemble where there are big markets. Key services, such as quality control of raw materials, quality control of goods, warehousing and storage, production administrations and other manufacturing supporting services enter the value chain at this stage.

4.2.2. *Post-manufacture Sales-related Activities*

After assembly of the machines is completed and products have passed quality control inspections, they will be dispatched to distributors. The company under study maintains three different channels of distribution around the globe. In mining economies such as Australia, Chile and South Africa, the company owns and operates distributors, which allows them to better meet the needs of large mining corporations, who are their major customers. In North America, Europe, and China, the company has regional headquarters which oversee manufacturing operations and sales activities by regional distributors. In the other areas, the company appoints local companies as distributors and sells the machines to them directly from its global production facilities. Market size and the nature of the customers determine the model of operations.

Sales arrangements can be customised for different buyers. Some customers may purchase the machines outright. Others may enter into various financial arrangements offered by the finance arm of the company. Across the range of these financial options, ownership may pass to the customer over time (e.g., through hire purchase), or the company may retain ownership.

4.2.3. *After-sales Services*

The value chain of the company does not end at product delivery. After-sales services and maintenance are important parts of the value chain and make a key contribution to the revenue of the company. Most mining and construction equipment manufactured by the company in recent years is equipped with remote monitoring systems. These systems record and transmit important operating information to the company's control centres through a satellite/mobile phone system. The company can support essential maintenance and support services to customers with up-to-date information, as well as key information for the optimisation of operations. The latter information addresses such matters as load factors, fuel consumption and machine operation status. The field information acquired through this continuous operational communication also helps the company to improve the design of its products.

4.2.4. *Re-manufacturing*

A special node in this value chain is re-manufacturing. In re-manufacturing plants, the company will dis-assemble old components, and restore and upgrade parts, with strict observance of the company's production standards. A warranty comparable to that on new parts will be offered to customers who buy these re-manufactured components at a discounted price. Re-manufactured products are distributed through the companies' normal distribution channels. Re-manufacturing is not only good commercial business, it also saves finite natural resources and benefits the environment.

4.2.5. *Back Office Functions and Group-wide Offerings*

In a complicated value chain, centrally supplied back office services are essential to the efficient functioning of the value chain. One of the key services in this context, provided at the pre-manufacturing stage of this value chain, is research & development (R&D). The company spends more than USD 500 million per year on improving existing models and developing new ones with new

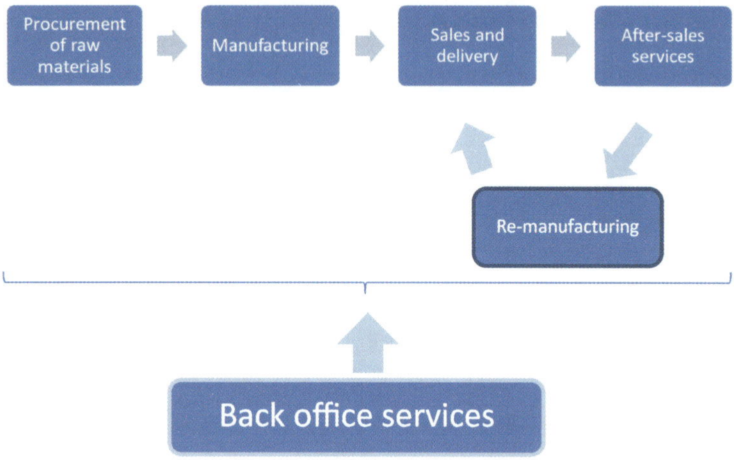

Figure 4.2: Brief Overview of the Value Chain.

Source: Authors.

technology. The Global Positioning System (GPS) satellite monitoring system described above is an example of group R&D.

Other back office functions listed in Table A.5 include finance, managerial services, legal services, advertising and the provision of utilities. These services are supplied at various and different stages of the value chain.

4.3. SERVICES, OUTSOURCING, AND BUNDLING

The tables in Appendix attempt to disentangle the value chain into six stages, namely pre-establishment, pre-manufacturing, manufacturing, post-manufacturing/delivery, back office and after-sales. A non-exhaustive list estimates that no fewer than 74 services are supplied as inputs onto the value chain. As can be observed in Tables A.1–A.6 (third column), these 74 services can be further disaggregated into around 90 service categories.

Among the 74 product categories, back office services account for one-third of the total, the manufacturing stage for one-fifth, the pre-manufacturing stage and post-sales stages for 14 per cent each, and the establishment and the post-manufacturing stages for 9 per cent each. These shares do not give us any sure indication of which value chain segment is the most costly in terms of services inputs, but one of the most costly is likely to be the pre-production segment, as it includes R&D. The importance of the back office services segment in terms of the number of services involved attests to the complexity of an internationalised capital goods global value chain.

4.3.1. *Outsourcing*

Outsourcing is a key aspect of the value chain. Among the 74 identified services, it is estimated that more than half of these services (52) are outsourced or partially outsourced to third parties outside the group. There are typically three reasons for outsourcing in this value chain.

First, a lack of expertise or specific equipment is one of the reasons. For example, maintaining a satellite system and transmission through ground radio or orbit are hardly tasks in which a company manufacturing construction and mining equipment would be specialized. This represents a fixed cost argument for outsourcing. Even as the company monitors around 400,000 machines around the world with satellite/mobile phone connections, the initial investment and fixed costs of owning and operating the systems may still be too high for in-group provision. This may be characterized as an internal economy of scale consideration.

A second reason for outsourcing relates to external economies of scale, sometimes referred to as economies of agglomeration. Some services, such as security services for factories or information technology (IT) specialist services, are commonly outsourced because third-party firms dedicated to these skill sets are located in the proximity of the lead firm. Cost savings and efficiency gains can accrue to the lead firm from taking advantage of the specialisation characteristics of third-party suppliers.

Third, some outsourcing services are required by laws and regulations derived from them. Services in this category include auditing and conformity assessment or inspection procedures associated with government standards (technical regulations).

These standards cover a range of public policy objectives, such as health, safety and the environment. These government services confer social value and may be considered additive with respect to private sources of value. But if the services are not provided in an efficient manner, they may also impose a negative value (cost) on firms. The company in the present case study did not report particular difficulties at the interface with public policy.

4.3.2. *Bundling*

The bundling phenomenon, where firms put together a packet of products as a composite market offering is commonplace along value chains. Bundling can achieve three major purposes (Cheung *et al.*, 2014). The first is related to efficiency gains associated with the joint supply of a variety of products, be they goods or services. Frequently, when an enterprise buys what is billed as a single product, it is in effect buying a suite of bundled products, many of which are likely to be services. The end product of after-sales services in this case, for example, is the repair and maintenance of machinery over its useful life. But a range of goods (replacement parts and lubricants etc.) and services (transport, insurance, financing and many others) go into the provision of the end-product.

Second, firms may bundle their offerings in the marketplace to acquire more value added and distinguish what they supply from what competing suppliers might offer. Successful product differentiation of this kind increases profits by creating temporary barriers to entry in a competitive market. The product differentiation engineered though bundling is also often a source of innovation and productivity growth.

Third, a by-product of bundling services with goods is that if some of those services were to be supplied separately, they could not be traded. Packaging services, for example, only enter trade if the goods that are packed in the supply of the services enter trade. The package services alone cannot travel across frontiers. This means that with bundling, an economy's trade potential can be increased.

The company in this study puts bundling at the centre of its business model. The company offers customers not only machinery and equipment, but many goods and services besides. These include, for example, after-sales services including warranty, maintenance, genuine spare parts and consumables, tools, information services, technical support and sometimes financial arrangements for hire purchases. The contracts for the sale of these products are often bundled with these services even if the company is transferring ownership of the machinery by selling it outright. The firm's bundling strategy embodies features of all three of the considerations mentioned above.

These bundled services constitute a significant part of the company's revenue. The value of after-sales services along the 20–30 years service life of the machines may be higher than the price of the machines themselves. When all is said and done, according to one executive of the company, services might count for over 50 per cent of costs, and therefore of added market value.

4.4. THE IMPACT OF POLICY ON THE VALUE CHAIN

In principle, all services (and goods) entering the value chain will be affected, however, tangentially, by the policy frameworks of the jurisdictions in which they work. Attempting a comprehensive coverage of these interfaces would be a near impossible task. For this reason, only a selection of issues is covered in each of the case studies and the selection is based on what firms in the case studies believe to be the most worthwhile issues to raise. In this case ownership and investment restrictions, labour market restraints, trade costs associated ports and customs, and issues involving re-manufacturing activities are covered (Table 4.1).

4.4.1. *Foreign Ownership Restrictions and Investment Constraints*

Investment constraints restrict foreign capital participation in many sectors. Although manufacturing is often open to majority foreign ownership, service sectors are usually more restricted. This pattern is well observed in the present value chain and consequently the company provides many of its services needs off-shore, often relying on in-group enterprises.

In many of the East Asian economies, manufacturing sectors are opened to full foreign equity ownership. However, investment constraints are rather widespread in wholesale and other distribution services, as well as in a range of professional services. For example, the newly enacted residential Regulation No. 39 of 2014 of Indonesia restricts foreign equity control in wholesale and distribution sectors to 33 per cent of the equity. The company also reported that the Thai government restricts foreign equity ownership in their distribution subsidiary to 40 per cent.

In some economies, there might be no formal restriction on foreign equity ownership but informal restrictions and barriers to establishing, owning and operating enterprises hinder value chain operations. Excessive documentation requirements, uncertain and unpredictable regulatory frameworks, and protective policies in local governments are reported to occur with relative frequency in several developing economies. These limitations cost companies time and resources.

In order to comply with the investment constraints, the company is sometimes obliged to seek local partners when entering new markets. Much depends on being able to find reliable and committed partners. When this is possible firms are often happy to enter into joint ventures as local partners can add considerable value to the company's value chain by their local expertise, knowledge, linguistic capabilities and connections.

4.4.2. *Foreign Labour Restrictions*

Foreign labour restrictions are common in many economies. While there is doubtless a case for protecting local employment and nurturing skills, barriers to work permits for foreign technical staff may actually hinder these objectives. This can occur for a number of reasons. First, foreign companies may invest less in the face of what they may consider an undue risk in terms of cost and reputation if they are unable to work out the optimal mix of local and foreign labour market participants at any point in time.

Second, where firms are already established, the inefficiency implicit in relying on inexperienced or under-qualified labour may reduce output, damage sophisticated machinery that requires key skills to output, and crimp employment opportunities. Third, on-the-job training for local employees who could certainly fill the jobs in question will be in shorter supply.

Similar restrictions also exist in distribution, sales and after-sales services in some Southeast Asian economies. In Thailand, for every foreign employee, the company has to employ four locals. In addition, not more than 10 foreigners can be employed unless the company has paid three million Baht or more in corporation tax in the previous fiscal year, operates an export business and brought in more than 30 million Baht the previous years, or has 100 or more Thai employees (Mayer Brown, 2008).

The company is active in adapting to labour requirements. In addition to sending engineers from Japan headquarters for setting up new plants, overseeing factory operations and providing technical guidance, the company has set up local training centres to nurture local engineers, sales and support staff. The company has also promoted locals to top management, including in China, Indonesia and Thailand. Companies often point to the relative cost of hiring foreigners as opposed to locals and point to this as a significant incentive to go local where possible.

4.4.3. *The Customs Regime and Efficiency*

A large body of research has demonstrated what the immediate costs and adverse knock-on effects are of inefficiency in entry ports and customs services (see, e.g., Engman, 2005; Wilson *et al.*, 2005). Complicated and inefficient customs clearance procedures have caused delay to the company's daily operations from time to time. The company often needs to ship spare parts from one economy to another for the maintenance of machines. Customs clearance services in some Southeast Asian economies may take few days, if not weeks, before the parts can arrive. One experience cited from one economy was that it took 3 days for an urgent component to pass through customs, causing a delay in the delivery of vital services to customers.

Another issue that arises in many economies in the APEC region is that the authorities sometimes insist that customs agents or brokers are officially licensed and they must be used to manage customs clearance procedures for firms wishing to import goods necessary for their operations. Where licenses are restricted to a few operators, this will tend to raise the costs of dealing with customs. On the other hand, some companies may find it convenient to outsource customs administration procedures in order to facilitate the transactions.

The recently agreed Agreement on Trade Facilitation in the World Trade Organization requires that "[W]ithout prejudice to the important policy concerns of some Members that currently maintain a special role for customs brokers, from the entry into force of this agreement Members shall not introduce the mandatory use of customs brokers." (Article 10.6:1). In Section II of the Agreement dealing with special and differential treatment, specific provisions are laid out for the timing and conditions of compliance with various provisions in Section I.

4.4.4. *Trade Restrictions in Re-manufacturing Goods*

Re-manufacturing goods are often subject to heavy regulation and restrictions on trade. USITC (2012) highlighted common trade barriers encountered by heavy duty off-road (HDOR) equipment manufacturers. Some of these difficulties include the lack of international standards to distinguish between new and used manufacturing products, limitations and even bans on trade in re-manufactured HDOR equipment and parts, a cumbersome registration system for re-manufacturing businesses, complex paper-work, unfriendly customs regulations, and a preference for (higher-priced) local production.

Table 4.1: Summary of Certain Policies Affecting the Value Chain.

Government policies/ services	Authority(ies) in charge	Details	How the policy affects services in the value chain
Customs procedures • Inefficiency in customs clearance • Compulsory customs agents	Customs	The company is required to clear the parts and goods with customs	Inefficient customs procedures raise costs and cause delays
Foreign ownership restrictions — formal restrictions • limitation on share of foreign equity	Ministry of Commerce/ Ministry of Investment	Regulations limit the share of equity owned by foreign companies to a certain proportion. Some economies prohibit foreign ownership in specific sectors	The company is restricted by regulations on foreign equity in wholesale and distribution sectors. Local partners are found to get around this issue
Foreign ownership restrictions — informal barriers • Uncertainty in policy regimes • Excessive requirements in documents	Various government departments	Uncertainty as to the content and implementation procedures of regulations, especially at local government level. Over-complicated requirements on documents adds cost and time to value chain	Extra cost and time on dealing with government policies
Foreign labour restrictions • Quota on work permits • Forbidden sectors for foreign workers	Ministry of Labour	Work permits for foreign labour may be restricted by quantity or through other means. Some sectors may only be open to local employees	May experience difficulty in acquiring enough skilled labour. The company in this case gets around the issue by training local staff
Trade on re-manufacturing goods	Ministry of Commerce/ Ministry of Trade	Importation of re-manufactured parts/ products is prohibited	The company in this case may not be able to provide re-manufactured products which are more cost efficient to their customers and environmental friendly, thus sales may be affected

Source: Authors.

The company in this case study has experienced some of these difficulties in their re-manufacturing business in Asia. One Asian economy, for example, prohibits the importation of parts from used engines and cars for re-manufacturing and has imposed very strict regulations on imports of used and re-manufactured machines. Li (2012) suggested the policy intentions were to protect domestic customers from low quality products, and prevent the export of electronic waste to the economy under the guise of used products. However, these policies frustrate the worldwide operation of the company's re-manufacturing business, and oblige the company to sell its remanufacturing parts in fewer economies. Considering the potential resource and environmental advantages of re-using materials and machines that still possess a useful life, this is an issue that merits closer attention and discussion.

APPENDIX

Table A.1: Establishment Phase.

Stage/category	Service	Central product classification (CPC) ver. 2 code	Outsourced or in-house/in-group provision
1. Government services (licensing etc.)	1. Business liaison services (e.g., services from trade development centres or SME centres)	9113 — Public administrative services related to the more efficient operation of business	Outsourced. Local consultants, business partners and trade facilitation organisations provide business liaison services
	2. Company registration and licensing services	91138 — Public administrative services related to general economic, commercial and labour affairs	Outsourced. Government services
	3. Information and statistical services	9113 - Public administrative services related to the more efficient operation of business	Outsourced. Business information usually provided by trade facilitation organisations and national statistical organisations
	4. Visa and immigration services for foreign investors/employees	91290 — Public administrative services related to other public order and safety affairs	Possibly outsourced. Many companies hire local agents to help with foreign work permit issues
2. Other services (professional etc.)	5. Business consultant services	8311 — Management consulting and management services 83129 — Other business consulting services	Outsourced. Consultants are required in some economies for new market
	6. Banking and finance services	71121 — Deposit services to corporate and institutional depositors 71135 — Non-mortgage loan services for business purposes	Partially outsourced and partially in-group. Banking services would be required for setting up new business. In-group finance company may also provide financial arrangements
	7. Legal services	82130 — Legal documentation and certification services	Possibly both in-group and outsourced. The company has in-group legal department but may possibly need local legal expertise

Source: Authors.

Table A.2: Raw Materials, Input/Pre-production Stage.

Stage/category	Service	CPC ver. 2 code	Outsourced or in-house/in-group provision
1. Procurement of raw materials	8. Procurement agent for raw material sourcing	85999 — Other support services n.e.c.	Possibly in-house. The company manages direct business relationship with suppliers
	9. Customs-related services for raw materials imported	85999 — Other support services n.e.c.	Outsourced. Customs brokers must be third-party entities in some economies
	10. Quality assurance services (of raw materials)	83441 — Composition and purity testing and analysis services	Mostly in-group
2. Logistics	11. Freight transportation services (of raw materials) by road, rail, sea or air	65 — Freight transport services	Partially outsourced and partially in-group. The company has an affiliate in logistics.
	12. Repair and maintenance for fleets	87143 — Maintenance and repair services of trailers, semitrailers and other motor vehicles n.e.c.	Possibly outsourced
3. Storage	13. Storage of raw materials — general storage	67290 — Other storage and warehousing services	Possibly In-house. Must be located near factories
	14. Storage of raw materials — tank farm	67220 — Bulk liquid or gas storage services	Possibly In-house. Must be located near factories
4. Research and Development	15. Conception and design of product	83920 — Design originals	In-house. The company houses industrial design and engineering centres
	16. Patent acquisition	83960 — Trademarks and franchises n/a	In-house. The company houses industrial design and engineering centres
	17. Product development/R&D for new technology	81129 — Research and experimental development services in other engineering and technology 81400 — Research and development originals	In-house and outsourced. In addition to in-house development teams, the company work with research institutes and universities in research

Source: Authors.

Table A.3: Manufacturing Stage.

Stage/category	Service	CPC ver. 2 code	Outsourced or in-house/in-group provision
1. Production administration	18. Production Administration — Production management	83115 — Operations management consulting services	Possibly in-house. Subsidiary in each economy has authority in production planning
	19. Production Administration — Quality assurance and compliance with ISO	83441 — Composition and purity testing and analysis services	Possibly in-house. Subsidiary in each economy has authority in production planning
2. Services supporting factory daily operations	20. Cleaning services of factory	85330 — General cleaning services	Possibly in-house
	21. Engineering Services	83310 — Engineering advisory services 8332 — Engineering services for specific projects	In-house. The company's engineer team provides engineering services
	22. Gardening services for factory lawn	85970 — Landscape care and maintenance services	Possibly in-house
	23. Security guards for factory and warehouses	85250 — Guard services	Possibly outsourced. Specialized security firms enjoy network effects
	24. Sewage water treatment services	94110 — Sewerage and sewage treatment services	Outsourced. Infrastructures are provided by host economies
	25. Repair and maintenance services of machines and equipments in the factory	87156 — Maintenance and repair services of commercial and industrial machinery	Both in-house and outsourced. Outsourced maintenance is required when in-house engineers cannot solve the issues arising
	26. Waste collection and recycling services	9421 — Collection services of hazardous waste 94229 — Collection services of non-hazardous recyclable materials, other 94239 — General waste collection services, other	Outsourced. Infrastructures are provided by host economies

(Continued)

Table A.3: (*Continued*)

Stage/category	Service	CPC ver. 2 code	Outsourced or in-house/in-group provision
3. Services from government regulation requirements	27. Government inspections on fire prevention, health hazards, environmental protection and other aspects	91133 — Public administrative services related to mining and mineral resources, manufacturing and construction 91290 — Public administrative services related to other public order and safety affairs	Possibly in-house and outsourced Government provides inspection services. Company may have to prepare documentation for inspection from time to time
4. Worker-related services	28. Catering services for workers	63393 — Other contract food services	Possibly in-group and outsourced. In some economies (such as United Kingdom) the company runs catering subsidiary. It is common to outsource canteen services
	29. Dormitory for factory workers	63220 — Room or unit accommodation services for workers in workers hostels or camps	Possibly in-group. Japanese large companies may offer dormitory for factory staff.
	30. Medical services	93121 — General medical services	Possibly outsourced
	31. Recreational facilities and services for workers	96520 — Sports and recreational sports facility operation services	In-house. The company provides sports facilities for workers on site
	32. Transportation services for employees	64114 — Local special-purpose scheduled road transport services of passengers	Possibly outsourced to local transportation companies

Source: Authors.

Table A.4: Delivery and Sales/Post-manufacturing Stage.

Stage/category	Service	CPC ver. 2 code	Outsourced or in-house/in-group provision
1. Delivery to wholesaler/retailer	33. Land transport of vehicles to distributors	65112 — Road transport services of freight by semi-trailers	Partially in-group and partially outsourced. In-group logistics arm arrange deliveries
	34. Sea transport of vehicles to distributors	65219 — Other coastal and transoceanic water transport services of other freight	Partially in-group and partially outsourced In-group logistics arm arrange deliveries
2. Sales services	35. Retail trade services — By distributors	621 — Non-specialised store retail trade services 622 — Specialised store retail trade services	Partially in-group and partially outsourced. Distributors in major markets are subsidiaries. Some are third parties
	36. Retail services on a fee or contract bases	625 — Retail trade services on a fee or contract basis	Partially in-group and partially outsourced
	37. Retail Administration Operation management	83115 — Operations management consulting services 83116 — Supply chain and other management consulting services	Partially in-group and partially outsourced
	38. Storage and warehousing services for machineries at distributors	67290 — Other storage and warehousing services	Partially in-group and partially outsourced. Distributors stores machineries
	39. Hire purchases and financing services for customers	71140 — Financial leasing services	Partially in-group and partially outsourced. In-group finance company provides financial arrangements in some economies. Third-party distributors also help customers of the company in financing

Source: Authors.

Table A.5: Back Office, Utilities and General Services.

Stage/category	Service	CPC ver. 2 code	Outsourced or in-house/in-group provision
1. Finance	40. Auditing on financial accounts	822 — Accounting, auditing and bookkeeping services	Outsourced as required by legislations. In-company accounting department assist in preparing documents for auditing
	41. Financial services	71121 — Deposit services to corporate and institutional depositors 71313 — Group pension services 71701 — Services of holding equity of subsidiary companies	Partially outsourced and partially in-group. Financial institutions and in-group financial company may both provide financial services to the company
	42. Internal auditing (including audits of financial accounts and corporate governance)	822 — Accounting, auditing and bookkeeping services 83118 — Head office services	In-group
	43. Insurance services for machinery used in factory	71334 — Other property insurance services	Partially in-group and partially outsourced. In-group company assisted in insurance policy
	44. Insurance for staff	71321 — Accidental death and dismemberment insurance services 71322 — Health insurance services	Partially in-group and partially outsourced. In-group company assisted in insurance policy
2. General Management	45. Business and management consultancy services	8311 — Management consulting and management services	In-house
	46. Company secretary services	83990 — All other professional, technical and business services, n.e.c.	In-house. As listed company, it employs its own company secretary

47. Corporate communications, marketing and public relationship	83114 — Marketing management consulting services 83121 — Public relations services	In-group corporate communications department responsible for corporate communications and marketing
48. Courier, postal and local delivery services	681 — Postal and courier services	Outsourced
49. Estate management	72112 — Rental or leasing services involving own or leased non-residential property	Possibly in-group and outsourced
50. Human resources management	83113 — Human resources management consulting services	In-group human resources department
51. Human resources management — Personnel search and referral services	8511 — Personnel search and referral services	In-group human resources department. Agents might be needed to recruit locally
52. IT and information system management, consulting and support, with webpage development	8313 — IT consulting and support services 8314 — IT design and development services 8316 — IT infrastructure and network management services	Possibly in-group and outsourced. In-group IT department maintains the computer system. Possibly outsource some of the tasks to third parties
53. Safety and security services	85230 — Security systems services 85250 — Guard services	Possibly outsourced. Local security firms enjoy network effects and connection with legal enforcement

(*Continued*)

Table A.5: *(Continued)*

Stage/category	Service	CPC ver. 2 code	Outsourced or in-house/in-group provision
	54. Telecommunication services	841 — Telephony and other telecommunications services 84210 — Internet backbone services 84221 — Narrowband Internet access services 84222 — Broadband Internet access services	Outsourced. Host economies provide telecommunication services
	55. Uniform —Design and alternation	83919 — Other specialty design services	Possibly outsourced
	56. Uniform —Laundry	97130 — Other textile cleaning services	Possibly outsourced
3. Legal	57. Legal services	82120 — Legal advisory and representation services concerning other fields of law 82130 — Legal documentation and certification services	In-group and possibly outsourced. Legal department exist within corporate structure. Possibly outsourced to local legal experts in some economies
4. Advertisement	58. Retail Administration — Advertising	83611 — Full service advertising	Partly in-group and partly outsourced. In-group company handles publication and publicity. Outsourced advertisement agencies are often required for production of advertisement materials
	59. Advertisement—advertisement agencies services	83611 — Full service advertising 83620 — Purchase or sale of advertising space or time on commission	

	60. Advertisement — designers' services	83611 — Full service advertising	
	61. Advertisement — cameramen, directors, photographers and models.	83611 — Full service advertising	
5. Utilities	62. Electricity supply	8631 — Support services to electricity transmission and distribution	Outsourced. Supplied by host economies
	63. Gas supply	86320 — Gas distribution services through mains (on a fee or contract basis)	Outsourced. Supplied by host economies
	64. Water supply	86330 — Water distribution services through mains (on a fee or contract basis)	Outsourced. Supplied by host economies

Source: Authors.

Table A.6: Post-sales Services.

Stage/category	Service	CPC ver. 2 code	Outsourced to third parties
1. Sales/After-sales customer services	65. Customer services — Complaints and compliments handling	85931 — Telephone call centre services	Partly outsourced. Both the company owned and third-party distributors resolve customer complaints
	66. Customer services —Technical support	85931 — Telephone call centre services	Partly outsourced. Both the company owned and third-party distributors resolve customer complaints
2. After-sales information support	67. Information services for customers	8315 — Hosting and IT infrastructure provisioning services	In-house. IT department maintain the information service system which customer can access by Internet
	68. Satellite services for remote monitoring system	84140 — Private network services (provision of satellite facilities for the exclusive use of the customer)	Possibly outsourced
	69. Data centre services for remote monitoring system	83159 — Other hosting and IT infrastructure provisioning services	Outsourced to MNC telecommunication services providers
	70. Ground radio communication services for remote monitoring system	84131 — Mobile telecommunications services — access and use	Outsourced to MNC telecommunication services providers
	71. GPS services	67990 — Other supporting transport services n.e.c. (radio navigational aid locating services, such as GPS provision)	Outsourced. GPS satellite system provided by US government

3. Repair and maintenance	72. Repair and maintenance of machines for customers	87156 — Maintenance and repair services of commercial and industrial machinery	In-house and outsourced. Distributors (may or may not be subsidiary, differed economywise) may help customers to repair machines. The company also sends in-house engineers when required
4. Remanufacturing services	73. Buy back services to buy used parts and machines from customers	62587 — Retail trade services on a fee or contract basis, of mining, construction and civil engineering equipment and components	In-house and outsourced. Distributors, both in-house and outsourced also handles buying back from customers
	74. Remanufacturing and refurbished services	4446 — Parts for the goods of classes 4441 to 4444 (Machinery and equipment for special purposes)	In-house. The company operates 10 re-manufacturing factories across the globe

Source: Authors.

REFERENCES

Businessvibes (2013). Global Construction Machinery Industry. https://www.businessvibes.com/blog/global-construction-machinery-industry-overview [15 March 2015].

Cheung, D, P Low and D Sit (2014). *Hong Kong-based Bakery Chain: The Short Life of a Loaf of Bread*. Hong Kong: Fung Global Institute Case Study.

Engman, M (2005). The economic impact of trade facilitation. *OECD Trade Policy Papers*, No. 21. Paris: OECD Publishing. DOI:http://dx.doi.org/10.1787/861403066656.

Li, YK (2012). 中国汽车零部件再制造产业发展现状 及政策环境. http://www.jc-web.or.jp/jcbase/publics/download/?file=/files/content_type/type019/204/201506121939324204.pdf [15 March 2015].

Mayer Brown, JSM. (2008). Guide to employment law in Thailand. http://www.mayerbrown.com/files/Publication/1ed36ca3-a467-4da1-8a05-9264088c9334/Presentation/PublicationAttachment/f1805434-c19c-465e-8a5d-555c2bec7ad1/jsm_thailand_employment_may2008.PDF [15 March 2015].

United States International Trade Commission (2012). Remanufactured goods: An overview of the U.S. and global industries, markets, and trade. USITC Publication 4356. http://www.usitc.gov/publications/332/pub4356.pdf [15 March 2015].

Wilson, JS, CL Mann and T Otsuki (2005). Assessing the benefits of trade facilitation: A global perspective. *World Economy*, 28, 841–871. DOI:http://dx.doi.org/10.1111/j.1467-9701.2005.00709.x.

World Trade Organization (2013). Agreement on trade facilitation. https://www.wto.org/english/thewto_e/minist_e/mc9_e/desci36_e.htm [15 March 2015].

Chapter 5

MANUFACTURING OF COMPUTER SERVERS

Yuhua Zhang

5.1. COMPANY BACKGROUND

The company represented in this case study is a leading global manufacturer of servers based in Chinese Taipei. It has grown over the years into a conglomerate, and has been listed on the Chinese Taipei Stock Exchange and has worldwide operations. The firm has manufacturing plants, research and development (R&D) centres, logistics and distribution centres, and after-sales service centres on different continents. The company has built a competitive edge in server production. The case study will focus on the server global value chain of this company.

5.2. THE SERVER INDUSTRY

Technically, a server is a computer or device on a network that manages network resources. It performs tasks on behalf of clients and facilitates the sharing of data, information, and hardware and software resources. In multiple operating systems, a server could refer to the program that is managing resources rather than the entire computer. Servers can be classified into different categories. Depending on the functionality, there are communication servers, database servers, media servers, web servers, application servers, etc. Depending on their size and sophistication, there are rack servers, tower servers, miniature servers, blade servers, ultradense servers, super servers, etc.

From a configuration point of view, a server is quite similar to a computer, because both have components such as a bare board, a processor, memory, hard drives, network connections and power supply. Yet, a server is different from a desktop or laptop computer in three major ways:

- First, a server is designed to carry out heavy-duty back-end tasks, which support desktop or laptop applications. For example, a server can be dedicated to host databases, while a computer only runs the interface of the databases and retrieves data.
- Second, the most important component for both a server and a computer is the central processing unit (CPU), which has a small cache to store frequently used data. A desktop or laptop may have one or two sets of CPU, but the size of cache is quite limited; while a server has larger and more varied caches, which increases the processing power substantially.
- Third, a desktop or laptop often has only one hard drive, while a server has several hard drives configured into a single disk, called redundant array of inexpensive disks (RAIDs). In this case, failure of one hard drive does not mean the failure of a server, as other hard drives back up the system.

The server emerged on the global information technology (IT) stage in 1981, when International Business Machines Corporation (IBM) made the LISTSERV, the first email server, to enable group email collaboration. From then on, web servers, rack-mountable servers, open search servers and blade servers gradually established their presence with the advance in technology. More recently, cloud computing has become widely popular. Virtualisation has led to more web-based applications and requires the establishment of mega data centres, which created enormous demand for cloud servers (iWeb Technologies, 2012).

As businesses become more reliant on advanced connectivity, and new uses of digital technology emerge, demand grows for data processing capacity. Intensified internet traffic and the growing need for real-time analytics in turn drive the development of IT.

The global server market has been expanding steadily since early 2000s thanks to increasing infrastructure investments (third platform infrastructure[1]) and the setup of hyper-scale datacentres, especially in the Asia and Pacific region. The server market is dominated by big brand vendors, such as IBM, Hewlett-Packard (HP), Dell, Oracle, Cisco, Lenovo and Fujitsu. Data obtained from Statista on the market share of major server vendors from 2009 to 2014 indicated that IBM, HP and Dell accounted for an overwhelming portion of server market, although their combined share had been declining. IBM led the market at the very beginning in the 1980s. In the last two years, however, it had been overtaken by HP. In 2014, IBM experienced a sharp decline in market share because it partially sold its server business to Lenovo. This also explained the quick catch up of Lenovo in 2014. In addition, the market share of original design manufacturing (ODM) direct[2] has been growing fast, reflecting intense competition in the market place.

Like in many manufacturing industries, in the server sector of the IT industry, vendors are not necessarily the manufacturers of the product; they are only the brand owners. Server manufacturing is outsourced to first tier suppliers, usually the contract manufacturers who supply several brand owners simultaneously. To understand what components make a server and how a server is manufactured, it is essential to examine the server global value chain.

5.3. DESCRIPTION OF THE VALUE CHAIN

Server global value chains (GVCs) are long and sophisticated, involving many participants. Two key participants are the brand owner (vendor) and the contract manufacturer. Brand owners pass

---

[1]The 3rd platform is a computing platform of technologies that emerged after 2010. According to International Date Corporation (IDC), the 3rd platform is built on mobile devices, cloud services, social technologies, and big data (IDC, 2013).

[2]ODM direct refers to the sales of servers by the server manufacturer not the brand owners.

the order to the contract manufacturers with requirements and specifications for the servers. Contract manufacturers usually manage the server GVC, and their capacity determines the structure of the chain. Notably, many server contract manufacturers are based in Chinese Taipei — an economy that has developed a reputation for the competitiveness of its IT sector over the years (The Economist, 2013).

Figure 5.1 presents the server GVC of the interviewed company. There are seven major stages of the chain, namely research and design, procurement, production and testing, logistics, assembly and package, distribution and after-sales service, and repair and recycling. These stages also reflect the major services involved. The path of physical components and products is depicted in the middle part of the figure, which shows the complexity of server production. Key players are specified on the top segment of the figure, where the coordinating role of contract manufacturer as the value chain manager clearly stands out.

5.3.1. *Research and Design*

A server GVC starts with the specific needs and requirements for a server on the part of the brand owner. The brand owner can either design the server in-house or work with the contractor manufacturer to design it. Depending on the level of engagement of contract manufacturers, there are two different ways that a server is designed:

- Original equipment manufacturing refers to the brand owner designing the server and passing it to the contract manufacturer to build it.
- ODM refers to the brand owner asking the contract manufacturer to design and produce the server.

Once a server is designed, the contract manufacturer will carry out three important testing:

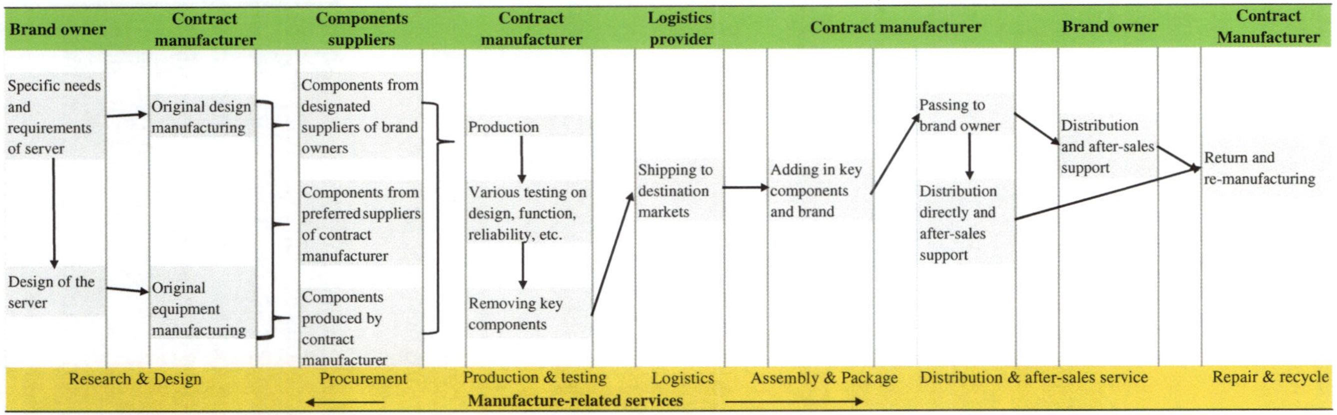

Figure 5.1: A Server Global Value Chain.

Note: Participants in a server global value chain are depicted in the upper range of the figure. They are brand owner, contract manufacturer, component suppliers and logistics providers. The middle range shows the value creation process along a server GVC, from conception to return and re-manufacturing. The lower range shows the key services involved in a server GVC at each stage and manufacture-related services cover several stages.

Source: Author.

(i) Engineering verification test (EVT). It assesses whether the design is free from any engineering bugs and all product functions conform to the needs of brand owner.

(ii) Design verification test. Once the product passed EVT tests, a pilot run will build several units with parts out of hard tooling. These units are subjected to various reliability and compatibility tests to ensure products can meet brand owner's requirements in compatibility and reliability.

(iii) Production process verification test (PVT). In PVT, higher volume is subjected to pilot run process and is fined tune until yield rate meet customer requests.

5.3.2. *Procurement*

The next stage of a server GVC is procurement. The contract manufacturer will purchase components to mass build servers following the prototype. There are three sources of components:

- Certain components are sourced from designated suppliers of the brand owners.
- If brand owners do not have designated suppliers for certain components, the contract manufacturer can source components from its own preferred suppliers.
- Contract manufacturers could also produce certain components that meet the design requirements of a server.

In general, 70 per cent of the server components are sourced from external suppliers, which account for a substantial part of the value generated in the chain. For server production, key components are controlled by Intel, Samsung and Seagate. As an industry practice, if the purchasing volume is high, the contract manufacturer will purchase directly from components suppliers; however, if the purchasing volume is low, the contract manufacturer could turn to an agent, who will collect and combine the orders from many contract manufacturers and purchase in bulk.

5.3.3. *Production and Testing*

After procurement, the contract manufacturer will conduct an incoming quality control inspection to ensure that the components meet the requirements of server design and functionality. Various testing procedures are required for quality assurance. Servers need to meet industry standards and/or the standards of brand owners, which are usually more stringent than industry standards. Figure 5.2 sketches the production process and the various testing procedures involved. From sourced components to finished products, there are eight testing and inspection steps to ensure quality and functionality.

Towards the end of production, a final testing on overall functionality and the quality of the product is carried out. However, this testing is not done in the destination markets, it is performed in Chinese Taipei. Once the server has passed the final testing process, it is considered finished and ready for shipping to destination markets. But the contract manufacturer will not ship the finished products as a whole. It will remove the key components from the finished products, leaving only the barebones, which represents only around 30 per cent of the value of a server.

The key components of a server are the CPU, the hard disk drive and the memory unit. All these components contain elements

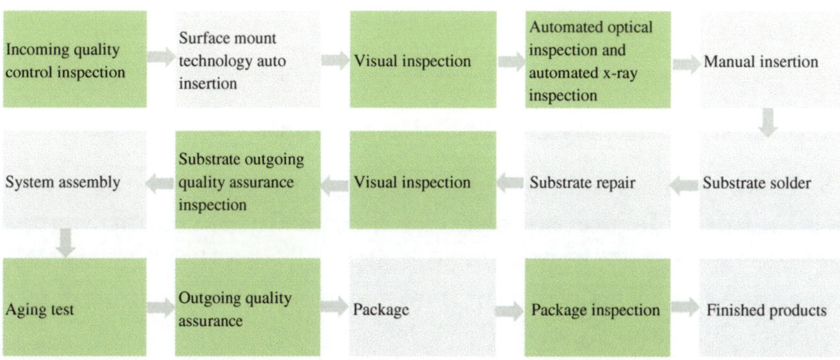

Figure 5.2: Server Testing Requirements at Different Stages of Production.

Note: Dark colour highlights various testing in the production process.

Source: 2013 annual report of the interviewed server manufacturer.

whose prices can fluctuate considerably. The price trend, however, tends to be in a downwards direction because of rapid technological change in this industry.

5.3.4. *Logistics*

The barebones are then shipped to destination markets, in this case the Americas, by logistics providers. Depending on the destination, shipping requires a significant amount of time, especially if the contract manufacturer is located in Asia and major markets are in North America or Europe, which would require over a month. To avoid the risk of losing too much value on the key components due to a drop in price, contract manufacturers choose to ship the barebones and then source the key components right before the barebones arrive.

5.3.5. *Assembly and Package*

In the destination markets, the barebones and newly purchased key components are assembled together. At this stage, the servers are ready to use. The next step is packaging, for example, the United States, the contract manufacturer will put the brand of the brand owner or its own brand. After that, a server is finalized and readied for distribution.

5.3.6. *Distribution and After-Sales Service*

The brand owners sell the servers through their distribution channels and provide after-sales service directly. If the contract manufacturer has its own brand, it could also sell directly to customers, categorized as ODM direct. In this case, the contract manufacturer will be responsible for the after-sales service.

5.3.6.1. *Repair and recycle*

Once a server is put into use and runs toward the end of its life cycle, usually five years, it then will be returned and remanufactured, which marks the end of this server GVC. The repair and remanufacturing can be done by the original contract manufacturer or companies that are specialised in repair and recycle.

5.4. SERVICES ALONG THE SERVER GLOBAL VALUE CHAIN

A total of 54 main services categories have been identified in the server GVC, which can be disaggregated into the 96 separate services recorded in Appendix. Information is not available on the share of these services in overall value creation. Research and design are usually high value-added services. According to the company, the value of manufacture-related services is not as high as physical components.[3] This is due to the nature of the sector — key components and physical materials hold the predominant share of value in a server.

The 54 aggregated categories of services have been listed in Table 5.1 as horizontal services and value chain specific services. Under horizontal services, 25 items are included, covering management and operation-related services, which are required at all stages of the value chain. The value chain specific services include 29 items and can be categorised by stages in which services are prominently involved. In order to avoid duplication in Table 5.1, services that are required in the previous stage are not shown twice.

5.4.1. *Services in the Value Chain*

The 96 services categories are listed in the second column of Table A.1 according to the United Nations Central Product Classification (CPC) list. The disaggregation underlines the sophistication

---

[3]Though the value creation of services in the server GVC is not high, it is understated. There are implicit services in production of the key components, as these require R&D, and specialised IT and engineering services, which are not counted separately in the server GVC.

Table 5.1: Main Service Categories in a Server Global Value Chain.

Horizontal services	Value chain specific services	
Company registration and licensing services	Research and design	Market research, customer needs assessment
Set up offices, computers		Conception of the product
Financial services (include Initial public offering (IPO) related services)		Design the product
Service charges for listed companies	Procurement	Customs-related services (for import and export)
Accounting, auditing and bookkeeping services		Procurement service
Legal consulting		Freight transportation services (raw materials)
Human resources		Storage of raw materials — general storage
Telecom services	Production and testing	Installation of production equipment
IT services		Utilities (electricity, water and gas)
Insurance service		Production administration/management
Public relations services		Technical testing
Business hospitality		Manufacture
Social insurance for factory workers		Meeting international standards
Accommodation for workers and managers		Maintenance and repair production equipment
Canteen service		Sewage water treatment services

Training services (for workers)	Logistics	Product testing (for export market)
Security services		Freight transportation
Cleaning services		Freight insurance
Waste collection services	Assembly and package	Design of packages
Government inspections on work environment		Packaging services
Travel services	Distribution and after-sales services	Advertising
Visa and immigration services		Marketing (trade fairs and conventions)
Patent and trademark protection		Warehousing
Industry association membership		Retail trade services
Interaction with Trade Union		Market research/consulting services
		News and information service
		Maintenance and after sales services
	Repair and recycle	Recycling services
		Disposal services

Source: Author.

of services in the chain as well as the varied sources of value created by services.

The 25 horizontal services in Table 5.1 add up to 49 more detailed categories that appear in Table A.1. Financial services and insurance services involve a relatively large number of different services. Under financial services, eight different services included are investment banking services, mergers and acquisition services, corporate finance and venture capital services, deposit services, credit-granting services, financial consultancy services, foreign exchange services and corporate tax consulting and preparation services. Included under insurance services are life insurance and pension services, accident and health insurance services, other property insurance services and general liability insurance services. The company has to use various financial services because it has operations worldwide and is listed on the stock exchange. Insurance services are required by law and regulations.

The 29 value chain specific services in Table 5.1 break out into 47 different services in Column 2 of Table A.1. In the seven value chain stages listed in Table 5.1, the eight broad categories in the production and testing stage contain a total of 15 distinct services. The next highest spread of different services is in the distribution and after-sales category, where the comparable numbers are seven and 12.

5.4.2. *In-house Services and Outsourced Services*

Among the 96 services listed in Column 2 of Table A.1, some are produced in-house and some are outsourced either to affiliates of the company or to a third party. In certain instances, one service can be supplied both internally and externally. Detailed listings on how services enter the value chain are provided in Table A.1. Table 5.2 indicates how services have entered the value chain.

As mentioned previously, 49 horizontal services are related to the company's management and operations. There are only five services sourced exclusively in-house and 13 services are supplied exclusively by third parties. These are investment banking,

Table 5.2: Number of Services by Source.

Supplier of services	Horizontal services	Value chain specific services
Company only	5	0
Affiliates only	0	0
Third parties only	13	13
Company and affiliates	1	4
Company and third parties	13	4
Affiliates and third parties	0	0
Company, affiliates and third parties	17	26
Total number of services	49	47

Source: Author.

corporate finance and venture capital, credit-granting, financial consultancy, internet access, insurance, waste collection and industry association membership services. The company and affiliates jointly provide one service and the company and third parties jointly provide 13 services. Seventeen services are partly sourced in-house and partly outsourced to affiliates or third parties.

In-house services are essential services that lay the foundation for daily operations, such as financial services; accounting, auditing and bookkeeping and human resources. When the firm sources services from its affiliates, it is mainly for resource sharing purposes so as to achieve economies of scale. Services are sourced from third parties for many reasons, such as to leverage the expertise lacking in-house, to conform to laws and regulations, to achieve economies of scale, to ensure access to the best services, and/or to build a stronger relationship with related stakeholders. In some cases, using services provided by third parties is simply due to a lack of feasibility to supply them in-house. There are also services provided by the government, for example, company registration services, licensing, visas, environmental inspection, etc.

As for value chain specific services, 47 services have been identified along the chain. Among these services, third parties exclusively provide 13 services, including freight transportation, water distribution, electricity and gas, technical testing for obtaining certificates for export markets, freight insurance, retail trade, recycling and disposal services. The company jointly with its affiliates provides four services, and with third parties provides four services. Twenty-six services are simultaneously supplied by the company, affiliates and third parties.

The company controls the majority of value chain services because they involve proprietary technology or are necessary to ensure product quality. Outsourced services are considered mainly as non-core for the company's business and the company does not have the equipment, the expertise or the capacity to provide such services.

5.5. POLICIES AFFECTING THE VALUE CHAIN

One objective of this case study is to analyse the impact of policies and public and private standards on the server GVC, and based on that, to identify measures that government can adopt to facilitate growth and contribution of services sectors to manufacturing value chains. The firm's view was that no significant policy-related barriers hindered the expansion of its value chain. The company has managed multinational operations for many years and has gained valuable experiences and financial resources in adapting to different policy environments and regulations. This did not mean, however, that policies never added unnecessary costs across the value chain.

5.5.1. *Trade Policies*

As the company needs to source components from overseas and ship barebones to destination markets, trade policies play a role at procurement and assembly and package stages of the value chain. Tariffs can be a potential hindrance to server import or export, but the tariff rates on IT products are already low. Many components, such as CPUs, enjoy a zero tariff rate in international trade.

Therefore, the company does not view tariffs as a big barrier to its business.

Globally, trade in many IT products are governed by Information Technology Agreement (ITA) signed in 1997. Today, some of the goods covered by the original ITA are obsolete, such as word-processing machines, accounting machines, magnetic tapes and so on. There are also other products that are being phased out, such as electronic calculators, analogue or hybrid computers and input or output units. By contrast, many new products developed since the mid-1990s are not part of the agreement, such as digital video discs (DVDs), media player 4 (MP4) and liquid-crystal displays. (APEC PSU, 2013). Signatories to the ITA are currently negotiating an expansion of the ITA to cover newly emerged products. A modernized ITA could make a valuable contribution to the growing server market.

5.5.2. *Public and Private Standards*

In the IT industry, a company has to meet numerous standards to be a qualified supplier, manufacturer or distributor. There are both public and private standards, though private standards are more dominant in this industry. Public standards, under rules and regulations, mainly concern public health, environment, consumer safety and safety at work place. The difficulty in complying with public standards lies in cross-border trade, when different economies have different public standards that companies are required to follow. Substantial resources are needed to ensure compliance with these public standards.

Private standards are more diverse, detailed and specific, and cover the following three main areas:

(i) Operation and product-specific standards: These are standards set up by industry associations, private companies who have an influential position in the industry, or the brand owners who have their own standards for a product. The TL9000, for example, is an international standard on quality management systems to meet the supply chain quality requirements of the global information and communication technologies industry.

(ii) Brand specific standards: These standards are implicit rather than explicit, in the sense that the quality characteristics of the product are signalled by the branding. They reflect the market position of the component suppliers or the identity of the brand owner. For example, in the server sector, Intel and Advanced Micro Devices Inc. are the two leading manufacturers of CPU, and the barebones of a server have to be designed for a CPU from one of the two. During production, contract manufacturers need to meet specific standards of the brand owner, who uses these standards to differentiate their brands from others.

(iii) Private standards with a public nature: These are standards established by international alliances or entities to meet social objectives, such as a safe and healthy work environment, and environmental sustainability. Examples are ISO 14001 Environment Management System, OHSAS 18001 Occupational Health and Safety Management System,[4] Electronic Industry Code of Conduct and Design for Environment.

The company devotes considerable resources to meeting these standards, such as using special equipment, assigning dedicated personnel, putting in place specific procedures and so on. Often, the company has to pay third parties who established these standards to certify its products or operations. The incremental costs from all the certifications add up to a significant amount.

5.5.3. *Labour Supply and Mobility*

Although the company has not experienced any issue in accessing labour supply, retaining staff or labour mobility, it is an area where

[4]OHSAS is a British Standard designed to work alongside UK health and safety legislation. It has been tweaked to confirm to ISO standards and hence it is a higher standard.

potential difficulties may arise. The company's main production sites are in China, where labour supply is in abundance. But recently, the wages of Chinese workers have been rising, which increases production costs. In the meantime, prices of servers are declining, squeezing the profit margin of server manufacturers.

The company runs R&D centres in Chinese Taipei. An emerging trend is for more and more well-trained researchers and engineers of Chinese Taipei-based companies to be headhunted by Chinese companies, who offer better compensation packages.

Due to its multinational nature, the company needs to send personnel to various plants to oversee or supervise operations. So far, the company has not faced any problem in sending staff to its affiliates in China, Europe or the United States. But the company is aware that if it expands operations in other locations in the future, labour mobility may arise as an issue.

5.5.4. *Intellectual Property Protection*

In general, as a multinational IT firm, the company has taken effective measures to protect its intellectual property (IP). But as mentioned in the last section, it is a growing concern when researchers and engineers leave companies based in Chinese Taipei to join Chinese companies because researchers and engineers carry the product and process knowledge with them and it is highly possible that the Chinese companies they join are competitors.

5.6. CONCLUSIONS AND POLICY RECOMMENDATIONS

This case study presents the background of server market and production, examines the server GVC, identifies the services along the chain, and studies the policies affecting the development of the value chain. The server global value chain is complex and sophisticated. A contract manufacturer needs to have multinational capacity to organise and manage it. Various policies and standards shape the development of the chain and impact the competitiveness of the server manufacturer.

Looking ahead, a successful expansion of the ITA would add momentum to the development of the global IT industry. Although tariffs on many IT products are quite low, there is still scope to reduce them further or to eliminate them in some markets. The IT industry is the backbone of modern economies and it is critical to ensure open trade in IT products. An extended coverage of quality IT infrastructure is essential to develop a knowledge-based economy.

Public and private standards are necessary to ensure public well-being and product quality. But too many standards complicate production and operations and incur significant costs for the company. Mutual recognition agreements can help reduce the burden on firms.

Labour shortage is a looming issue and could be solved partially through smart manufacturing, which integrates information along value chains, optimises manufacturing intelligence, operates industrial plants though innovation, and reduces production costs. Smart manufacturing improves operational efficiency and labour productivity and helps to ease the constraint of labour shortage. Labour mobility is another area which can help the smooth functioning of businesses. Visa facilitation and more open labour markets are critical for the survival and growth of many companies in the region.

Protection of IP through stricter implementation is another key area for public policy. Where researchers and engineers possess patented knowledge and join competitor companies, more stringent rules in the contract may be required on maintaining trade secrets. In such cases, governments could provide guidance on reasonable ways to protect trade secrets through the design of employment contracts in the private sector.

Last but not least, facilitating the development of services catering to GVCs involves multiple government agencies. Proper coordinating mechanisms to ensure corresponding measures are carried out across different agencies to promote value chain development should be put in place.

APPENDIX

Table A.1: Services in Server Global Value Chain.

Services	Corresponding CPC ver. 2 code	In-house	If outsourced: To affiliates; reasons	If outsourced: To third parties; reasons	Bundled
Value chain specific services					
Stage 1: Research and design					
1. Market research, customer needs assessment	837 — Market research and public opinion polling services	X	X: resource sharing	X: certain industry trend, new application, etc.	
2. Conception of the product	8112 — Research and experimental development services in engineering and technology	X	X: resource sharing	X: leverage the expertise	
3. Design the product	8314 — IT design and development services	X	X: resource sharing	X: leverage the expertise	
	8391 — Specialty design services	X	X: resource sharing	X: leverage the expertise	
	8392 — Design originals	X	X: resource sharing	X: leverage the expertise	

(*Continued*)

Table A.1: (*Continued*)

Services	Corresponding CPC ver. 2 code	In-house	If outsourced: To affiliates; reasons	If outsourced: To third parties; reasons	Bundled
Stage 2: Procurement					
4. Customs-related services (for import)	85999 — Other support services n.e.c.	X	X: resource sharing	X: brokers for clearance, etc.	
5. Procurement service	85999 — Other support services n.e.c.	X	X	X: brokers/vendor	
6. Freight transportation services (of raw materials) by road, rail, sea or air	Division: 65 — Freight transport services			X: not feasible to supply in house	
7. Storage of raw materials — general storage	67290 — Other storage and warehousing services	X	X: resource sharing	X: transition	

Stage 3: Production and testing				
8. Installation of production equipment	87320 — Installation services of industrial, manufacturing and service industry machinery and equipment	X	X: resource sharing	X: certain type of equipment
	87360 — Installation services of electrical machinery and apparatus n.e.c.	X	X: resource sharing	X: certain type of equipment
9. Utilities (electricity, water and gas)	691 — Electricity and gas distribution (on own account)	X	X: resource sharing	X: lack of expertise
	692 — Water distribution (on own account)			X: lack of expertise
	863 — Support services to electricity, gas and water distribution			X: required by law and regulations
10. Production administration — Production management	83115 — Operations management consulting services	X	X: resource sharing	X: for better management
	83116 — Supply chain and other management consulting services	X	X: resource sharing	X: for better management
	83117 — Business process management services	X	X: resource sharing	
	83118 — Head office services	X	X: resource sharing	

(Continued)

Table A.1: (*Continued*)

Services	Corresponding CPC ver. 2 code	In-house	If outsourced: To affiliates; reasons	If outsourced: To third parties; reasons	Bundled
11. Technical testing (International Quality Consultant (IQC) inspection, visual inspection, Automatic Optical Inspection / Automatic X-Ray Inspection (AOI/ AXI) test, substrate OQA inspection, package inspection)	8344 — Technical testing and analysis services	X	X: resource sharing		
12. Manufacture	8874 — Computer, electronic and optical product manufacturing services	X	X: resource sharing	X: economic scale	
	8875 — Electrical equipment manufacturing services	X	X: resource sharing	X: economic scale	
13. Meeting international standards	990 — Services provided by extraterritorial organisations and bodies	X	X: resource sharing	X: strong relationship with the related stakeholder	

14. Maintenance and repair of production equipment	87156 — Maintenance and repair services of commercial and industrial machinery	X	X: resource sharing	X: lack of expertise and for better service
15. Sewage water treatment services	94110 — Sewerage and sewage treatment services	X		X: required by law and regulations
Stage 4: Logistics				
16. Product testing (for obtaining certification at the export market)	8344 — Technical testing and analysis services			X: strong relationship with the related stakeholder
17. Freight transportation (domestic and cross-border)	651 — Land transport services of freight			X: not the core business
	652 — Water transport services of freight			X: not the core business
	653 — Air and space transport services of freight			X: not the core business
	67910 — Freight transport agency services and other freight transport services			X: not the core business
18. Freight insurance	71333 — Freight insurance services			X: not the core business
Stage 5: Assembly and package				
19. Design of packages	83919 — Other specialty design services	X		X: leverage the expertise

(Continued)

Table A.1: (*Continued*)

Services	Corresponding CPC ver. 2 code	In-house	If outsourced: To affiliates; reasons	If outsourced: To third parties; reasons	Bundled
20. Packaging services	85400 — Packaging services	X	X: resource sharing	X: economic scale	
Stage 6: Distribution and after-sales service					
21. Advertising	8361 — Advertising services	X	X: resource sharing	X: leverage the expertise	
	83611 — Full service advertising	X	X: resource sharing	X: leverage the expertise	
	83612 — Direct marketing and direct mail services	X		X: overseas partner	
	83619 — Other advertising services				
22. Marketing (attending trade fairs and conventions)	8596 — Convention and trade show assistance and organisation services	X	X: resource sharing	X: leverage the expertise	
	83114 — Marketing management consulting services	X	X: resource sharing	X: leverage the expertise	
23. Warehousing	672 — Storage and warehousing services	X	X: resource sharing	X: EU/US partner	

24.	Retail trade services	624 — Other non store retail trade services			X: better coverage
25.	Market research/ consulting services	8370 — Market research and public opinion polling services	X		X: leverage the expertise
		8311 — Management consulting and management services			X: better service coverage
26.	News and information service (PRNewswire/ Thomson Reuters)	84410 — News agency services to newspapers and periodicals	X	X: resource sharing	X: strong relationship with the related stakeholder
27.	Maintenance and after sales services	8713 — Maintenance and repair services of computers and peripheral equipment	X	X: resource sharing	X: certain maintenances and service agreement
		8715 — Maintenance and repair services of other machinery and equipment (including electrical machinery and apparatus, telecommunication equipment and apparatus, commercial and industrial machinery, etc.)	X	X: resource sharing	X: certain maintenances

(Continued)

Table A.1: (*Continued*)

Services	Corresponding CPC ver. 2 code	In-house	If outsourced: To affiliates; reasons	If outsourced: To third parties; reasons	Bundled
Stage 7: Repair and recycle					
28. Recycling services	894 — Material recovery (recycling) services, on a fee or contract basis			X: required by laws and regulations	
29. Disposal services	943 — Waste treatment and disposal services			X: required by laws and regulations	
Horizontal services					
30. Company registration and licensing services	91138 — Public administrative services related to general economic, commercial and labour affairs	X	X: resource sharing	X: government service	
31. Set up offices, computers	8733 — Installation services of office and accounting machinery and computers	X	X: resource sharing	X: government service	

32. Financial services (including IPO related services)	7120 Investment banking services		X: leverage the expertise
	71511 — Mergers and acquisition services	X	X: leverage the expertise
	71512 — Corporate finance and venture capital services	X	
	7112 Deposit services		
	7113 — Credit-granting services		X: not the core business
	71591 — Financial consultancy services	X	X: required by laws and regulations
	71592 — Foreign exchange services		X: leverage the expertise
	8231 — Corporate tax consulting and preparation services	X	X: required by laws and regulations
33. Service charges for listed companies in the securities market	71552 — Financial market regulatory services	X	

(*Continued*)

Table A.1: (*Continued*)

Services	Corresponding CPC ver. 2 code	In-house	If outsourced: To affiliates; reasons	If outsourced: To third parties; reasons	Bundled
34. Accounting, auditing and bookkeeping services	8221 — Financial auditing services	X	X: resource sharing	X: required by laws and regulations	
	8222 — Accounting and bookkeeping services	X	X: resource sharing	X: required by laws and regulations	
35. Legal consulting, including legal compliance, dispute settlement and arbitration	8212 — Legal advisory and representation services concerning other fields of law	X	X: resource sharing	X: required by laws and regulations	
	8213 — Legal documentation and certification services	X	X: resource sharing	X: required by laws and regulations	
	82191 — Arbitration and conciliation services	X	X: resource sharing	X: required by laws and regulations	
36. Human resources	8512 — Labour supply services	X		X: better coverage	
	8511 — Personnel search and referral services	X		X: better coverage	
	83113 — Human resources management consulting services	X			

37.	Telecom services	841 — Telephony and other telecommunications services	X		X: better coverage
		8422 — Internet access services			X: better coverage
38.	IT services	83132 IT support services	X	X: resource sharing	X: better coverage
		83151 — Website hosting services	X		X: better service
		8422 — Internet access services			X: better service
39.	Insurance service (commercial life and accident/health insurance, property insurance for the factory compound, product quality insurance, management liability insurance)	7131 — Life insurance and pension services			X: required by laws and regulations
		7132 — Accident and health insurance services			X: required by laws and regulations
		71334 — Other property insurance services			X: required by laws and regulations
		71335 — General liability insurance services			X: required by laws and regulations
40.	Public relations services	83121 — Public relations services	X		X: leverage the expertise
41.	Business hospitality	63 Accommodation, food and beverage services	X		X: better coverage

(*Continued*)

Table A.1: (*Continued*)

Services	Corresponding CPC ver. 2 code	In-house	If outsourced: To affiliates; reasons	If outsourced: To third parties; reasons	Bundled
42. Social insurance for factory workers	91320 — Administrative services related to government employee pension schemes; old-age disability or survivors' benefit schemes, other than for government employees	X		X: required by laws and regulations	
	91330 — Administrative services related to unemployment compensation benefit schemes	X			
43. Accommodation for workers and managers	63220 — Room or unit accommodation services for workers in workers hostels or camps	X	X: resource sharing		
	72111 — Rental or leasing services involving own or leased residential property	X	X: resource sharing	X: required by law and regulations	
44. Canteen service	63393 — Other contract food services	X	X: resource sharing	X: economics scale	

45.	Training services (for workers)	9291 Other education and training services	X	X: resource sharing	X: certain professional training
46.	Security services	8523 — Security systems services	X		X: for better service
		8525 — Guard services	X		X: for better service
47.	Cleaning services	85330 — General cleaning services	X	X: resource sharing	X: economic scale
48.	Waste collection services	942 — Waste collection services			X: required by law and regulations
49.	Government inspections on fire prevention, health hazards, environmental protection and other aspects	91133 — Public administrative services related to mining and mineral resources, manufacturing and construction	X	X: resource sharing	X: required by law and regulations
		91290 — Public administrative services related to other public order and safety affairs	X	X: resource sharing	X: required by law and regulations
50.	Travel services	642 — Long-distance transport services of passengers	X		X: better coverage
		641 — Local transport and sightseeing transportation services of passengers	X	X: resource sharing	X: better coverage

(*Continued*)

Table A.1: (*Continued*)

Services	Corresponding CPC ver. 2 code	In-house	If outsourced: To affiliates; reasons	If outsourced: To third parties; reasons	Bundled
51. Visa and immigration services for foreign investors/employees (travel between headquarter and offices in different economies)	91290 — Public administrative services related to other public order and safety affairs	X	X: resource sharing	X: required by laws and regulations	
52. Patent and trademark protection	8396 — Trademark and franchises	X	X: resource sharing	X: required by laws and regulations	
53. Industry association membership	9511 — Services furnished by business and employers organisations			X: strong relationship	
	9512 — Services furnished by professional organisations			X: strong relationship	
54. Interaction with Trade Union	9520 — Services furnished by trade unions	X		X: strong relationship	

Source: Author.

REFERENCES

Gens, F (2013). IDC Predictions 2013 competing on the 3rd platform [Powerpoint Slides]. http://jetroplatforms.com/wp-content/uploads/2013/04/IDC Predictions-2013.pdf [15 March 2015]

Kuriyama, C and A Ogazon (2013). Expanding the Information Technology Agreement. Singapore: APEC Policy Support Unit Policy Brief No. 6, May 06.

iWeb Technologies (2012). History of Servers in Pictures from 1981 to today. http://blog.iweb.com/en/2012/01/history-of-servers-in-pictures-from-1981-to-today/1959.html [15 March 2015].

The Economist (2013). The Server Market Shifting Sands. http://www.economist.com/news/business/21578678-uphearal-less-visible-end-computer-industry-shifting-sands [15 March 2015].

United Nations (2013). United Nations Central Product Classification, Version 2. http://unstats.un.org/unsd/registry/cpc-2.asp [15 March 2015].

Chapter 6

WASTEWATER TREATMENT SERVICES

Arian Hassani and Andre Wirjo

6.1. INDUSTRY OVERVIEW

Driven by various issues, such as rapid urbanisation, growing populations, the importance of sustainability and the need to mitigate climate change, issues related to water have gained significant prominence in recent years. The 2015 United Nations World Water Development Report noted that the planet is expected to face a 40 per cent shortfall in water supply by 2030 unless the management of this resource is dramatically improved.

With such information widely available, it is no surprise that the water Design, Build, Operate (DBO) market is estimated to grow at a compound annual growth rate (CAGR) of 8.6 per cent between 2010 and 2020 (Royan, 2012). However, it is worthwhile to mention that the global expansion is led by not only the emerging needs of developing economies but also the upgrading requirements of plants in developed economies. Indeed, while the CAGR for the Asia-Pacific region is forecast to grow at 10.4 per cent between 2010 and 2020, the CAGR for North America is also growing healthily at 7.8 per cent over the same period.

As an industry, water treatment evolves continuously with time. The technological, logistical and regulatory complexities associated with water treatment make bundled design, construction and operation service packages highly profitable. Water infrastructure, industrial needs and uses, in addition to environmental norms and regulations, make the water DBO market highly sensitive to local conditions and know-how. In line with the idea of sustainability, the trend for water treatment has also moved towards sustainable water management solutions.[1] Among newly employed technologies are the use of membrane bioreactors and organisms to treat wastewater.

The key to competitive positioning in water DBO market appears to lie in the provision of fast, low-cost and environmental-friendly services that embody a "plug-and-play" component. This is especially true in markets where the labour pool tends to be made up of low-skilled workers because it minimises knowledge-specific/specialised local interventions. To maintain a competitive edge, industry players need to carve out "clear geographic priorities, balanced business models, and state-of-art technologies so as to constantly stay ahead of environmental challenges and client needs" (Royan, 2012).

6.2. BACKGROUND INFORMATION OF FIRM[2]

The firm depicted in this case study provides end-to-end waste water treatment and management services, from data gathering and analysis to the design and construction of water treatment plants, and the manufacturing and sales of water treatment products and chemicals. It is headquartered in Japan and currently has operations in China, Chinese Taipei, Thailand as well as Viet Nam. It has plans to expand to Indonesia and other economies in Southeast Asia.

The firm's client base generally stems from the automotive and food and beverage industries. To meet customers' diverse wastewater treatment needs, the firm determines the optimum processing conditions and proposes the use of its proprietary treatment procedures. The firm does not only deal with the establishment of

---

[1] In terms of water reuse and recycling, in addition to adopting more sophisticated technologies and reducing the carbon footprint.

[2] All information about the firm in this case study is from its corporate brochure as well as from the interview.

Table 6.1: Comparison Between Firm's and Other Physico-Chemical Treatment Process.

Firm's treatment process	Other physico-chemical treatment process
Automated operation requires fewer staff	Manual operation requires more staff
Minimal sludge and other by-products	Abundant sludge and other by-products
Utilise less chemicals	Utilise more chemicals
Treatment plants occupy smaller area	Treatment plants occupy larger area
Treatment cost for different industry and type of wastewater ranges between ¥1 million and ¥20 million	Treatment cost for different industry and type of wastewater ranges between ¥18 million and ¥80 million

Source: Authors and courtesy of firm.

new water treatment plants but also meets the modification and enhancement needs of already-existing plants.

Central to the firm's business model is its proprietary biological chemical which has been developed in-house and used in virtually all of the firm's recommended treatment procedures. This chemical contains living microbes/bacteria that can decompose various kinds of organic matter commonly found in wastewater discharged by manufacturing plants. More importantly and in the context of environmental friendliness, water and carbon dioxide are the two main by-products generated by the firm's treatment procedures. In contrast, other wastewater treatment procedures usually lead to the formation of larger quantity of sludge (see Table 6.1).

As sludge treatment costs continue to rise worldwide, many companies are starting to avoid processes that will lead to the production of large quantities of sludge. Depending on the industry and type of wastewater, firm's technology can reduce the cost of its clients' wastewater treatment process to between 1/3 and 1/25 of the original cost. Even in Thailand, where sludge treatment has remained affordable, the firm continues to grow its customer base. Other advantages of firm's treatment process vis-à-vis other treatment process include automation, less chemical utilisation and more compact treatment plants (see Fig. 6.1).

Figure 6.1: Layout of Wastewater Treatment Facility of a Client Before and After Installation of Firm's Wastewater Treatment Plant.

Source: Courtesy of the firm.

6.3. DESCRIPTION OF THE VALUE CHAIN

This case study will focus on the firm's Thailand subsidiary and examines a particular water DBO value chain that the subsidiary provides to its automotive client in Thailand.

Although the case study has focused on a value chain specific to a client to facilitate the mapping process of services needed, it should be noted that a similar value chain generally applies to clients from different industries. Variations, if any, are usually observed in the type of biological chemicals used to treat the

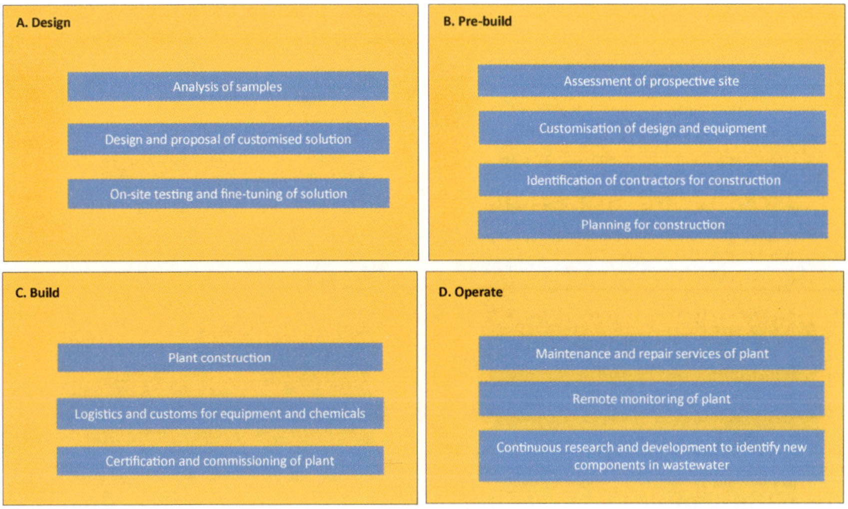

Figure 6.2: Dimension of the Value Chain Covered by Case Study.

Source: Authors.

wastewater considering that different wastewater are made up of different components. As an illustration, wastewater produced by sugar manufacturing plant may be different in content from that produced by automotive manufacturing plant. In addition, firm shared that wastewater produced by certain industries such as automotive need to be pre-treated prior to treatment with its biological chemical while that from industries such as food do not need pre-treatment.

For the purpose of this case study, the value chain begins when the firm receives a request from its client to provide water treatment DBO services and ends when the firm operates the treatment plant on behalf of its client, which includes services such as monitoring and maintenance and disposal of treatment by-products (see Fig. 6.2)

6.3.1. *Design*

Upon receiving a request from its client in the automotive industry to provide water DBO services to its manufacturing facility in

Thailand, the firm flies in engineers from its headquarters in Japan to collect, analyse and run tests on samples from the client's existing water treatment plant. The entire process has to be done in Thailand because of the instability of the contents within the wastewater.

Based on the test results, the engineers design and suggest a customised treatment solution that is tested on-site and fine-tuned as the pilot proceeds. Past experience in working with the client on the same project for its facility in Japan also helped the firm to quickly identify the client's needs.

6.3.2. *Pre-building*

Once the samples from its initial pilot are further assessed and its client is satisfied with the proposed solution, the firm moves to the pre-building stage of the treatment plant where among others, it carries out an assessment of the prospective site for the plant and fine-tunes the design and customisation of its equipment.

In addition, the firm has to obtain construction permits from relevant government agencies and identify contractors to outsource construction activities such as welding and piping work, which very often have to be local firms.

6.3.3. *Building*

The construction period for a treatment plant usually takes approximately six months. During this time, the firm will also be importing core equipment and chemicals from Japan so that they can be installed when the plant is ready. The firm takes care of the certification and commissioning services of the equipment as well.

6.3.4. *Operation*

As part of its strategy to capture more from its value chain, the firm never supplies its proprietary biological chemical on its own. Instead, it provides them together with wastewater treatment services. In order to make sure its proprietary technology is well protected, it essentially undertakes all operational activities of the treatment plant, including regular testing of treated water, remote monitoring of the plant, maintenance and repair of installed equipment, as well as continuous research and development to identify new chemicals that need to be removed from the water prior to discharge into public area. Firm also takes care of arranging logistics firm to collect the oil by-products produced by its treatment process.

6.4. SERVICES ALONG THE VALUE CHAIN

6.4.1. *Services Identification and Value Contribution*

A total of at least 118 services can be identified in this value chain and have been listed in Appendix A under 54 more aggregated categories. The services have been grouped according to the various stages within the chain: (i) services provided during the design stage, (ii) pre-building services, (iii) services provided during the building stage, (iv) services during the operation stage and (v) back office services (see Fig. 6.3).

Although there is limited information on which services inputs are the most important in terms of value added, it can be surmised that engineering services are perhaps among the most critical inputs because they are pervasive throughout the entire DBO chain, with the exception of back office support. The substantial number of services entering the value chain described in this case study shows that services are ubiquitous and crucial for the value chain to function properly.

The important role of services in the value chain can also be seen in the number of staff involved. The complement of five staff in the firm's Thailand subsidiary is generally providing services to clients and none are directly involved in the chemicals and equipment manufacturing processes. Among the services activities provided are engineering services throughout the entire DBO value chain, remote monitoring services of installed equipment, as well

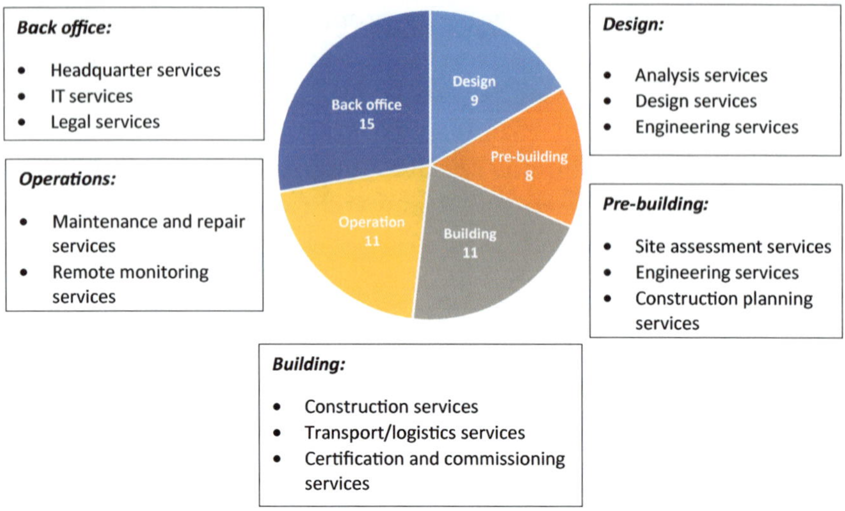

Figure 6.3: Breakdown of Services by Stage and Examples of Key Services.

Source: Authors.

as sales and marketing services for its product-cum-services offerings.

From a cost perspective, the firm indicated that the cost of the chemicals and equipment is relatively low compared to the costs of services identified in Fig. 6.3. Moreover, the cost of chemicals and equipment also embodies a number of services inputs such as research and development activities undertaken to formulate the chemicals and design services for the equipment.

In the near future, the share of services in the value chain is expected to fall slightly as the firm is currently building a chemical manufacturing and centralised wastewater treatment facility in Chonburi Province. The facility is expected to become operational by mid-2015. The services share in total value will nevertheless remain significant because the facility intends to expand its customer base to include clients who prefer to have their wastewater treated at a central location. Among the additional services that will be required for this new offering is transport/logistics services to transport wastewater from clients' plants to the firm's facility. The firm's expansion plans in Southeast Asia also suggest that the

Thailand office may have a regional role to play, such as the provision of regional headquarters services.

6.4.2. *Outsourcing, Bundling and other Aspects of Services Supply*

Among the 118 services identified in this value chain, 54 are supplied in-house, 3 are partially outsourced and the remaining 61 services are fully outsourced (see Appendix A).[3]

Services provided entirely in-house are those that the firm considers core services activities such as project-specific design and engineering services, analysis services of samples, installation and commissioning services as well as remote monitoring of treatment plants. These services are provided in-house to minimise the risk of proprietary information leakage. In fact, the firm not only keeps most of its core services in-house (i.e., within its Thailand operations), but it actually obtains them from its headquarters in Japan. From the collection and analysis of a client's wastewater, to the testing, design and implementation of treatment systems, in-house talent is solicited from Japan. Even the core components of a treatment plant are produced in-house in Japan and shipped to Thailand for assembly. Besides ensuring the security of its proprietary information, these arrangements allow the firm to overcome the lack of skilled labour locally. Furthermore, firm representatives noted that

[3] The reasons identified by the firm for outsourcing depend on the services involved but can generally be grouped into the following explanation categories: (i) government services, such as company licensing and issuance of construction permits for plants; (ii) services required by laws and regulations, such as by-product collection services following wastewater treatment and auditing services; (iii) lack of feasibility to supply service in-house, such as utilities and freight insurance services; (iv) lack of expertise such as site assessment and preparation services; (v) external economies of scale such as transport/shipping services of chemical solutions and equipment; (vi) need for strong relationships with government agencies, such as customs clearance services and (vii) network economies such as personnel search and referral services.

it takes years of on-the-job training to reach the required level of expertise.

In addition to its stringent outsourcing ethos, the firm also relies upon a strategic bundling of services to ensure the sustainability of its cash flow generation model. As indicated earlier, the firm's core asset is the proprietary biological chemical which it manufactures in-house in Japan. If this product were sold by itself in the market, clients would simply integrate it into their wastewater treatment procedures and firm would not be able to obtain more value from its value chain. The biological chemical is not complicated to produce, and when patents on the product run out, the firm will face heavy competition and significant pressure to capture market share.

The firm realised that it could not depend on this core proprietary product alone to ensure the sustainability of its business model. Instead, it decided to offer this proprietary chemical as a bundled package of services that also includes proprietary treatment procedures. This particular addition makes the core service more difficult to mimic as a bundle. With a relatively small initial investment (when compared to ensuing cost savings) to build the water treatment plant, clients enter into what could virtually be referred to as a "lifelong" (the duration of the life of the water treatment plant) contractual agreement with the firm. As a result, the firm does not have to constantly seek new clients. Rather, it can focus on providing top-notch, cutting-edge services to a few major clients who are bound to the firm's services over the long term.

Several of the firm's offerings are provided as bundled services. One example of such bundled services is transport/shipping services of chemical solutions. The package offered by the third-party logistics firm includes customs clearance, a vehicle and driver (as well as vehicle repair and maintenance) if land transportation is involved, the unloading and transfer of goods at the warehouse, insurance and other financial services. Another important example in this value chain is construction services where in addition to the construction itself, the contractor would be responsible for a range of other services, including sourcing of workers and construction equipment, providing training to workers, purchasing their insurance and possibly housing them in temporary accommodation throughout the entire construction period.

6.5. POLICIES AFFECTING THE VALUE CHAIN

An important part of this study involves the analysis of how government policies affect the value chain (see Appendix B). The case study firm has constructed a business model that combines the manufacture of proprietary biological chemical and other products with customer contracts that embody a range of services. This multi-product business model is not uncommon in the world of global value chains and it makes enterprises sensitive to a range of possible policy risks, including in respect of services. A number of different policies could therefore undermine the firm's strategy and overall viability.

6.5.1. *Enforcement of Environmental Laws and Regulations*

The firm's product-cum-service offerings are competitive both in terms of cost and environmental friendliness. Its wastewater treatment technology reduces cost by between 1/3 and 1/25 of the previous cost. The firm's proprietary process also utilises a lesser quantity of chemicals and produces less sludge compared to other treatment procedures. An increasing awareness among individuals and firms of the need to minimise environmental degradation might have served to increase the firm's client base. But interest in Thailand and a few neighbouring economies in the company's offerings has seemingly been driven more by cost considerations. The environmental-friendliness aspect comes either: (1) as a bonus; (2) because it comprises part of the corporate social responsibility (CSR) policy of its clients or (3) because it is a requirement of clients of firms supplied by the case study company.

One explanation for a weak response on the environmental side of things could be a lack of enforcement of environmental laws and regulations, notwithstanding strict laws on the maximum allowable limit for treated wastewater discharged in public domains (chemical oxygen demand (COD) of less than 120 mg/L).[4] This leads to situations where treatment of industrial wastewater is more process-driven to adhere to Section 70 of the Enhancement and Conservation of National Environmental Quality Act B.E. 2535[5]; despite not observing the stipulated limits for industrial discharge, firms try to build water treatment facilities as required in order to minimise the risk of being found non-compliant.

The issue is further compounded when even Section 70 as well as related Sections 71 and 72[6] are not fully enforced, leading to many firms not installing wastewater treatment operations on-site or sending their wastewater to centralised wastewater treatment plants. Indeed, government regulators rarely check the quality of industrial discharge into public domains such as rivers and appear, for now, to leave compliance to individual firms' responsibility. This leads to a situation where only firms that care about the reputational risk of a lack of enforcement in their own economies, or have their own CSR, to undertake proper wastewater treatment processes.

When government regulators come for inspection visits, they sometimes appear to have limited knowledge of the technology

---

[4] COD up to 400 mg/L is allowed for food furnishing factories, animal food factories, textile factories and pulp and paper factories (http://www.pcd.go.th/info_serv/en_reg_std_water04.html).

[5] Section 70 of the Enhancement and Conservation of National Environmental Quality Act B.E. 2535 (1992) states that as determined by the pollution control authorities, the owner or possessor of the point of source of pollution may have the duty to construct, install or bring into operation an on-site facility for wastewater treatment or waste disposal. For this purpose, the firm concerned may be required to commission a Monitoring Control Operator to control wastewater treatment or a waste disposal facility that shall be constructed, installed or brought into operation accordingly. More details can be obtained at: http://www.pcd.go.th/info_serv/en_reg_envi.html.

[6] Sections 71 and 72 of the Enhancement and Conservation of National Environmental Quality Act B.E. 2535 (1992) require the owner or possessor of any point source of pollution to send its wastewater to the centralised treatment plant if the operation concerned does not have on-site facility. More details can be obtained at: http://www.pcd.go.th/info_serv/en_reg_envi.html.

behind wastewater treatment. In such cases they may sign off on reports with whose contents they are not entirely familiar. This will tend to make site visits more of a cost than a benefit.

6.5.2. *Intellectual Property (IP) Regime*

As a developing, middle-income economy, the economic prospects of Thailand are attractive to businesses. Besides hoping to reach out to more clients via its current offerings, the firm is currently constructing a chemical manufacturing and centralised wastewater treatment facility in Chonburi Province. The facility, when operational in mid-2015, will treat wastewater from clients that do not have their own treatment facilities, or prefer to treat them in a centralised facility.

One important consideration among businesses, including the firm when executing its expansion plan, pertains to the host economy's IP regime. Depending on its perception of the current IP regime, businesses may exercise more caution about transferring technology and knowledge to their local operations.

Acknowledging the importance of IP protection, the Thai government has taken steps to enhance its IP regime so that it is more in line with international standards of protection and enforcement.[7] It established the Central Intellectual Property and International Trade Court (CIPITC) in 1997 to adjudicate cases involving IP.[8] It also signed the Paris Convention for the Protection of Industrial Property and the Patent Cooperation Treaty, in 2008 and 2009, respectively, to align its national standards better with those set out in the WTO Agreement on Trade Related Aspects of International Property Rights agreement. More recently, it has established the National IPR Centre of Enforcement to coordinate efforts to prevent IP violations.[9]

---

[7]More information about Thailand's IP regime can be obtained at: http://www.ipthailand.go.th/en/.

[8]More information about the Central Intellectual Property and International Trade Court (CIPITC) can be obtained at: http://www.ipitc.coj.go.th/?co=en.

[9]http://www.aseanbriefing.com/news/2014/08/21/thailands-intellectual-property-regime.html.

Yet, the firm continues to express concerns about transferring technology and technical know-how to its local operations. This is not a surprising finding because despite the steps mentioned above, Thailand has been in the Watch List or Priority Watch List of the US Trade Representative Special 301 Report since its inception in 1989.[10] The annual report provides a review of the state of IP protection and enforcement by US trading partners.

6.5.3. *International Labour Mobility*

One option utilised by firm to alleviate its IP concerns is to entrust staff from its headquarters in Japan to carry out tasks requiring higher technical expertise in Thailand and potentially other economies such as Indonesia. While the firm faces no significant issues in obtaining temporary-stay visa for its staff in Thailand, it indicated that it may face challenges in obtaining temporary-stay visas or work permits for its international staff in Indonesia and other economies where it plans to have operations.

The main issue is likely related to whether the economy of interest has put in place some form of quota system, or requires a labour market test before it will issue a temporary-stay visa or work permit for intra-corporate transferees. Even if these do not pose barriers to firm, the permitted length of stay may be an issue. The OECD Services Trade Restrictiveness Index (STRI) regulatory database for cross-border provision of engineering services in economies where the firm has or plans to have operations, and where data are available (China and Indonesia), show that both economies require a labour market test for intra-corporate transferees (see Table 6.2).

Besides limiting the stays of intra-corporate transferees between 12 and 36 months, the Ministry of Manpower of Indonesia is reported to be working on an Indonesian language proficiency test

[10] https://ustr.gov/sites/default/files/USTR%202014%20Special%20301%20Report%20to%20Congress%20FINAL.pdf and http://www.iipa.com/pdf/2014SPEC301HISTORICALCHART.pdf.

Table 6.2: Policies Affecting Intra-Corporate Transferees in China and Indonesia.

	China	Indonesia
Quota for intra-corporate transferees	No	No
Labour market test for intra-corporate transferees	Yes	Yes
Limitation on stay for intra-corporate transferees	More than 36 months	Between 12 and 36 months

Source: OECD STRI Regulatory Database.

for both existing international staff who are already based in Indonesia as well as prospective international staff.[11,12] The provision, which reportedly requires foreigners (excluding commissioners, directors and those in temporary employment) to master the Indonesian language before they are able to obtain a work permit, would further strengthen the 2013 Indonesian Minister of Manpower Regulation No. 12.

6.5.4. *Local Third-Party Logistics*

The Foreign Business Act B.E. 2542 (1999) determines the extent of foreign business activities and limits the rights of foreigners to undertake certain business activities in Thailand. Activities that are controlled are categorised into three schedules and transportation/logistics industry can be found in schedule 2 of the Act.[13] This essentially means that the case study firm has to utilise locally licensed third-party logistics firms to transport imported equipment and chemicals as well as to collect the oil by-products produced by its treatment process. Such arrangements reduce

[11] http://www.thejakartapost.com/news/2015/03/05/minister-prepares-language-proficiency-test.html.

[12] There is an indication that the plan for language proficiency test will be withdrawn but this has not been announced officially (http://thejakartaglobe.beritasatu.com/news/indonesia-withdraw-local-language-plan-foreign-workers-sources/).

[13] http://bic.thaiembassy.sg/node/104,

competition in the logistics sector and are likely to raise the costs of these services.

The firm may also face a challenge in finding reliable logistics firms to transport wastewater from clients' premises to its new facility when this becomes operational. Of late, there have been news reports about the illegal disposal of waste, with detrimental effects on the environment and nearby communities.[14] The possibility of hiring a logistics company that does not comply with regulations represents a reputational risk for the firm.

6.5.5. *The Way Forward*

This case study has attempted to give a wastewater treatment firm's perspectives on the importance of services to its operations. Government policies clearly impact the firm's ability to access and provide services and hence, its overall viability. More importantly, policies may have unintended consequences on the economy as a whole when businesses find ways to go around them. Policymakers, therefore, have a significant role in supporting businesses to the benefit of their own economies.

Lack of enforcement of environmental laws and regulations, for example, leads to situations where the firm is unable to reach its full potential. By strengthening the enforcement of these laws, governments facilitate the operations of firms, bringing economic and social benefits, both in terms of employment and environmental improvements.

Similarly, regulations affecting the movement of natural persons such as stay limitations, labour market tests and language proficiency requirements may also cause firms to reconsider their present operations and future plans for the region. Governments may want to weigh the costs of such limitations, in terms of their

[14] http://www.nationmultimedia.com/national/Praksa-fire-leads-to-calls-for-nationwide-dumpsite-30230655.html and http://news.yahoo.com/thailand-totters-towards-waste-crisis-041321001.html.

effects on foreign direct investment (FDI) flows, against their perceived benefits.

Finally, despite improvements, the current IP regime in Thailand remains a concern to firms when considering the transfer of technology and know-how to their local operations. More technology and knowledge transfer will foster domestic innovation and ultimately greater upgrading opportunities in value chains. It may also encourage additional expansion by firms and increase output and employment opportunities.

APPENDIX A

Table A.1: Services during Design Stage.

	Service	Corresponding CPC ver. 2 code	Supplied in-house	Outsourced to affiliated companies and reasons	Outsourced to third-party suppliers/ government and reasons	Bundled
1	Travel services for engineers and other staff to collect and analyse samples	8551 — Reservation services for transportation	No	No	Yes, efficiency	
		85521 — Reservation services for accommodation	No	No	Yes, efficiency	
2	Visa and immigration services for staff	91290 — Public administrative services related to other public order and safety affairs	No	No	Yes, government services	
3	Analysis services of samples	8344 — Technical testing and analysis services	Yes	No	No	

4	Design services to support proposed solution	8314 — Information technology (IT) design and development services	Yes	No	No
		83920 — Design originals	Yes	No	No
		83912 — Industrial design services	Yes	No	No
		8393 — Scientific and technical consulting services n.e.c.	Yes	No	No
		83142 — IT design and development services for networks and systems	Yes	No	No
		83143 — Software originals	Yes	No	No

(*Continued*)

Table A.1: (*Continued*)

Service		Corresponding CPC ver. 2 code	Supplied in-house	Outsourced to affiliated companies and reasons	Outsourced to third-party suppliers/ government and reasons	Bundled
5	Engineering services to support proposed solution	83310 — Engineering advisory services	Yes	No	No	
		83326 — Engineering services for waste management projects (hazardous and non-hazardous)	Yes	No	No	
		83327 — Engineering services for water, sewerage and drainage projects	Yes	No	No	
6	Transport/shipping services of chemical solutions	651 — Land transport services of freight	No	No	Yes, efficiency; economies of scale	Bundled with transport, personnel, repair and insurance
		652 — Water transport services of freight	No	No	Yes, efficiency; economies of scale	Bundled with transport, personnel, repair and insurance

		6531 — Air transport services of freight	No	No	Yes, efficiency; economies of scale	Bundled with transport, personnel repair and insurance
		67910 — Freight transport agency services and other freight transport services	No	No	Yes, efficiency; economies of scale	Bundled with transport, personnel, repair and insurance
7	Freight insurance of chemical solutions	71333 — Freight insurance services	No	No	Yes, not possible to supply in-house service in-house	Bundled with transport
8	Customs clearance services and logistics of chemical solutions	67110 — Container handling services	No	No	Yes, efficiency; strong relationship with government agencies; government services	Bundled with transport, personnel, repair and insurance

(*Continued*)

Table A.1: (*Continued*)

Service	Corresponding CPC ver. 2 code	Supplied in-house	Outsourced to affiliated companies and reasons	Outsourced to third-party suppliers/ government and reasons	Bundled
	85999 — Other support services n.e.c.	No	No	Yes, efficiency; strong relationship with government agencies; government services	Bundled with transport, personnel, repair and insurance
9 Storage and warehousing services of chemicals solutions	67220 — Bulk liquid or gas storage services	Yes	No	No	
	67290 — Other storage and warehousing services	Yes	No	No	

Source: Authors.

Table A.2: Services during Pre-Build Stage.

Service		Corresponding CPC ver. 2 code	Supplied in-house	Outsourced to affiliated companies and reasons	Outsourced to third-party suppliers/ government and reasons	Bundled
10	Site assessment services	83411 — Geological and geophysical consulting services	No	No	Yes, lack of expertise	
		83412 — Geophysical services	No	No	Yes, lack of expertise	
		83421 — Surface surveying services	No	No	Yes, lack of expertise	
		83442 — Testing and analysis services of physical properties	No	No	Yes, lack of expertise	
11	Architectural services for plant	8321 — Architectural services and advisory services	No	No	Yes, lack of expertise	
		8323 — Landscape architectural services and advisory services	No	No	Yes, lack of expertise	

(*Continued*)

Table A.2: (*Continued*)

Service		Corresponding CPC ver. 2 code	Supplied in-house	Outsourced to affiliated companies and reasons	Outsourced to third-party suppliers/ government and reasons	Bundled
12	Design services for equipment as well as analysis and monitoring software	83912 — Industrial design services	Yes	No	No	
		83920 — Design originals	Yes	No	No	
		8393 — Scientific and technical consulting services n.e.c.	Yes	No	No	
		83141 — IT design and development services for applications	Yes	No	No	
		83142 — IT design and development services for networks and systems	Yes	No	No	
		83143 — Software originals	Yes	No	No	
13	Engineering services for equipment	83310 — Engineering advisory services	Yes	No	No	
		83326 — Engineering services for waste management projects (hazardous and non-hazardous)	Yes	No	No	
		83327 — Engineering services for water, sewerage and drainage projects	Yes	No	No	

14	Environmental consulting services	83931 — Environmental consulting services	No	No	Yes, lack of expertise
15	Product testing to obtain certification at export market	8344 — Technical testing and analysis services	No	No	Yes, government services
16	Government-related pre-building services (e.g., securing government permits for construction)	91133 — Public administrative services related to mining and mineral resources, manufacturing and construction	No	No	Yes, government services
17	Planning and management services	83223 — Project site master planning services	Yes	No	No
		83330 — Project management services for construction projects	Yes	No	No
		83990 — All other professional, technical and business services, n.e.c.	Yes	No	No

Source: Authors.

Table A.3: Services during Build Stage.

Service		Corresponding CPC ver. 2 code	Supplied in-house	Outsourced to affiliated companies and reasons	Outsourced to third-party suppliers/ government and reasons	Bundled
18	Engineering services during building/ construction	83310 — Engineering advisory services	Yes	No	No	
		83321 — Engineering services for building projects	Yes	No	No	
		83326 — Engineering services for waste management projects (hazardous and non-hazardous)	Yes	No	No	
		83327 — Engineering services for water, sewerage and drainage projects	Yes	No	No	
19	Site preparation services	54310 — Demolition services	No	No	Yes, lack of expertise	
		54320 — Site formation and clearance services	No	No	Yes, lack of expertise	
		54330 — Excavating and earthmoving services	No	No	Yes, lack of expertise	

20	Importation of core equipment and chemicals: Customs clearance and logistics	67110 — Container handling services	No	No	Yes, efficiency; strong relationship with government agencies; government services	Bundled with transport, personnel, repair and insurance
		85999 — Other support services n.e.c.	No	No	Yes, efficiency; strong relationship with government agencies; government services	Bundled with transport, personnel, repair and insurance
21	Importation of core equipment and chemicals: Freight transportation services	651 — Land transport services of freight	No	No	Yes, efficiency; economies of scale	Bundled with transport, personnel, repair and insurance
		652 — Water transport services of freight	No	No	Yes, efficiency; economies of scale	Bundled with transport, personnel, repair and insurance

(*Continued*)

Table A.3: (*Continued*)

	Service	Corresponding CPC ver. 2 code	Supplied in-house	Outsourced to affiliated companies and reasons	Outsourced to third-party suppliers/ government and reasons	Bundled
		6531 — Air transport services of freight	No	No	Yes, efficiency; economies of scale	Bundled with transport, personnel, repair and insurance
		67910 — Freight transport agency services and other freight transport services	No	No	Yes, efficiency; economies of scale	Bundled with transport, personnel, repair and insurance
22	Freight insurance of chemical solutions and equipment	71333 — Freight insurance services	No	No	Yes, not possible to supply in-house service in-house	Bundled with transport
23	Importation of core equipment and chemicals: Storage and warehousing services	67220 — Bulk liquid or gas storage services	Yes	No	No	
		67290 — Other storage and warehousing services	Yes	No	No	
24	Construction services	54 — Construction services	No	No	Yes, lack of expertise	

25	Manufacturing services provided by suppliers of activities such as welding, coating, etc.	886 — Basic metal manufacturing services	No	No	Yes, required by laws and regulations; lack of expertise
		887 — Fabricated metal product, machinery and equipment manufacturing services	No	No	Yes, required by laws and regulations; lack of expertise
26	Installation services for equipment and related components such as wiring	5461 — Electrical installation services	Yes	No	No
		873 — Installation services (other than construction)	Yes	No	No
27	Certification and commissioning services of plant and equipment by firm	8344 — Technical testing and analysis services	Yes	No	No

(*Continued*)

Table A.3: (*Continued*)

Service		Corresponding CPC ver. 2 code	Supplied in-house	Outsourced to affiliated companies and reasons	Outsourced to third-party suppliers/ government and reasons	Bundled
28	Government inspections on fire prevention, health hazards, environmental protection and other aspects	91133 — Public administrative services related to mining and mineral resources, manufacturing and construction	No	No	Yes, government services	
		9129 — Public administrative services related to other public order and safety affairs	No	No	Yes, government services	

Source: Authors.

Table A.4: Services during Operate Stage.

Service		Corresponding CPC ver. 2 code	Supplied in-house	Outsourced to affiliated companies and reasons	Outsourced to third-party suppliers/ government and reasons	Bundled
29	Production administration — Production management	83115 — Operations management consulting services	Yes	No	No	
30	Procurement services	83116 — Supply chain and other management consulting services	Yes	No	No	
		85999 — Other support services n.e.c.	Yes	No	No	
31	Utilities (electricity, gas and water supply)	691 — Electricity and gas distribution (on own account)	No	No	Yes, not possible to supply in-house	
		692 — Water distribution (on own account)	No	No	Yes, not possible to supply in-house	
32	Engineering services during operations	83310 — Engineering advisory services	Yes	No	No	

(*Continued*)

Table A.4: (*Continued*)

	Service	Corresponding CPC ver. 2 code	Supplied in-house	Outsourced to affiliated companies and reasons	Outsourced to third-party suppliers/ government and reasons	Bundled
		83326 — Engineering services for waste management projects (hazardous and non-hazardous)	Yes	No	No	
		83327 — Engineering services for water, sewerage and drainage projects	Yes	No	No	
33	Testing services for quality control and assurance	8344 — Technical testing and analysis services	Yes	No	No	
34	By-product collection services	942 — Waste collection services	No	No	Yes, required by laws and regulations	
35	IT services for on-site and remote monitoring of plant	8315 — Hosting and information technology (IT) infrastructure provisioning services	Yes	No	No	

		8316 — IT infrastructure and network management services	Yes	No	No
36	Telephone-based support services	85931 — Telephone call centre services	Yes	No	No
37	Diagnostic, inspection, maintenance and repair of equipment	87156 — Maintenance and repair services of commercial and industrial machinery	Yes	No	No
38	Training services for workers	9291 — Other education and training services	Yes	No	No
39	Research and development services for continuous upgrading	81112 — Research and experimental development services in chemistry and biology	Yes	No	No

Source: Authors.

Table A.5: Business Processes (Back Office Support).

Service		Corresponding CPC ver. 2 code	Supplied in-house	Outsourced to affiliated companies and reasons	Outsourced to third-party suppliers and reasons	Bundled
40	Company registration and licensing services (obtaining permit to operate)	91138 — Public administrative services related to general economic, commercial and labour affairs	No	No	Yes, government services	
41	Headquarter services	83118 — Head office services	No	Yes, economies of scale	No	
42	Management services	83111 — Strategic management consulting services	Yes	Yes, economies of scale	No	
43	Accounting, auditing and bookkeeping services	82210 — Financial auditing services	Yes	No	Yes, required by laws and regulations	
		8222 — Accounting and bookkeeping services	Yes	No	No	

44	Sales and marketing services	83612 — Direct marketing and direct mail services	Yes	No	No
		8596 — Convention and trade show assistance and organisation services	No	No	Yes, lack of expertise
		8370 — Market research and public opinion polling services	Yes	No	No
		8311 — Management consulting and management services	No	No	Yes, lack of expertise
45	Financial services	7112 — Deposit services	No	No	Yes, not possible to supply in-house
		7113 — Credit granting services	No	No	Yes, not possible to supply in-house
		71512 — Corporate finance and venture capital services	No	No	Yes, not possible to supply in-house
		71591 — Financial consultancy services	No	No	Yes, not possible to supply in-house
		71592 — Foreign exchange services	No	No	Yes, not possible to supply in-house

(Continued)

Table A.5: (*Continued*)

Service		Corresponding CPC ver. 2 code	Supplied in-house	Outsourced to affiliated companies and reasons	Outsourced to third-party suppliers and reasons	Bundled
		83112 — Financial management consulting services	No	No	Yes, not possible to supply in-house	
46	Legal services	82120 — Legal advisory and representation services concerning other fields of law	No	No	Yes, lack of expertise	
		82130 — Legal documentation and certification services	No	No	Yes, lack of expertise	
		82191 — Arbitration and conciliation services	No	No	Yes, lack of expertise	
		8396 — Trademark and franchises	No	No	Yes, lack of expertise	

47	Insurance services (commercial life and accident/health insurance, property insurance for the factory compound, product quality insurance, management liability insurance)	7131 — Life insurance and pension services	No	No	Yes, not possible to supply in-house
		7132 — Accident and health insurance services	No	No	Yes, not possible to supply in-house
		71334 — Other property insurance services	No	No	Yes, not possible to supply in-house
		71335 — General liability insurance services	No	No	Yes, not possible to supply in-house
48	Social insurance for staff	91320 — Administrative services related to government employee pension schemes; old-age disability or survivors' benefit schemes, other than for government employees	Yes	No	No
		91330 — Administrative services related to unemployment compensation benefit schemes	Yes	No	No

(Continued)

Table A.5: (*Continued*)

	Service	Corresponding CPC ver. 2 code	Supplied in-house	Outsourced to affiliated companies and reasons	Outsourced to third-party suppliers and reasons	Bundled
49	Corporate communications and public relationship	83121 — Public relations services	Yes	Yes, economies of scale	No	
50	Human resources management	8511 — Personnel search and referral services	Yes	No	Yes, network economies	
		83113 — Human resources management consulting services	No	No	Yes, lack of expertise	
51	Information technology services	83132 — IT support services	Yes	No	Yes, efficiency; lack of expertise	
		83151 — Website hosting services	No	No	Yes, efficiency; lack of expertise	

52	Telecommunication services	841 — Telephony and other telecommunications services	No	No	Yes, not possible to supply in-house
		8422 — Internet access services	No	No	Yes, not possible to supply in-house
53	Security services	85230 — Security systems services	No	No	Yes, lack of expertise
		85250 — Guard services	No	No	Yes, lack of expertise
54	Cleaning services	853 — Cleaning services	No	No	Yes, lack of expertise

Source: Authors.

APPENDIX B

Table B.1: Policies Affecting Services in the Value Chain.

Government policies/ services	Authority(ies) in charge	Details	How the policy affects services in the value chain
Environmental laws and regulations	Pollution Control Department of Thailand (PCD)	Weak enforcement of environmental laws and regulations which are strict on paper	It results in the firm competing only from the cost perspective and not on the environmental-friendliness of its offerings. It may also influence the firm's investment plans
Environmental, health and safety inspection	PCD	Low awareness of technology related to wastewater treatment among regulators	It results in inspection visits which are more of a cost than a benefit, considering the time and effort needed to prepare for the visit
IP regime	Department of Intellectual Property of Thailand The CIPITC	Absence of strong IP regime despite significant improvements recently	Firm is reluctant to transfer technology and knowledge to its local operations for fear that former staff will share its proprietary technology with competitors. This also hampers the involvement of local staff in tasks requiring higher technical expertise

Movement of natural persons	Various ASEAN economies	They limit stays by firms' foreign engineers and staff. There is also a requirement mooted for a proficiency test in the local language	Certain restrictions may limit the number of days that engineers can stay in the economy and hence affect the quality of services provided. Firms also become more cautious when about increasing FDIs because the role of foreign staff in ensuring the success of operations is very important
Local third-party logistics firms	Thailand Ministry of Tra nsport Thailand Ministry of Industry	Foreign Business Act B.E. 2542 (1999) restricts foreigners' participation in controlled industries such as transport/logistics	Firm has to engage the services of local third-party logistics firms to transport imported equipment and chemicals as well as to collect oil by-products produced by its treatment process. Restrictions on participation in the logistics industry restrict choice and raises costs In addition, the firm is worried about identifying unreliable logistics firms which may dispose wastewater illegally, potentially causing the firm and its clients to face reputational risk

Source: Authors.

REFERENCES

AFP (2014). Thailand totters towards waste crisis. *The Bangkok Post*. http://www.bangkokpost.com/news/local/429877/thailand-totters-towards-waste-crisis [9 April 2015].

ASEAN Briefing, Thailand's Intellectual property regime. http://www.aseanbriefing.com/news/2014/08/21/thailands-intellectual-property-regime.html [9 April 2015].

General Restrictions on Foreign Participation, the Foreign Business Act. Business Information Centre, Royal Thai Embassy. http://bic.thaiembassy.sg/node/104 [9 April 2015].

International Intellectual Property Alliance (IIPA) (2014). Appendix C: Chart of Countries' Special 301 Placement (1989–2013) and IIPA 2014 Special 301 Recommendations. Washington, D.C.: IIPA. http://www.iipa.com/pdf/2014SPEC301HISTORICALCHART.pdf [9 April 2015].

Jakarta Globe, Indonesia to withdraw local language plan for foreign workers, sources say. http://jakartaglobe.beritasatu.com/news/indonesia-withdraw-local-language-plan-foreign-workers-sources/ [9 April 2015].

OECD Services Trade Restrictiveness Index. http://www.oecd.org/tad/services-trade/services-trade-restrictiveness-index.htm [9 April 2015].

Office of the United States Trade Representative. Special 301. https://ustr.gov/issue-areas/intellectual-property/Special-301 [9 April 2015].

Parlina, I (2015). Minister prepares language proficiency test. *The Jakarta Post*. http://www.thejakartapost.com/news/2015/03/05/minister-prepares-language-proficiency-test.html [9 April 2015].

Pongrai, J (2014). Praksa fire leads to calls for nationwide dumpsite controls. *The Nation*. http://www.nationmultimedia.com/national/Praksa-fire-leads-to-calls-for-nationwide-dumpsite-30230655.html [9 April 2015].

Royan, F (2012). *Sustainable Water Treatment Technologies in the 2020 Global Water Market*. Mountain View: Frost & Sullivan. http://www.forbury-invest.com/fileuploads/Frost%20&%20Sullivan.pdf [9 April 2015].

Thailand Central Intellectual Property and International Trade Court. http://www.ipitc.coj.go.th/?co=en [9 April 2015].

Thailand Department of Intellectual Property. http://www.ipthailand.go.th/en/ [9 April 2015].

Thailand Pollution Control Department. http://www.pcd.go.th/index-Eng.cfm [9 April 2015].

UNESCO (2015). *The United Nations World Water Development Report 2015 — Water For A Sustainable World*. Paris: UNESCO. http://unesdoc.unesco.org/images/0023/002318/231823E.pdf [9 April 2015]

Chapter 7

MANUFACTURING OF AUTOMOTIVE COMPONENTS IN THE ASEAN REGION

Denise Cheung

7.1. BACKGROUND INFORMATION

7.1.1. *Company Overview*

The company represented in this case study is a leading supplier of advanced automotive technology, systems and components for major automakers. Based in Japan, the company has operations in over 30 economies, including eight Association of Southeast Asian Nations (ASEAN) member states, of which six are APEC members.

Global turnover is about JPY 4 trillion (US$30 billion) on a consolidated basis in FY2013-14. Japan is the company's largest market, accounting for half of its revenues. North America and Asia Oceania each accounts for one-fifth of the company's global turnover, and Europe around a tenth. The company employs approximately 140,000 persons globally, 27 per cent of whom are located in Japan and 15 per cent in ASEAN economies.

7.1.2. *Products*

The company has two main product lines, namely original equipment manufacturer (OEM) automotive parts and aftermarket automotive-related products. The former group of products is sold to automakers as intermediate inputs and will not bear the company's brand. These include powertrains, climate control equipment, body electronics (e.g., windshield wipers, horns), safety components, and information and communications equipment (e.g., navigation systems, toll collection systems). The latter product line — after-market products — consists of automotive parts and accessories and navigation apps targeted directly at consumers.

7.1.3. *ASEAN Regional Network*

The company has factories for OEM components in seven ASEAN economies — Thailand, Indonesia, Myanmar, Malaysia, the Philippines, Viet Nam and Cambodia (Table 7.1). Within the ASEAN region, the company has two regional headquarters (RHQ), in Singapore and Thailand. The company's choice of Singapore as a location for RHQ can be explained by the economy's well-developed physical and financial infrastructure. In contrast, the locational choice of Thailand is driven by a large scale of supporting industry, as many major automakers/suppliers have production hubs in Thailand.

Research and development (R&D) centres have been established in Thailand. The company has designated the R&D facility in Thailand as the core technical centre for the ASEAN region.

There are sales offices in Thailand and Indonesia to sell and distribute in the domestic markets.

Table 7.1: Operations in ASEAN.

	Regional headquarters	Production sites	R&D centres	Sales offices
Cambodia		✓		
Indonesia		✓		✓
Malaysia		✓		
Myanmar		✓		
The Philippines		✓		
Singapore	✓			
Thailand	✓	✓	✓	✓
Viet Nam		✓		

Source: Author.

7.2. DESCRIPTION OF THE VALUE CHAINS

This case study considers two types of value chains — decentralised and centralised — for OEM automotive components. In general, the company adopts centralised value chains for smaller-sized and higher value-added parts and components. On the other hand, decentralised value chains are adopted for bulky parts and components.

In a decentralised value chain, production of the component is allocated among various locations within the region. This is the case for the assembly of heat, ventilation and air conditioning units.

7.3. SERVICES ALONG THE VALUE CHAIN

This section highlights some of the services that are particularly important in these OEM automotive component value chains. Table 7.2 shows a list of services inferred from the interview with company's management.

7.3.1. *Production Planning and Logistics*

With a large turnover, multiple production sites and multiple markets, careful planning of production volume, inventory control and efficient transportation of materials and products among economies can be significant cost savers.

7.3.2. *Research and Development*

The automotive industry continuously undergoes technological changes. To keep up with demand from automakers and consumers as well as changes in regulatory requirements, the company spends approximately 10 per cent of its revenues on R&D.

R&D services are supplied in-house, mainly from the Japanese headquarters but increasingly from the European and North American headquarters. The company creates roadmaps to track the progress of its R&D activities and updates them annually to incorporate market and regulatory changes. The roadmaps are

Table 7.2: Services Entering the Value Chains of OEM Automotive Components.

Services	Corresponding CPC rev. 2 code
Raw materials, input/pre-production stage	
1. Industrial design	83912 — Industrial design services
2. Procurement agent for raw material sourcing	85999 — Other support services n.e.c.
3. Customs-related services for imported raw materials	85999 — Other support services n.e.c.
4. Quality assurance services (of raw materials)	83441 — Composition and purity testing and analysis services
5. Freight transportation services (of raw materials) by road, rail, sea or air	65 — Freight transport services
6. Storage of raw materials — general storage	67290 — Other storage and warehousing services
In factory stage	
7. Production management	83115 — Operations management consulting services
8. Quality assurance and compliance with International Organization for Standardization (ISO)	83441 — Composition and purity testing and analysis services
9. Warehousing services for intermediate goods	67290 — Other storage and warehousing services
10. Cleaning services for factory	85330 — General cleaning services
11. Security guards for factory and warehouses	85250 — Guard services
12. Sewage water treatment services	94110 — Sewerage and sewage treatment services
13. Industrial waste disposal	94212 — Collection services of industrial hazardous waste (except medical and other biohazardous waste) 94339 — Other non-hazardous waste treatment and disposal services
14. Repair and maintenance services of machines and equipment	87156 — Maintenance and repair services of commercial and industrial machinery

(*Continued*)

Table 7.2: (*Continued*)

Services	Corresponding CPC rev. 2 code
15. Government inspections on fire prevention, health hazards, environmental protection and other aspects	91133 — Public administrative services related to mining and mineral resources, manufacturing and construction 91290 — Public administrative services related to other public order and safety affairs
16. Catering services for workers	63393 — Other contract food services
17. Personnel search and referral services — Recruitment of factory workers	85112 — Permanent placement services, other than executive search services
18. Social insurance for factory workers	91320 — Administrative services related to government employee pension schemes; old-age disability or survivors' benefit schemes, other than for government employees 91330 — Administrative services related to unemployment compensation benefit schemes
Delivery and sales stage	
19. Design of packages	83919 — Other specialty design services
20. Packaging services	85400 — Packaging services
21. Freight transportation services (of finished goods) by road, rail, sea or air	65 — Freight transport services
22. Storage and warehousing services for finished goods	67290 — Other storage and warehousing services
Post-sales stage	
23. Customer services — technical support, complains handling and enquiry related to recalls	85931 — Telephone call centre services
24. Repair and maintenance	87141 — Maintenance and repair services of motor vehicles
Operation and management related services	
25. Auditing on financial accounts	822 — Accounting, auditing and bookkeeping services

(*Continued*)

Table 7.2: (*Continued*)

Services	Corresponding CPC rev. 2 code
26. Auditing with respect to ISO certification	822 — Accounting, auditing and bookkeeping services
27. Financial services	71121 — Deposit services to corporate and institutional depositors
28. Internal auditing (including audits of financial accounts and corporate governance)	822 — Accounting, auditing and bookkeeping services 83118 — Head office services
29. Insurance services for machinery	71334 — Other property insurance services
30. Corporate communications, marketing and public relationship	83114 — Marketing management consulting services 83121 — Public relations services
31. Real estate services	72112 — Rental or leasing services involving own or leased non-residential property
32. Human resources management	83113 — Human resources management consulting services
33. Human resources management: Personnel search and referral services for back office staff	8511 — Personnel search and referral services
34. Information technology (IT) and information system management	8314 — IT design and development services 8316 — IT infrastructure and network management services
35. Legal services	82120 — Legal advisory and representation services concerning other fields of law 82130 — Legal documentation and certification services
36. Product development/R&D	81129 — Research and experimental development services in other engineering and technology
37. Visa and immigration services for foreign investors/employees	91290 — Public administrative services related to other public order and safety affairs

Source: Author.

shared among various divisions which are responsible for different stages throughout the value chain. Decisions about which projects to start, continue, modify or terminate are made based on feedback from different divisions in the company.

7.3.3. *Testing and Certification*

For automotive components, expectations regarding product safety and quality are high compared to many other goods. In addition, customers and regulators have in recent years become concerned about the environmental impact of automotive components. These factors have increased the importance of testing and certification in recent years, pushing up the value contribution of these services to the supply chain.

Testing and certification services are required throughout the value chain. Inspection and sample testing are conducted on the firm's suppliers. Design testing is done to confirm the feasibility of product ideas. Product testing is conducted on protocols and samples of finished products. In addition, the company has adopted the ISO 26262 certification standard for the functional safety of the design process.

The company has also acquired the ISO/TS16949 certification standard for quality management systems. The certification is valid for three years and has to be confirmed annually based on external audit results. On the environmental front, most of the company's subsidiaries have obtained ISO 14001 certification and implemented environmental management systems.

7.3.4. *Supplier Management*

With a diverse supplier base, the company pays considerable attention to supplier management. The company holds regular meetings to communicate its procurement policy to major suppliers. In addition, it runs an assessment program for suppliers, which assesses supplier performance based on quality, cost, delivery, management, and environmental and social compliance.

The company realises that without cooperation throughout the value chain, it cannot fulfil its social responsibilities such as safety and product quality, legal compliance, labour protection and environmental protection. For example, the company has to ensure that its suppliers refrain from using "conflict materials" (raw materials mined from conflict zones whose proceeds end up financing armed forces) in their automotive components. The company requires that its suppliers maintain detailed records of materials used, and this information is audited by the lead firm.

The company is in the process of developing corporate social responsibility (CSR) guidelines for its suppliers. Suppliers are asked to self-diagnose based on the guidelines. In the future, the company plans to inspect its suppliers to verify the self-diagnostic results and hold meetings with suppliers to explore ways to improve.

7.3.5. *After-sales Services*

After-sales services are important elements of the automotive component value chain. The company has offices that provide basic repair and maintenance services as well as high-tech service facilities that undertake more advanced diagnostic and repair functions. The latter facilities are equipped with the company's proprietary assets and are staffed by experts who have received specialised internal training and certification. In addition, the company has a recall mechanism to repossess problematic or defective products. This mechanism includes making announcements via newspaper advertisements and the company website, and responding to enquiries via a dedicated toll-free phone number. Finally, after-sales services staff also communicates with customers and auto-dealers for feedback on potential market needs.

7.4. POLICIES AFFECTING SERVICES IN THE VALUE CHAIN

With value chains spanning most of the ASEAN economies, the company constantly faces regulatory hurdles emanating from different policy backdrops. This section discusses some examples of these challenges.

7.4.1. *Foreign Investment Restrictions*

In 2014, one ASEAN economy government revised its Negative Investment List for foreign investment, which indicates restrictions placed on foreign investment. The new Negative Investment List explicitly spelled out restrictions on foreign investment in retail services. In addition, the new list tightened foreign ownership in distribution and warehousing services. The tightened restrictions affect some of the value chain's activities in the economy concerned, but offers grandfathering of existing arrangements as long as the subsidiary involved does not undergo any legal changes.

7.4.2. *Operational Restrictions*

When establishing a new subsidiary in this ASEAN economy, the company originally intended the new subsidiary to share certain back office functions with the group's existing company in the same jurisdiction for cost and efficiency purposes. However, this arrangement was not approved by the government because the law requires any registered company to be completely equipped with all back office functions and a stand-alone office address. If two companies within the same group were to share certain functions, formal transactions would need to take place, meaning that formal contracts would be required. From the company's perspective, this regulation does nothing but augments costs and lowers efficiency.

7.4.3. *Labour Regulations*

The company used to hire term employees to perform labour-intensive tasks. The company believed this was a win-win transaction for the company and the workers. The company could enjoy more flexibility in response to fluctuations in demand, while the

Table 7.3: Policies Affecting Services in the Value Chain.

Government policies/ services	Authority(ies) in charge	Details	How the policy affects services in the value chain
Foreign investment restrictions	Ministry of Industry	The Negative List 2014 specifies restrictions on foreign ownership in certain industries	The list specifies foreign ownership restrictions on retail and wholesale services
Requirements on established companies	Ministry of Industry and Trade	Registered companies are required to be fully equipped with back office management functions	The company cannot share its back office operations among its subsidiaries in the same jurisdiction
Labour regulations	Department of Labour Protection and Welfare	Legislation prohibits discrimination in terms of pay and benefits to different workers performing the same task	Reduces flexibility in adjusting workforce in response to seasonal fluctuations in workload
Intellectual property concerns	Customs department	Counterfeit products are commonplace in the region. The company depends on customs departments' enforcement to combat counterfeiting activities	The company proactively seeks cooperation with customs departments in its major markets

Source: Author.

term employees who were mostly rural labour could earn additional income, including during their seasonal downtime.

However, with labour law amendments that came into effect in 2008 in this ASEAN economy, companies are required to provide the same pay and benefits to all workers who undertake the same task. In other words, the firm was no longer allowed to hire employees on different employment terms (i.e., full-time versus casual employment) for the same duty. This has reduced the company's flexibility in adjusting manpower according to workload, leading to seasonal labour shortages and reduced output.

7.4.4. *Intellectual Property Concerns*

Counterfeit products are commonplace in the region and have become a threat to the company's business. However, they mainly affect the aftermarket product lines and replacement parts, which account for a minority share of the company's turnover. The OEM product lines have not been much affected because the company deals with automakers directly.

To combat counterfeit products, the company has been cooperating with customs departments in its major markets. For example, it provides samples of its products to customs departments so that customs officers can compare imported goods to the genuine product. However, the company thinks that it is also important to root out "production source" of counterfeit products because it is quite difficult to detect and exterminate smuggling (Table 7.3).

Chapter 8

MANUFACTURING OF OIL AND GAS INDUSTRY EQUIPMENT IN SINGAPORE

Andre Wirjo and Gloria O. Pasadilla

8.1. INDUSTRY OVERVIEW

Energy consumption is projected to increase by 41 per cent between 2012 and 2035 (British Petroleum, 2014), which means that the oil and gas industry will retain the importance it has held since machines were invented. The Asia-Pacific region will contribute to meeting this increased demand; its share of global energy production over the same period is estimated to increase to 47 per cent. For the oil and gas industry as a whole, compound annual growth is forecasted to be around 6 per cent between 2014 and 2017 (GE, 2014).

The oil and gas industry can generally be divided into three stages, each requiring specialised equipment and infrastructure (see Fig. 8.1). The upstream stage covers the exploration and production of crude oil and gas from both onshore and offshore fields. Firms in the midstream stage then collect the crude oil and gas prior to processing, storing, transporting and distributing the products as feedstock to the petrochemical plants and refineries. Following further processing and refining, downstream firms market and distribute the end product to the gas stations as well as final consumers.

Within the upstream stage of the oil and gas industry, subsea[1] appears to be the segment with most promising growth prospects;

[1]Subsea systems are equipment used to explore, drill and develop oil and gas fields in underwater locations.

Figure 8.1: Simplified Oil and Gas Value Chain.

the capital expenditure for subsea systems is forecasted to increase by about 9 per cent annually between 2014 and 2017 (GE, 2014). Oil and gas exploration firms now have to look for resources in more remote and harsher environments to meet global energy demand. Over the last decade, more than half of the main discoveries have been offshore and in the past five years, more than half of these discoveries have been in deep water. In fact, approximately half of the production of international oil companies now comes from deepwater reserves (GE, 2014).

The needs of the oil and gas industry are continuously evolving in line with newer technologies, and manufacturers of equipment used for activities such as exploration and extraction are reacting to the changing needs through various strategies. First, manufacturers have moved to modularisation and standardisation as opposed to customisation so that they are able to provide equipment that is capital-efficient to their customers. Second, manufacturers have incorporated remote monitoring technology in their equipment, which together with their service agreements, will ensure full life-cycle productivity of the equipment supplied to their customers. Moreover, manufacturers have to regularly extend the physical boundaries that their equipment can operate in such as newer depths and newer parameters of pressure requirements.

8.2. BACKGROUND INFORMATION OF FIRM[2]

The firm represented in this case study has operations across six continents. The equipment that it supplies to the oil and gas

[2]All information about the firm in this case study is from its corporate website as well as from the interview.

industry can be used in upstream applications as well as downstream processing plants.

Besides selling products, the firm has a business focusing on services. It has over 40 service centres and workshops in many of the world's main oil and gas extraction and production regions. Specifically, it provides upgrades as well as maintenance and repair services to customers' equipment so as to extend their production capability and performance.

8.3. DESCRIPTION OF THE VALUE CHAIN

This case study will focus on the firm's Singapore facility, which is one of its main manufacturing and services facilities of products for subsea systems such as subsea trees, controls and wellheads. For tractability, this case study focuses on one product — subsea trees — whose primary function is to control and monitor the flow of oil and gas out of the well. These products are usually built on the sea floor and constitute part of an entire underwater/sea bed production system (see Fig. 8.2). The technology system includes other products or structures such as manifolds and work over systems, hydraulic power units, subsea compression system, riser

Figure 8.2: Overview of Subsea Systems and the Position of Subsea Trees within the System.

Source: Courtesy of the firm.

base, tie-in and connection systems, hydraulic power units and others. The technology system's function is to extract oil or gas.

While the intent of the case study is to focus on a single product, the same value chain or process described in what follows applies to other products of the subsea systems such as wellheads and in fact, likewise, to surface systems. Differences are usually in the specifications and life-cycle of the products, since they operate under variable environments such as different pressure and temperature.

For the purpose of this case study, the value chain begins with a request for quotation (RFQ) from the customer and ends with the provision of after-sales services to the customer (see Fig. 8.3).

8.3.1. *From Customer Specification to Product Engineering and Design*

In the RFQ, the customer lists the requirements of the product(s), such as the ability to withstand a certain temperature range, water pressure, and depth. The firm then works with its design and engineering departments as well as its sales team to provide the best possible quote for the RFQ.

The firm's strategy in responding to the RFQ could be simply summarised as standardisation first, customisation later. Specifically, the firm makes use of its design and engineering departments to determine whether its existing product range can meet the requirements. If not, the firm has to work out how customised products could be manufactured to meet them. Customised products are often needed because either the customer wants exact duplicates of existing assets which they can store and use interchangeably, or the environment in which the products operate is distinct from other situations where the existing range of available products can be used.

The quotation/sales department works in parallel to identify the most cost-competitive way of manufacturing the products. This involves exploring where the most competitively priced inputs could be sourced and to whom certain manufacturing processes could be outsourced. However, it should be noted that while the

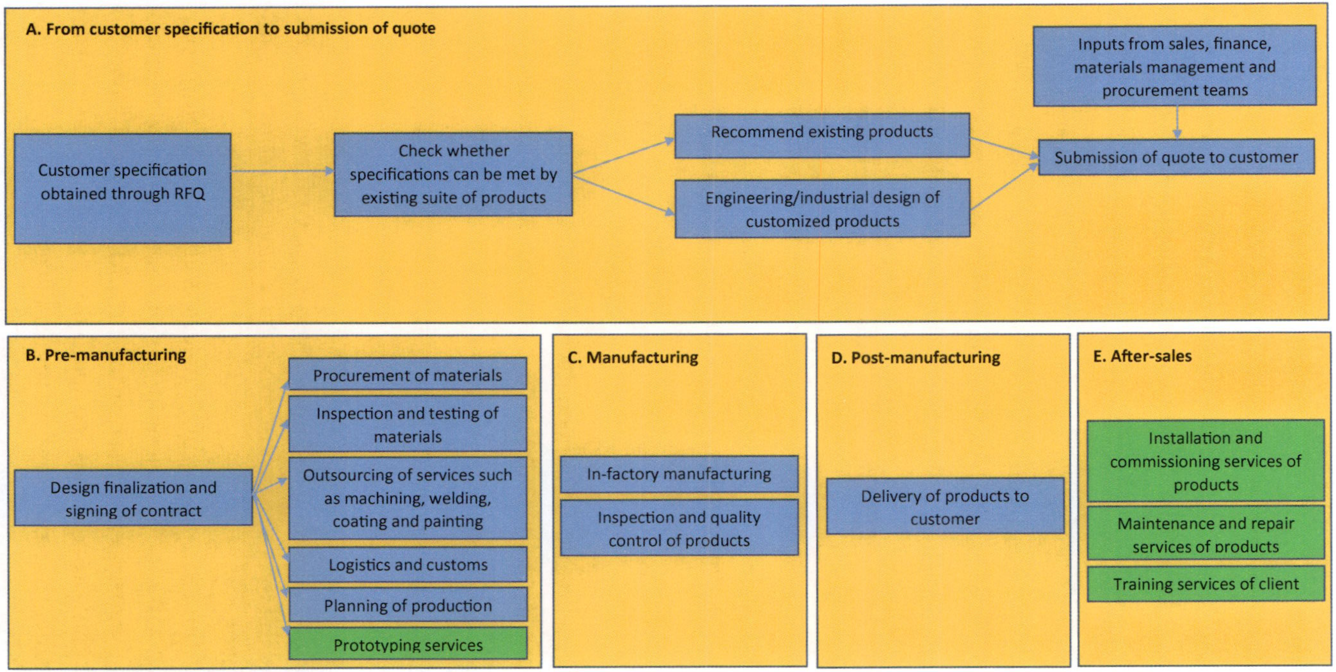

Figure 8.3: Dimension of the Value Chain Covered by Case Study.

Note: Optional activities in the value chain are indicated by green boxes.

Source: Authors.

firm has a wide range of options for minimizing cost, it may be constrained by a customer's requirements that some inputs and processes are (or are not) sourced from certain locations. This could be for various reasons, including previous experience with specific suppliers, or close supplier relationships.

8.3.2. *Manufacturing*: *Critical Role of Industry Standards*

The manufacturing process for each subsea tree is generally similar although the specifics, such as resistance to varying degrees of underwater pressure or temperature, may differ among projects. The design and engineering departments would look at the agreed blueprint and finalise it (incorporate additional specs from the customer, if necessary). The relevant department verifies the different inputs as well as processes needed to manufacture the product, which would then be sourced from various channels. The role of its material management team is critical to ensure that all inputs and processes adhere to the contract requirements such as local content and sourcing of inputs from specific suppliers. Depending on the source of inputs, brokerage and customs clearance services would also be needed. For an advanced technology product, such as a subsea tree, most of the expensive material inputs are specially engineered products which are either produced in-house or by an affiliate, or outsourced to third-party suppliers. It is worth noting that these engineered materials themselves contain significant services value which is not tracked in this study.

The firm noted that in-house activities can only be carried out after the necessary equipment is reserved in advance via the internal shop ordering service. Prototyping may also be required under certain contracts.

Crucial to the manufacturing process is the role of third-party inspectors, who must be present at various stages of the manufacturing process to inspect the equipment and products as well as sign off that the process has adhered to stipulated regulations. A schedule indicating the various manufacturing stages is agreed between the firm and the customer in the presence of inspectors and listed clearly in the signed contract. The inspection is carried out by a third party and is either sourced by the firm or by the customer. The inspection may be carried out at the firm's premises or at the suppliers' premises.

The presence of third-party inspectors is a norm agreed by industry players who are members of the American Petroleum Institute (API)[3] to ensure that products will be able to perform reliably in a challenging environment, as well as adhere to health, safety and environment standards. It is worth noting that the oil and gas industry is a highly regulated industry because of the potential risks it poses, particularly for the environment.

Final products are tested in-house. The more challenging the operating environment, the more stringent the requirements are in terms of materials and testing.

8.3.3. *Post-Manufacturing (Packing, Warehousing, Logistics and Commissioning)*

Products are packed and stored in a warehouse before they are delivered to customers. The mode of delivery depends on the customer's preference. Customers can collect the product from the factory or the firm delivers to an agreed location. In the latter case, the contract may entail installation services. This could involve installation or the supervision of installation and is an additional source of revenue. The provision of these commissioning services avoids a blame game that may occur if a third party is contracted to undertake installation, and is especially important if the product is still under warranty.

8.3.4. *After-Sales Services*

Maintenance and repair services are provided by the firm and are usually provided for between 5 and 20 years, depending on the contractual services agreements which is tailored to suit the needs

[3] The American Petroleum Institute is the largest US trade association for the oil and natural gas industry. Among its important functions is the establishment and certification of industry and technical standards.

of customers. Post-sales services agreements may include parts supply; customer property management by fielding service and resident engineers; routine, planned and unplanned maintenance; logistics and inventory management. More recently, the firm has also been able to remotely access data pertaining to installed products, allowing for round-the-clock monitoring and support to ensure full lifecycle productivity.

In order to ensure quick turn-around and minimise losses caused by breakdowns, the firm has established service facilities close to where its customers are located. Localisation shortens transport times and avoids customs and immigration procedures that are commonly associated with crossing borders. Services agreements also sometimes include training services for staff of the customers.

Firm indicates that as original equipment manufacturer, its after-sales business has been growing since investigation of the Deepwater Horizon accident in the Gulf of Mexico in 2010[4] revealed that layers of subcontracting could have caused the incident. Although exact figures are not provided, publicly available data of the firm showed that the oil and gas subsidiary contributed to the increase in the sales of product services[5] for the corporations as a whole.

8.4. SERVICES ALONG THE VALUE CHAIN

8.4.1. *Services Identification and Value Contribution*

A total of 55 services are identified in this value chain and they have been grouped according to the various stages within the chain: (i) services in responding to RFQ from customer, (ii) services during pre-manufacturing and manufacturing,

---

[4]http://www.theguardian.com/environment/2010/jul/18/deepwater-horizon-blow-out-preventer-china and http://www.telegraph.co.uk/finance/newsbysector/energy/oilandgas/10345677/BP-failed-to-plan-for-Deepwater-Horizon-disaster.html.

[5]The firm defines sales of product services as sales under product services agreements and sales of both goods (such as spare parts and equipment upgrades) and related services (such as monitoring, maintenance and repairs).

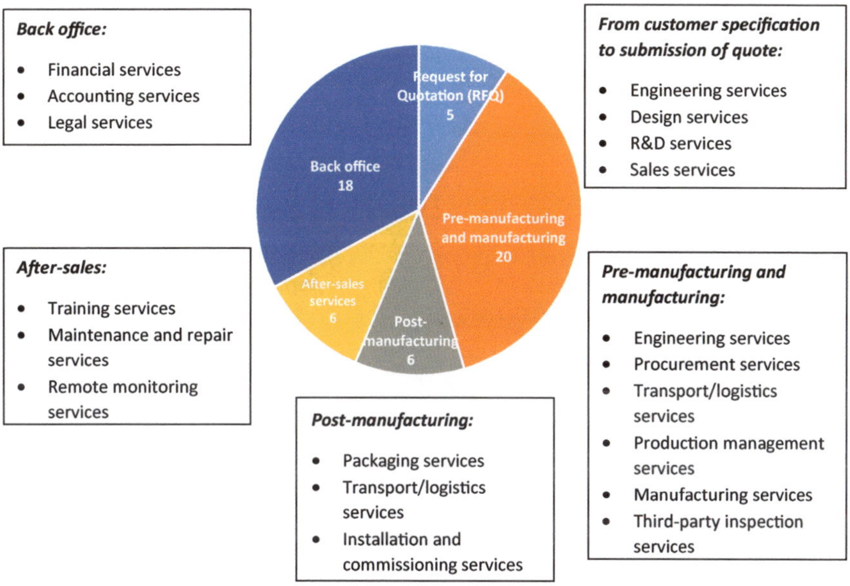

Figure 8.4: Breakdown of Services by Stage and Examples of Key Services.
Source: Authors.

(iii) post-manufacturing services, (iv) after-sales services and (v) back office services (see Fig. 8.4).

The 55 services can be further disaggregated into at least 94 services if one is to count individually the sub-services that made up a single services activity.[6] While we have limited specific information on which services inputs are the most important in terms of added value, it can be surmised that engineering and design services are among these. Engineering services, in particular, are required in the design and manufacturing stages (including quality control) as well as in the after-sales phase. The large number of services entering the value chain of subsea tree production

[6]For instance, utilities can be separated into water as well as electricity and gas distribution while legal services can be broken down into legal advisory and representation services, legal documentation and certification services as well as arbitration and conciliation services. Transport/shipping services of raw materials and products can also be broken down into land, water and air transport.

validates the view that services are ubiquitous, even in the case of firms that are putatively manufacturing enterprises.

While the services contribution is not easily quantified, it is surely considerable. If data were available for the cost of material and manufacturing-specific labour involved in subsea tree production, we could assume that the remaining portion of what is considered "production" cost can, in fact, be attributed to the various services identified in Tables 8.1–8.5 and Fig. 8.4.

The cost of material inputs is likely to be high in this industry because many of them are engineered materials. However, if one considers that these high-value material inputs have themselves a large component of services value-added, the total services contribution in subsea tree production is magnified. Another difficulty in estimating the services contribution in a particular product is that expenditures on research and development — a service activity, though significant, is difficult to attribute to specific product lines. The firm's parent company has a strong research and development services range, from materials research (such as identifying new types of coating that will allow the materials to be resistant to extreme conditions) to other types of basic research which may not (yet) have any immediate application in the oil and gas industry.

Significantly, among the 600 employees in the firm's Singapore facility, two-thirds are involved in the provision of the identified services in Fig. 8.4 and Tables 8.1–8.5. Only one-third is in manufacturing positions and this includes supervisors, managers and engineers who provide management and production planning services. Therefore, "services employees" may well make up more than two-thirds of the facility's total employees.

Looking ahead, the share of services in value addition is likely to remain significant because the firm's long-term operating results depend substantially on its ability to: (1) continually develop, introduce, and market new and innovative products and services; (2) modify existing products and services; (3) customise products and services and (4) respond to technological change. Many of these functions require services at their core or are at least heavily reliant on services.

Table 8.1: Services in Responding to RFQ from Customer.

	Service	Corresponding CPC ver. 2 code	Supplied in-house	Outsourced to affiliated companies and reasons	Outsourced to third-party suppliers/and reasons	Bundled
1	Project-specific research and development services to support quotation	8111 — Research and experimental development services in natural sciences	Yes	No	No	
		8112 — Research and experimental development services in engineering and technology	Yes	No	No	
2	Project-specific design services to support quotation	8314 — Information technology (IT) design and development services	Yes	No	No	
		83920 — Design originals	Yes	No	No	
		83912 — Industrial design services	Yes	No	No	
3	Project-specific engineering services to support quotation	83310 — Engineering advisory services	Yes	No	No	
		8332 — Engineering services for specific projects	Yes	No	No	

(Continued)

Table 8.1: (*Continued*)

	Service	Corresponding CPC ver. 2 code	Supplied in-house	Outsourced to affiliated companies and reasons	Outsourced to third-party suppliers/and reasons	Bundled
4	Procurement and quotation services	83116 — Supply chain and other management consulting services	Yes	No	No	
		85999 — Other support services n.e.c.	Yes	No	No	
5	Sales and marketing services	8596 — Convention and trade show assistance and organisation services	No	No	Yes, lack of expertise	
		8370 — Market research and public opinion polling services	No	No	Yes, lack of expertise; need to access the best services	
		8311 — Management consulting and management services (may include customer relationship management)	Yes	No	Yes, lack of expertise; need to access the best services	

Source: Authors.

Table 8.2: Services during Pre-Manufacturing and Manufacturing Process.

	Service	Corresponding CPC ver. 2 code	Supplied in-house	Outsourced to affiliated companies and reasons	Outsourced to third-party suppliers/government and reasons	Bundled
6	Project-specific research and development services prior to manufacturing	8111 — Research and experimental development services in natural sciences	Yes	No	No	
		8112 — Research and experimental development services in engineering and technology	Yes	No	No	
7	Project-specific design services prior to manufacturing	8314 — IT design and development services	Yes	No	Yes, lack of expertise; need to access the best services	
		83920 — Design originals	Yes	No	No	
		83912 — Industrial design services	Yes	No	No	

(*Continued*)

Table 8.2: (*Continued*)

	Service	Corresponding CPC ver. 2 code	Supplied in-house	Outsourced to affiliated companies and reasons	Outsourced to third-party suppliers/ government and reasons	Bundled
8	Project-specific engineering services prior and during manufacturing	83310 — Engineering advisory services	Yes	No	No	
		8332 — Engineering services for specific projects	Yes	No	No	
9	Government licensing and inspections on fire prevention, health hazards, environmental protection and other aspects)	91133 — Public administrative services related to mining and mineral resources, manufacturing and construction	No	No	Yes, government services	
		91290 — Public administrative services related to other public order and safety affairs	No	No	Yes, government services	
10	Prototyping services	887 — Fabricated metal product, machinery and equipment manufacturing services	Yes	No	No	

11	Procurement services	83116 — Supply chain and other management consulting services	Yes	No	No	
		85999 — Other support services n.e.c.	Yes	No	No	
12	Training services for staffs	92919 — Other education and training services, n.e.c.	Yes	No	Yes, professional certification may be required by laws and regulations	
13	Transport services of raw materials	651 — Land transport services of freight	No	No	Yes, efficiency; economies of scale	Bundled with transport, driver, repair and insurance
		652 — Water transport services of freight	No	No	Yes, efficiency; economies of scale	Bundled with transport, driver, repair and insurance
		6531 — Air transport services of freight	No	No	Yes, efficiency; economies of scale	Bundled with transport, driver, repair and insurance

(*Continued*)

Table 8.2: (*Continued*)

Service		Corresponding CPC ver. 2 code	Supplied in-house	Outsourced to affiliated companies and reasons	Outsourced to third-party suppliers/ government and reasons	Bundled
		67910 — Freight transport agency services and other freight transport services	No	No	Yes, efficiency; economies of scale	Bundled with transport, driver, repair and insurance
14	Freight insurance of raw materials	71333 — Freight insurance services	No	No	Yes, not possible to supply in-house	Bundled with transport, driver, repair and insurance
15	Customs clearance services and logistics of raw materials	67110 — Container handling services	No	No	Yes, efficiency; strong relationship with government agencies	Bundled with transport, driver, repair and insurance
		85999 — Other support services n.e.c.	No	No	Yes, efficiency; strong relationship with government agencies	Bundled with transport, driver, repair and insurance

16	Storage and warehousing services of raw materials	67290 — Other storage and warehousing services	Yes	No	No
17	Technical testing of raw materials	83441 — Composition and purity testing and analysis services	Yes	No	Yes, required by laws and regulations
18	Production administration — Production management and shop ordering services	83115 — Operations management consulting services	Yes	No	No
19	Maintenance and repair of production equipment	87156 — Maintenance and repair services of commercial and industrial machinery	Yes	No	Yes, lack of expertise; by machine vendor
20	Utilities (electricity, gas and water supply)	691 — Electricity and gas distribution (on own account)	No	No	Yes, not possible to supply in-house
		692 — Water distribution (on own account)	No	No	Yes, not possible to supply in-house

(*Continued*)

Table 8.2: (*Continued*)

	Service	Corresponding CPC ver. 2 code	Supplied in-house	Outsourced to affiliated companies and reasons	Outsourced to third-party suppliers/ government and reasons	Bundled
21	Manufacturing services provided in-house and by suppliers of activities such as welding, coating, etc.	886 — Basic metal manufacturing services	No	No	Yes, efficiency; need to access the best services	Bundled with transport, driver, repair and insurance
		887 — Fabricated metal product, machinery and equipment manufacturing services	No	No	Yes, efficiency; need to access the best services	Bundled with transport, driver, repair and insurance
22	Testing services for quality control and assurance	8344 — Technical testing and analysis services	Yes	No	No	
23	Third-party inspection services	990 — Services provided by extraterritorial organisations and bodies	No	No	Yes, required by laws and regulations	

24	Product testing to obtain certification at export market	8344 — Technical testing and analysis services	No	No	Yes, required by laws and regulations
25	Sewage and waste treatment and collection services	94 — Sewage and waste collection, treatment and disposal and other environmental protection services	No	No	Yes, efficiency; not possible to supply in-house

Source: Authors.

Table 8.3: Post-manufacturing Services.

Service		Corresponding CPC ver. 2 code	Supplied in-house	Outsourced to affiliated companies and reasons	Outsourced to third-party suppliers and reasons	Bundled
26	Packaging services	83919 — Other specialty design services	Yes	No	No	
		85400 — Packaging services	Yes	No	No	
27	Storage and warehousing services of products	67290 — Other storage and warehousing services	Yes	No	No	
28	Transport services of products	651 — Land transport services of freight	No	No	Yes, efficiency; economies of scale	Bundled with transport, driver, repair and insurance
		652 — Water transport services of freight	No	No	Yes, efficiency; economies of scale	Bundled with transport, driver, repair and insurance
		6531 — Air transport services of freight	No	No	Yes, efficiency; economies of scale	Bundled with transport, driver, repair and insurance

		67910 — Freight transport agency services and other freight transport services	No	No	Yes, efficiency; economies of scale	Bundled with transport, driver, repair and insurance
29	Freight insurance	71333 — Freight insurance services	No	No	Yes, not possible to supply in-house	Bundled with transport, driver, repair and insurance
30	Installation services for equipment and related wiring	5461 — Electrical installation services	Yes	No	No	
		873 — Installation services (other than construction)	Yes	No	No	
31	Certification and commissioning services of equipment	8344 — Technical testing and analysis services	Yes	No	No	

Source: Authors.

Table 8.4: After-sales Services.

Service		Corresponding CPC ver. 2 code	Supplied in-house	Outsourced to affiliated companies and reasons	Outsourced to third-party suppliers and reasons	Bundled
32	Training services for workers of customers	9291 — Other education and training services	Yes	No	No	
33	Maintenance and repair services	87156 — Maintenance and repair services of commercial and industrial machinery	Yes	No	No	
34	Telephone-based support services	85931 — Telephone call centre services	Yes	No	No	
35	IT services for on-site and remote monitoring of equipment	8315 — Hosting and IT infrastructure provisioning services	Yes	No	No	
		8316 — IT infrastructure and network management services	Yes	No	No	

36	Travel services for engineers and other staffs pertaining to after-sales services	8551 — Reservation services for transportation	No	No	Yes, lack of expertise
		85521 — Reservation services for accommodation	No	No	Yes, lack of expertise
37	Visa and immigration services for staffs	91290 — Public administrative services related to other public order and safety affairs	No	No	Yes, government services

Source: Authors.

Table 8.5: Business Processes (Back Office Support).

	Service	Corresponding CPC ver. 2 code	Supplied in-house	Outsourced to affiliated companies and reasons	Outsourced to third-party suppliers and reasons	Bundled
38	Company registration and licensing services (obtaining permit to operate)	91138 — Public administrative services related to general economic, commercial and labour affairs	No	No	Yes, government services	
39	Headquarter services	83118 — Head office services	No	Yes, economies of scale	No	
40	Management services	83111 — Strategic management consulting services	Yes	Yes, economies of scale	No	
41	Accounting, auditing and bookkeeping services	82210 — Financial auditing services	Yes	No	Yes, required by laws and regulations	
		8222 — Accounting and bookkeeping services	Yes	No	No	

42	Financial services	7113 — Credit granting services	No	Yes, economies of scale; not possible to supply in-house	Yes, economies of scale; not possible to supply in-house
		71512 — Corporate finance and venture capital services	Yes	No	Yes, not possible to supply in-house
43	Legal services	82120 — Legal advisory and representation services concerning other fields of law	No	No	Yes, lack of expertise
		82130 — Legal documentation and certification services	No	No	Yes, lack of expertise
		82191 — Arbitration and conciliation services	No	No	Yes, lack of expertise

(*Continued*)

Table 8.5: (*Continued*)

Service		Corresponding CPC ver. 2 code	Supplied in-house	Outsourced to affiliated companies and reasons	Outsourced to third-party suppliers and reasons	Bundled
44	Insurance services (commercial life and accident/health insurance, property insurance for the factory compound, product quality insurance, management liability insurance)	7131 — Life insurance and pension services	No	No	Yes, not possible to supply in-house	
		7132 — Accident and health insurance services	No	No	Yes, not possible to supply in-house	
		71334 — Other property insurance services	No	No	Yes, not possible to supply in-house	
		71335 — General liability insurance services	No	No	Yes, not possible to supply in-house	
45	Social insurance for staffs	91320 — Administrative services related to government employee pension schemes; old-age disability or survivors' benefit schemes, other than for government employees	Yes	No	No	
		91330 — Administrative services related to unemployment compensation benefit schemes	Yes	No	No	

46	Corporate communications and public relationship	83121 — Public relations services	Yes	Yes, economies of scale	No
47	Human resources management	8511 — Personnel search and referral services	Yes	No	Yes, network economies
		8512 — Labour supply services	No	No	Yes, network economies
		83113 — Human resources management consulting services	Yes	No	Yes, lack of expertise
48	IT services	83132 — IT support services	Yes	No	Yes, efficiency; lack of expertise
		83151 — Website hosting services	No	No	Yes, efficiency; lack of expertise
49	Telecommunication services	841 — Telephony and other telecommunications services	No	No	Yes, not possible to supply in-house
		8422 — Internet access services	No	No	Yes, not possible to supply in-house

(Continued)

Table 8.5: (*Continued*)

	Service	Corresponding CPC ver. 2 code	Supplied in-house	Outsourced to affiliated companies and reasons	Outsourced to third-party suppliers and reasons	Bundled
50	Transport services for staffs	641 — Local transport and sightseeing transportation services of passengers	No	No	Yes, efficiency; not possible to supply in-house	
51	Property management services	72212 — Non-residential property management services on a fee or contract basis	No	No	Yes, efficiency; lack of expertise	
52	Medical services	93121 — General medical services	No	No	Yes, lack of expertise	
53	Catering services	63393 — Other contract food services	No	No	Yes, lack of expertise	
54	Security services	85230 — Security systems services	No	No	Yes, lack of expertise	
		85250 — Guard services	No	No	Yes, lack of expertise	
55	Cleaning services	853 — Cleaning services	No	No	Yes, lack of expertise	

Source: Authors.

8.4.2. *Outsourcing, Bundling and other Aspects of Services Supply*

Among the 94 services categories identified in this value chain, our analysis estimates 38 services are supplied in-house, 11 are partially outsourced, while 45 are fully outsourced (see Tables 8.1–8.5).

The reasons for the firm to outsource services vary but can broadly be grouped into the following: (i) government services, such as company licensing, visa and immigration services and inspections pertaining to environmental, health and safety; (ii) services required by laws and regulations, such as the submission of RFQ by local sales agents in partnership or joint ventures with the firm, manufacturing training for staff for professional certification, and third-party inspection services provided by inspectors certified by the API; (iii) a lack of expertise or specialisation in-house to provide certain services, such as consulting services, legal services, medical services, catering services and cleaning services; (iv) the need to access to the best services, such as market research services and certain types of manufacturing services including basic metal and fabrication services; (v) a lack of feasibility in supplying services in-house, such as freight insurance and utilities services; (vi) economies of scale, such as transport services; (vii) network economies, such as recruitment services and (viii) the need for strong relationships with government agencies, such as customs clearance services.

Services provided in-house are generally those that can be considered as core services such as project-specific design and engineering services, procurement and quotation services, prototyping services, production management, quality control and assurance services, installation and commissioning services, remote monitoring services, and maintenance and repair services. These services are provided in-house because they either involve proprietary technology or are essential to the quality of products. Other non-core services, such as packaging services, and storage and warehousing services, are also provided in-house because these are continuing, repeated services for which the firm has established the requisite infrastructure to ensure low-cost provision.

Interestingly, the firm does not always base outsourcing decisions on its own assessment of who are the best suppliers because customers sometimes specify their own specific preferred suppliers. This may be result of a customer's own previous experience with a supplier, or its strong relationship with the supplier. Some SMEs are among the firm's suppliers. Often, however, SMEs find it very costly to upgrade their processes and own equipment for products that need to satisfy the exacting standards demanded by the oil and gas industry.[7]

Outsourcing decisions may also be motivated by the need to satisfy local content requirements, particularly if the customer is a state-owned-enterprise.

Activities such as specialised coating, threading and painting are usually outsourced to different external parties because some may involve proprietary technology and the firm does not have the equipment to do them in-house, or it is more cost-efficient to outsource to external parties.

Regarding bundled services, several examples can be found in the value chain. One example pertains to the high-value/engineered inputs purchased by the firm. While the firm views the price it paid as the material cost of these inputs, this price actually has a certain percentage of services inputs embedded in its value, such as the underlying research and development and engineering services carried out by the suppliers in order to produce these inputs.

Another example is the bundling of transport services together with manufacturing processes that are outsourced. Providers of processes (such as machining, welding, and coating) also take care of transportation and logistics to move products from one location to another for further processing. In other words, the price paid by the firm to these suppliers has the cost of transportation and other

---

[7]For example, API has standards for manufactured products such as precision thread gauges, plain plug and ring gauges, thread measuring systems, metrology and industrial supplies, measuring instruments, custom gauges, precision machining and grinding.

services imbedded in it. Indeed, bundling can be observed in transport services as well, where the package offered by third-party providers typically includes the driver, insurance, loading of goods at the warehouse, customs clearance and the delivery of goods to customers.

Other examples of bundled services include catering services, where the caterer presumably takes care of recruiting workers and training them, as well as transporting food from the central kitchen to the firm's facility. It is also highly likely that the same bank and insurance firm provide in a single bundle the whole range of financial and insurance services needed by the firm.

8.5. POLICIES AFFECTING THE VALUE CHAIN

An important component of this study is the analysis of how policies, both government and private, impact the value chain discussed here, and hopefully, identify areas of improvements (see Appendix). This firm faces few policy challenges, for example, in trade policy (customs and logistics) and even in labour policies. This is due to the fact that it is located in Singapore which has consistently ranked very high in its business environment quality. But importantly, part of the reasons why it does not face major policy problems is that it is large, it is a subsidiary of a major multinational firm, and possesses the necessary capacity to comply with regulations.

8.5.1. *Government Incentives*

The Singapore government has always been proactive in providing assistance to the private sector to ensure that businesses remain competitive in the global economy. It gives tax incentives for productivity-enhancing investments in activities such as: (i) training of employees; (ii) purchase/leasing of specific IT and automation equipment; (iii) acquisition/licensing of intellectual property; (iv) registration of intellectual property; (v) research and development and (vi) approved design projects.

The government also alleviates the rising business costs to firms in the last few years by providing a corporate income tax rebate[8] or by co-funding 40 per cent of wage increases given to Singaporean employees earning a gross monthly wage of S$ 4000 and below over the period 2013–2015.[9] This policy is intended to give businesses more time to adjust to the tight labour market and concurrently, free up resources to undertake productivity investment activities.

8.5.2. *International Labour Mobility and Equity Ownership*

While the government provides incentives to firms to improve productivity, its labour policies result in rising costs to businesses. One such measure is the maximum number of foreign workers that a firm can hire, depending on the sector.[10] For the manufacturing sector, the quota for foreign workers is set at 60 per cent, which means that a firm is only allowed to hire up to 1.5 foreign workers for every local full time employee that it employs. Once a firm has reached its quota for foreign workers, it has to employ more local workers before being allowed to hire more foreign workers. Not only does this measure raise the manpower costs of businesses, manufacturing firms also face the problem of meeting their target output and deadlines due to the shortage of manpower.

Through its fair consideration framework (FCF), the government has also mandated that firms hiring foreigners as Professionals, Managers and Executives (PMEs) after 1 August 2014 need to consider local candidates fairly prior to submitting new Employment

---

[8]More information about the Corporate Income Tax Rebate can be obtained at: http://www.iras.gov.sg/irasHome/page04.aspx?id=14564

[9]More information about the WCS can be obtained at: http://www.iras.gov.sg/irashome/WCS.aspx. Both measures have been extended by two more years (i.e., up to 2017) in the recent Budget 2015 announcement although at a lower rate or cap.

[10]More information about the quota for foreign workers can be obtained at: http://www.mom.gov.sg/Documents/services-forms/passes/Guidelines_on_Levy_Bill_Computation.pdf

Pass (EP) applications to the Ministry of Manpower (MOM). Firms now need to advertise job vacancies on a new jobs data bank administered by the Singapore Workforce Development Agency (WDA) for at least 14 calendar days before making an EP application.[11] This measure lengthens firms' hiring processes and can potentially lead to firms losing ideal candidates for the position because of their nationalities.

8.5.2.1. *Travel outside singapore*

The firm's after-sales services, such as installation services, commissioning services, repairs and maintenance, and training services, make up a significant share of the firm's business. To provide this service to customers, the firm needs its staff, particularly engineers, to travel seamlessly from one location to another. However, the firm often encounters challenges in obtaining visas, resulting in situations where the most qualified staff are unable to provide the requisite services. Some economies also restrict the number of days that foreign professionals can go offshore where many of the oil and gas extraction facilities are located. In addition, foreign professionals are subjected to withholding tax if they stay in some economies beyond a certain number of days. To address this, as well as to provide more rapid service, the firm has established local service facilities in close proximity to its customers.

8.5.2.2. *Investment restrictions*

In the area of investment, several economies appear to have restrictions on foreign ownership, particularly when it is a service industry. The firm gave an example of an economy which requires the local subsidiary providing after-sales services to its state-owned enterprises to be majority owned by locals, which sometimes complicates decision-making by management.

---

[11] More information about the FCF can be obtained at: http://www.mom.gov.sg/newsroom/Pages/PressReleasesDetail.aspx?listid=523

8.5.3. *Local Content Requirements*

On the manufacturing side, local content requirements feature in several projects, particularly when they bear some relation to the government or a state-owned enterprise. These requirements increase operational costs since the firm has to track the procurement of inputs by source in order to ensure compliance with local content requirements. In addition, local suppliers may not necessarily be the ones giving the most competitive quotes or providing the best service. Despite this, the additional cost is not such a heavy burden because the firm can always pass on the additional cost to the customer. The more difficult part in complying with local content requirements is the absence of local capacities, the lack of local suppliers who can meet the stringent criteria and standards of the firm and industry.

Related to the issue of local content is the regulations in certain jurisdictions which requires a firm to partner with a local agent if it wants to respond to a RFQ. Depending on the local market, this requirement may lead to an increase in the firm's costs, which have to be passed on in the final quote for the products, potentially making its price bid less competitive. However, in some cases, the firm itself prefers to work with a local agent because the agent has specific knowledge of the customer and local market conditions.

8.5.4. *Other Policies*

Other policies identified by the firm as affecting services in manufacturing include periodic visits by government inspectors to ensure that the facility adheres to the stipulated guidelines on environment, health and safety, and the requirement for new staff to attend manufacturing orientation courses given by training agencies appointed by government. These policies are not considered burdensome by the firm.

8.5.5. *Private Standards: Requirement for Third-Party Certification*

Private standards are usually not strictly part of government policies but, for the oil and gas industry, are an important aspect of policy. In particular, the firm must adhere to the standards and requirements by the API for third-party inspection at various stages of the manufacturing process. API is a private trade association which has more than 500 standards and recommended practices. API's Monogram Program verifies that manufacturers of equipment are operating in compliance with these standards and its Witnessing Program provides witnesses to observe critical material and equipment testing and verification. Many manufacturers including the firm in this case study would like to have the API Monogram mark on their products because it serves as a form of quality assurance to their customers, who in effect often make this a requirement.

The firm has no significant issues meeting the requirements of the programs, considering that its facility already has certification and licenses from API. However, this may not be the case for its suppliers, particularly SMEs which also have to be certified if they are interested in being part of the firm's manufacturing value chain. The certification process for API Monogram[12] is not a simple process which can be met by firms with limited resources. Indeed, several of the firm's suppliers had determined that the process is too cumbersome and costly for them to do business with the firm if an API certification is required. Consequently, the firm may have to curtail its outsourcing strategy and will have access to a smaller pool of suppliers of inputs and processes.

As a means of supporting SME access to global value chains, the Singapore government has established the Partnerships for Capability Transformation (PACT) initiative. The initiative allows large organisations to identify and implement projects between itself and local SMEs in areas such as knowledge transfer, capability upgrading, and the development and testing of innovative solutions.[13]

---

[12] http://www.api.org/certification-programs/api-monogram-program-and-apiqr/certification-process.

[13] More information about the PACT can be obtained at: http://www.spring.gov.sg/Developing-Industries/Industry-Initiatives/Pages/partnerships-for-capability-transformation.aspx.

8.5.6. *The Way Forward*

This case study has attempted to give an oil and gas firm's perspectives on the importance of services to its operations. It is evident that policies impact the firm's access to services as well as its ability to provide them. These can have implications on the firm's overall competitiveness. Policymakers have a significant role to play in supporting businesses, particularly in creating the right regulatory environment to facilitate optimal management decisions.

Restrictions on the movement of natural persons impedes the ability of foreign firms to provide after-sales services. Likewise, the restrictions also hamper the access of its customers (such as state-owned enterprises) to the most qualified engineers to resolve the issues at hand. The case study has demonstrated that particularly in the provision of after-sales services, the different modes of supply are in fact complementary to one another. Mode 3 (in the case of establishment of local service facilities) and mode 4 (in the case of movement of engineers) go hand-in-hand in ensuring that customers are provided with the best and most efficient after-sales services.

The same can be said about local content requirements. While these certainly limit the pool of suppliers from whom the firm can source inputs and processes, they may also raise the total cost to customers and possibly diminish product quality where local suppliers do not have the right capabilities.

On the support that can be provided to local suppliers, especially SMEs, in promoting more engagement in global value chains, a good area to explore is how to help them meet global standards and pass various certification processes. This help may take the form of capacity building activities, or the provision of grants and incentives to obviate the costs of certification.

Looking deeper at the regulations mentioned in this case study, it should also be acknowledged that some of them are cross-agency in nature. For example, the restrictions on movement of foreign professionals and capital are likely to have been enacted by agencies whose mandates cover areas such as labour or foreign direct investment. For many developing economies, coordination among agencies, and awareness of the effects of policies, pose major challenges.

APPENDIX

Table A.1: Policies Affecting Services in the Value Chain.

Government and private policies/services	Authority(ies) in charge	Details	How the policy affects services in the value chain
Productivity and Innovation Credit Scheme	Inland Revenue Authority of Singapore (IRAS)	Scheme allows businesses to obtain tax savings in the form of 60 per cent cash payout and/or 400 per cent tax deduction on qualifying activities	It encourages businesses to invest in productivity improvement activities
Corporate Income Tax Rebate	IRAS	It allows firms to receive 30 per cent Corporate Income Tax Rebate in Years of Assessment (YA) 2013, 2014 and 2015 that is subject to a cap of S$30,000 per YA. For YA 2016 and 2017, the cap is S$20,000 per YA	It helps businesses to cope with rising costs
Wage Credit Scheme (WCS)	IRAS	Scheme enables government to co-fund 40 per cent of wage increases given in the period 2013 to 2015 to Singapore employees earning a gross monthly wage of up to S$4000. For 2016 and 2017, lower co-funding rate of 20 per cent applies	It gives businesses time to adjust to rising costs and concurrently, frees up resources to undertake productivity improvement activities

(Continued)

Table A.1: (*Continued*)

Government and private policies/services	Authority(ies) in charge	Details	How the policy affects services in the value chain
Quota for foreign workers	Singapore MOM	Manufacturing firms can only hire up to 1.5 foreign workers for every local full-time employee that it employs	It increases business costs to firms because local workers generally receive higher wages than foreign workers
FCF	Singapore MOM	It mandates firms to consider Singaporeans fairly for Professionals, Managers and Executives (PMEs) positions before applying for new EP for eligible foreigners by requiring firms to advertise job vacancy on a new jobs bank administered by the Singapore WDA for at least 14 calendar days	It lengthens the hiring process and may potentially lead to firms losing ideal candidates for a position because of their nationalities
PACT	Spring Singapore	It allows large organisations to identify and implement collaborative projects with local SMEs in areas of knowledge transfer, capability upgrading as well as development and testing of innovative solutions	It allows the firm to upgrade the capabilities of its suppliers and potentially widens the pool of suppliers to whom it can outsource inputs and outsource processes

Third-party inspection requirements as part of API's Monogram and Witnessing Programs	API	The challenging environment that many of its products have to operate in makes it imperative that third-party inspectors are on-site to monitor the design, engineering and manufacturing processes as and when necessary	The certification process is a challenge for some of the firm's suppliers, particularly SMEs, who elect not to do business with the firm if it requires them to be certified by API. This affects firm's outsourcing strategy and restricts the pool of suppliers the firm can access
Movement of natural persons	Various jurisdictions	They limit the stay by firm's foreign engineers and staff in the jurisdictions	The restriction may affect the quality of services provided, particularly if the engineers are the most qualified persons for the task
Withholding tax	Various jurisdictions	They require earnings made by foreign engineers and staff to be subjected to withholding tax if they provide services over a certain period of time	The restriction may limit the number of days that the engineers can stay in the economy and inadvertently affect the quality of services provided

(Continued)

Table A.1: (*Continued*)

Government and private policies/services	Authority(ies) in charge	Details	How the policy affects services in the value chain
Foreign equity restrictions	Various jurisdictions	They restrict the level of foreign equity that local maintenance and repair services facility could have	The restriction may limit the level of involvement of firms in the decisions and daily operations of the facility
Local content requirements	Various jurisdictions	They require the products sold to have certain percentage of local content	The requirements may increase prices and diminish quality because firm has to obtain quotations and secure parts/components and services only from providers who meet the criteria In addition, there may be limitations in local capacities
Use of local agent for submission of quotation and when securing order (if awarded)	Various jurisdictions	Some economies require that the firm's sales team work with local agent when making submission of quotation and when securing order	The need to involve a local agent may increase firm's operational costs

Source: Authors.

REFERENCES

American Petroleum Institute. http://www.api.org/ [23 March 2015].

British Petroleum (2014). *BP Energy Outlook 2035*. London: BP (January 2014). http://www.bp.com/content/dam/bp/pdf/Energy-economics/Energy-Outlook/Energy_Outlook_2035_booklet.pdf [23 March 2015].

Certification. American Petroleum Institute. http://www.api.org/certification-programs/api-monogram-program-and-apiqr/certification-process [23 March 2015].

GE Oil & Gas (2014). Investor meeting GE oil & gas for 10 September 2014. http://www.ge.com/sites/default/files/ge_webcast_presentation_09102014_final_0.pdf. [23 March 2015]

Inland Revenue Authority of Singapore. http://www.iras.gov.sg/irashome/default.aspx [23 March 2015].

Rushton, K (2013). BP 'failed to plan' for Deepwater Horizon disaster. *The Telegraph* (30 September 2013). http://www.telegraph.co.uk/finance/newsbysector/energy/oilandgas/10345677/BP-failed-to-plan-for-Deepwater-Horizon-disaster.html [23 March 2015].

Singapore Ministry of Manpower. http://www.mom.gov.sg/Pages/default.aspx [23 March 2015].

Spring Singapore. http://www.spring.gov.sg/Pages/Home.aspx [23 March 2015].

Webb, T (2010). BP oil spill: failed safety device on Deepwater Horizon rig was modified in China. *The Guardian* (18 July 2010). http://www.theguardian.com/environment/2010/jul/18/deepwater-horizon-blow-out-preventer-china [23 March 2015].

Chapter 9

CAR MANUFACTURING IN THE PHILIPPINES

Sherry Stephenson

9.1. FIRM BACKGROUND

This manufacturing company that produces cars in the Philippines is a large multinational producer. The company operates manufacturing operations in over 25 economies. In 2014, the company sold over 10 million vehicles in some 170 economies. The company's Head Office is in Japan, and the regional office for the ASEAN economies is located in Thailand. The case study firm has a vehicle line-up of 18 models in the Philippines and a sales distribution and service network composed of 44 outlets nationwide as of end 2014.

The Philippine automotive industry, of which is the case study company is part, comprises of two sectors: assembly operations and manufacture of parts (Aldaba, 2007). Imported cars are classified based on their level of completion: completely knocked down (CKD), referring to imported parts such as engines or transmission, which are combined with locally-produced parts to be assembled in the economy by different automotive firms; semi-knocked down, which are semi-assembled vehicles without tires or batteries; and completely built-up vehicles, which are completely assembled cars.

The firm's activities in the Philippines have been of considerable economic importance over the period 1988–2012 and can be summarised in the following figures taken from the company's website:

- Investments in the Philippines totalled Php \$35.5 billion (1988–2012).

- Procurement purchases from local suppliers totalled Php $95 billion (1989–2012).
- Vehicle production in the Philippines came to over 500,000 units (1989–2012), of which CKD were two-thirds, or 67 per cent.
- Total vehicle sales in the Philippines came to 750,000 units (1989–2012).
- Government revenues generated by the firm in the Philippines amounted to Php $136 billion (1988–2012).
- The firm's total value chain employment in the Philippines was over 40,000 employees as of end 2012.

The firm is located in a special economic zone (SEZ) south of Metro Manila, where it is permitted to undertake automobile assembly and production. The firm does not export vehicles and thus does not enjoy tax and duty-free benefits from the Philippine Economic Zone Authority (PEZA) as these are only accorded to exporting manufacturers located there. However, there are logistical advantages to being located in the zone close to other suppliers and the main plant. Additionally, the firm enjoys the advantage of being able to receive incentives from the Government in the form of duty-free importation of parts and income tax holiday for 4–6 years, depending on the specific guidelines of the particular incentive granted by government.

The plant's capacity is supposed to be 35,800 units per year. However, in 2014 the plant produced 41,000 units in response to increased demand, using adjusted production with two overtime shifts. The manufacturing assembly plant employed 1,600 regular employees as of December 2014.

The firm's manufacturing operation is supported by group company suppliers, many of whom are located within the SEZ and provide manufacturing inputs. These group suppliers may also export their intermediate inputs to other company locations overseas, or sell to other automobile manufacturers in the economy. The firm has an after-sale services office located in Santa Rosa, Laguna. The vehicle sales and customs relations departments are located in

Makati City. Several of the group suppliers export parts to other economy in the Asia Pacific region, especially Thailand, Viet Nam and Indonesia.

The firm produces two models in the Philippines. Both cars are produced for domestic sales only; at this point the Philippines does not export any fully finished automobiles, only parts. The firm's goal is to surpass production of 100,000 models per year in order to be able to export cars.

According to a research group in the Philippines, the firm had 39.4 per cent of the market share for sales in the economy at end 2014, with over 106,000 unit sales.[1] One of the firm's models was the best-selling vehicle in the economy and the market leader for the twelfth year in a row, representing nearly 10 per cent of the domestic car market.

9.2. THE VEHICLE VALUE CHAIN

The study's value chain is depicted in a simple, stylised form in Fig. 9.1.

The value chain for the car is fairly complex, involving many stages and many suppliers into the final assembly.

- The design and manufacturing blueprint for the car are provided by the firm's Head Office in Japan.
- The procurement of some of the raw materials required in the production process (steel, aluminium, rubber and many others) is sourced from various locations in the world. Sixty percent of the raw materials used in the assembly manufacture process are imported.
- Many of the manufactured inputs into the car production assembly are supplied by several companies that are not affiliated with the cases study firm but located within the SEZ, labelled as "locators" because they, as first-tier suppliers,

[1] http://focus2move.com.

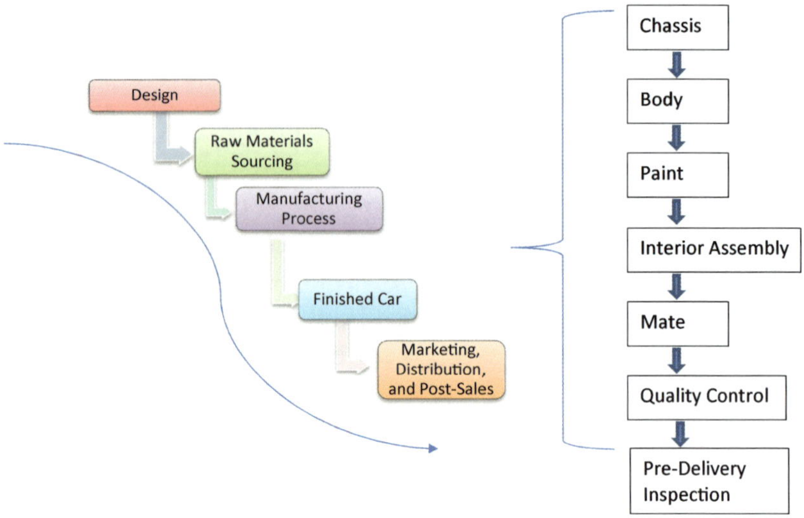

Figure 9.1: Value Chain for Car Production.

Source: Adaptation from firm's website.

followed the case study firm (lead firm) in locating to the Philippines. There are 44 local suppliers of parts and nine indirect suppliers for paints, chemicals and steel sheet that are locators in the SEZ. Their sales to the case study firm are considered domestic sales. Examples of these first tier supplies include:

- manual transmission
- switches
- machines parts
- transmissions and engine gears

- Inspection and quality control are carried out at different stages of the manufacturing process and at the end for each finished automobile according to the strict specifications provided by the Head Office.
- The final stage of the value chain for the production of the car is the post-production phase of marketing and distribution.

The activities in the production or manufacturing phase of the value chain shown in Fig. 9.1 carried out in the firm's assembly plant include the following:

- pressing and stamping (shaping metal)
- welding
- painting
- assembly
- mate[2]
- inspection and testing for quality control

Once the car is assembled and passes quality control, the unit is retained in the SEZ until it is distributed to the franchise outlet or dealership in various parts of the Philippines. As of the first quarter of 2015, there are 46 exclusive dealers in all of the Philippines for the car. The value chain studies ends with sales at one of the retail outlets and post-sales servicing.

9.3. SERVICES ALONG THE VALUE CHAIN

As has been emphasised in other mapping studies, services fulfil a range of functions along value chains. A classification of these is set out in Table 9.1. This clearly shows that for manufactured products such as an automobile, services enter the value chain at all stages, including the pre-production stage. They also accompany the final product in the post-manufacturing stage. This multi-functional role of services is the reason that their overall contribution to economic output is so large.

The number of services used in the assembly value chain for the car as identified in this study was 70 in total (see Appendix). This number includes services used in the pre-manufacturing, manufacturing and post-manufacturing phases of the value chain. The identified services have been broken down in Fig. 9.2

[2]The chassis assembly conveyor and the body shell conveyor meet at this stage of production.

Table 9.1: Multiple Functions of Services in the Manufacturing Value Chain.

Functions of services	Illustrative list of services by function
Government regulatory services	• Licensing and certification • Customs administration • Product safety standards • Environmental quality
Pre-manufacturing services	• Design • Prototyping • Intellectual property management
Manufacturing support and coordination services	• Management and administration • Personnel and human resources • Insurance • Financial services • Accounting • Legal services • Repairs and maintenance • Cleaning • Facilities services (water, electricity, gas and sewage) • Security services • Packaging and labelling
Linking services in value chains	• Transport • Communications
Post-manufacturing services	• Marketing • Advertising • Distribution • Retail • After-sales services

Source: Author.

as between pre-production stage, manufacturing stage, delivery and sales/post-manufacturing stage, back office services and post-sales services, with the largest number of services listed as back office support. As in the other case studies, the analysis treats both government and private services as additive sources of value.

A full listing of the services identified as sources of value along the chain is provided in Appendix, together with an indication of whether these services are sourced in-house or through first-tier suppliers (from group companies or otherwise), and the reasons for the choice of sourcing. This is also discussed in the section below.

9.3.1. *Outsourced Services*

Thirty-nine services are fully outsourced and nine are partially outsourced by the firm for its car production are listed below (Table 9.2), divided between the categories shown in Fig. 9.2. The reasons given for the outsourcing of services by the firm revolve around specialised skills, efficiency considerations, network economies, cost considerations and government regulatory requirements (such as the legal requirement to contract out the auditing services and the customs brokerage services at the port). While nearly all of the outsourced services are contracted to arm's length third parties within the Philippines, several outsourced services are obtained from the Head Office in Japan or other affiliated subsidiaries in Asia and therefore within the firm's own corporate structure. This includes design services and technical assistance services for various aspects of the production process when required. The reasons given for within-company outsourcing of certain services centre on the protection of proprietary information, as well as the maintenance of the corporation's worldwide product standards.

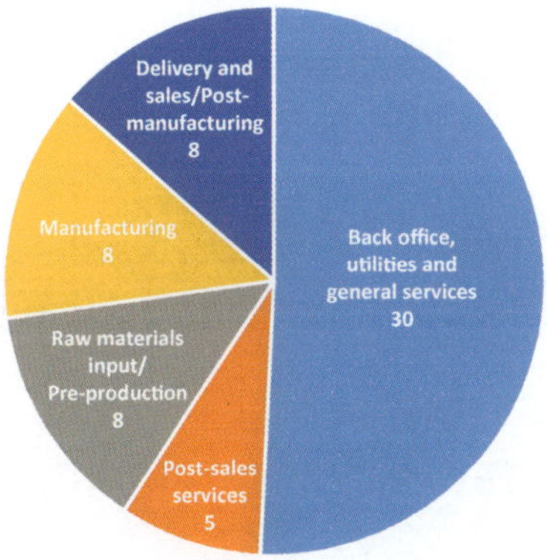

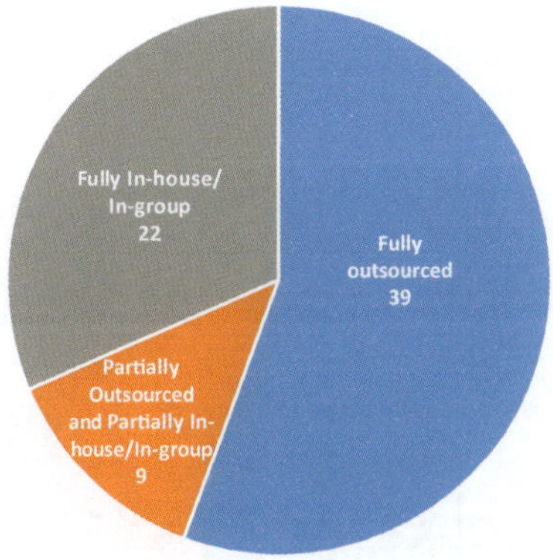

Figure 9.2: Services Used in the Car Value Chain by Function and Source.
Source: Author.

Table 9.2: List of Services Used in Car Manufacturing.

Raw materials, input/ pre-production	Manufacturing services	Delivery and sales/post-manufacturing services	Back office services	After-sales services
(1) Procurement agent for raw material sourcing	(7) Cleaning services of factory (SMEs)	(19) Land transport of vehicles to dealers	(27) Auditing services	45 Customer services — mystery shoppers
(2) Customs-related services for raw materials imported	(8) Gardening services for factory lawn	(20) Retail trade services — by store	(28) Insurance services for machinery and inventory[4]	46. Customer satisfaction surveys
(3) Quality assurance services (of raw materials)	(9) Security guards for factory and warehouses (SMEs)	(21) Retail trade services — by other non-store retail trade service	(29) Insurance for staff[5]	
(4) Freight transportation services (of raw materials) by road, rail, sea or air	(10) Sewage water treatment services (partly outsourced)	(22) Retail services on a fee or contract bases	(30) Courier, postal and local delivery services	
(5) Storage of raw materials — general storage (SMEs)	(11) Waste collection — general (SMEs)	(23) Retail administration — operation management	(31) IT and information system management, consulting and support, with webpage development	
	(12) Waste collection — scrap metal disposal (SMEs)		(32) Safety and security services (SMEs)	
	(13) Waste collection — sludge hauling services (SMEs)		(33) Telecommunication services	
			(34) Uniform — design and alternation	
			(35) Uniform — laundry	
			(36) Legal services	

[4]The firm has different insurance policies for the plants and for the vehicles produced and for the vehicles driven.

[5]Group life insurance, health insurance and hospitalisation insurance apply to workers. The latter two are not offered to workers after their retirement.

(6) Patent filing and acquisition (partially government services)	(14) Government inspections on fire prevention, health hazards, environmental protection and other aspects (partly in-house and partly government services) (15) Catering services for workers (SMEs) (16) Transportation services for employees (SMEs) (17) Moving cars from assembly line to storage parking yard (SMEs) (18) Repair and maintenance services for car carriers and busses (SMEs)	(24) Retail administration — site development for new shops (25) Storage and warehousing services for vehicles at retail outlets (26) Installation services for car accessories	(37) Retail administration — advertising (38) Advertisement — advertisement agencies services (39) Advertisement — designers' services (40) Advertisement — direct marketing and mail-in advertisement (41) Advertisement — cameramen, directors, photographers and models (42) Electricity supply (43) Gas supply (44) Water supply

Source: Author.

Many of the third parties to whom the various services tasks are outsourced within the Philippines are small and medium-sized firms (SMEs). This is the case for car carrier services, transport services for employees, hauling and disposal of scrap materials, security services, cleaning services and installation services for accessory parts, among several others. In addition, certain form of research and development, for example, testing if the addition of coconut oil in gasoline used in cars in the Philippines would have adverse effect on the car machines, is likewise outsourced to third parties. This demand thus generates additional employment and value-added in-economy. The firm specifically states that part of the reason for this outsourcing is to integrate into the domestic economy and build networks of domestic capability, especially among SMEs.[5]

9.3.2. *In-house Services*

Twenty-two in-house services are provided entirely by the firm or by in-group suppliers for the identified value chain. A listing of the in-house/in-group services is provided below. It is of note that the firm has chosen to supply itself those services that are the closest to the identity of the company (design and engineering) or to the reputation and safety of the product (quality assurance control services and repair and maintenance services), or those that involve sensitive financial information. Quality controls are carried out by the firm but are dictated to the company by the Head Office in Japan through its established system of quality controls. For the Philippines, oversight of these controls is supported by the regional office in Thailand. Compliance with government standards and policies are dealt with by concerned business units. This major responsibility of compliance is done in-house (Table 9.3).

---

[6]Around 40 per cent of parts and components used by the firm in the value chain are sourced locally, while 60 per cent are imported. The company did not have a similar figure for the services that it uses in its output.

Table 9.3: In-house Services.

Raw materials, input/pre-production	Manufacturing services	Delivery and sales/ post-manufacturing services	Back office services	After-sales services
(1) Conception and design of product (2) Industrial design	(3) Production Administration — Production management (4) Production Administration — Quality assurance and compliance with ISO (5) Engineering services (technical assistance service for special skills is provided to group companies) (6) Repair and maintenance services of machines and equipment (7) Personnel search and referral services — Recruitment of factory workers (8) Accommodations for dealers (9) Prototype testing services	None	(10) Financial services[7] (11) Internal auditing (including audits of financial accounts and corporate governance) (12) Business and management consultancy services (13) Company secretary services (14) Corporate communications, marketing and public relationship (15) Estate management (16) Human resources management (17) Human resources management —Personnel search and referral services for back office staff (18) Training of staff (19) Product development/R&D	(20) Customer services — Complains and compliments handling (21) Customer services — Technical supports (22) Customer services — Warranty and repair services

[7]Financial accounting is done in-house and all loans are obtained in-house, within the company group. The 1997 Asian crisis was the last time the firm took out a loan.

Source: Author.

9.3.3. *Bundling*

The researcher found only one case of services that were bundled when being contracted by the firm in its assembly production. These are information technology (IT)/computer offerings. A single IT vendor can provide both software development and interface services for other systems within the company. It can provide a package of services including website design and maintenance, software development and computer programming and computer repair services.

9.4. POLICIES AFFECTING SERVICES IN THE VALUE CHAIN

Several government policies and regulatory requirements in the Philippines affect the production of the supply chain — prior, during and after — its assembly. The firm indicated during the interview that no fewer than ten different government departments were involved in various aspects of vehicle assembly and subsequent sale in terms of granting incentives, giving permits of various types, implementing environmental and technical standards and determining and enforcing fuel standards, labour laws and vehicle registration, among others. Table 9.4 lists these agencies, together with their area of regulatory responsibilities.

In terms of the government policies that are the most imposing on the firm's costs and efficiency, the firm pointed to labour policies, taxation rates and inefficiencies in logistics (including infrastructure and customs procedures), each of which is discussed in turn. These policies impact heavily on the current cost structure of production for the car, which is 10 per cent higher in the Philippines than it is in Thailand. The cost of production of another model produced by the firm is 18 per cent higher in the Philippines than it currently is in Indonesia.

9.4.1. *Labour Policies*

Labour laws in the Philippines are stringent and offer several layers of legal protection to workers. Current labour law strictly

Table 9.4: Government Agencies in the Philippines Involved in Policy Interface with the Production of a Car.

Government department	Area of regulatory responsibility
PEZA	Grants incentives in the SEZ
Department of Trade and Industry and Department of Finance	Gives approval for these incentives including tax and duty rates that must then be passed by Congress into law
Department of Finance: Bureau of Internal Revenue and Bureau of Customs	Determines and collects tax remissions and taxes and duties
Department of Transportation and Communication and Land Transport Office	Responsible for registration of vehicles
Department of Land Franchising Registry Board	Approves applications for franchising and operation of public utility vehicles
Department of Energy	Establishes requirement for fuels to contain a certain percentage of coconut oil, affecting the standards for the engines
Department of Environment and Natural Resources	(i) Establishes and implements emission standards (ii) Grants certification for each model of automobile so that it can be put on the market for sale (iii) Grants permits for construction and disposal of waste (iv) Carries out annual inspections for painting and waste water treatment
National Biofuels Board (under Department of Energy)	Reviews and enforces fuel standards for automobiles
Metro Manila Development Authority	Responsible for all roads and traffic management
Department of Labour and Employment	Administers and enforces all labour laws

Source: Author.

prohibits labour-only contracting and there is a pending bill in Congress that states that only a maximum of 20 per cent of labour in a firm can be contractual; all others must be full-time employees.[7] Firms are not allowed to lower salaries or dispense with any type of benefits, unless it is possible to convince the labour unions to accept these changes. To date, there are no specific guidelines for part-time workers, which reduce the flexibility of the labour market and the firm's ability to deal with fluctuations in demand and the production cycle. Worker lay-offs can only be made through a notification to the Department of Labour 30 days in advance, and after having presented the plan to the labour union. Strikes are allowed anywhere in the economy except within the SEZ. Labour costs were indicated as one of the factors inhibiting expansion of motor vehicle production.

9.4.2. *Taxation/Subsidy Policies*

The firm indicated that tax rates in the Philippines are high, with 30 per cent corporate tax rate, even for SMEs. Employee allowances are also taxed (the increment). Currently the cost of production of the case study car is not directly linked to its sales price, which is determined by competition in the domestic market but subject to final approval by the regional office. There is a roadmap pending in Government offices to provide a $1,000 incentive per qualified automobile unit to help fill in the $2,000 gap in the cost of domestic production; however, this roadmap has not been approved and acted upon in 2 years. The car manufacturing industry has been requesting a more stable automobile policy framework from the Government that would allow the company to be more competitive and to be able to increase its domestic production base and capacity in order to be able to exports cars.

---

[8]The firm stated that in Thailand, one of the major automobile manufacturers in ASEAN, the labour laws are very different, and over 50 per cent of the labour force can be short term.

9.4.3. *Logistics Policies*

Many aspects of logistics also impact on the production of the car, adding significantly to cost and time delays.[8] There is a widely recognised, urgent need to upgrade port and road infrastructure so that port services and transport services can be more efficient in coping with current and future needs. Delays in receiving imported raw materials, parts and components through customs procedures and out of the port in Manila were cited by the firm as a major reason for cost increases, given the congestion of the port, deteriorated infrastructure and slow processing times.

A mandatory inspection must be carried out for all imported parts and components of the car by Customs/Bureau of Finance and Internal Revenue before they are unloaded, but due to the port congestion, this requirement cannot be adequately implemented. Previously, this inspection was done by a pre-shipment inspection company.

In addition to port congestion, there was a local government ban on trucks in Manila where the port is located since February 2014, leading to a backlog of goods needing to be transported out of Customs. Due to these restrictions, truckers often double their charges and work overtime in order to deliver. The trucking ban has significantly added to production costs for the firm and its suppliers.[9] Also, because of the congestion and delays, the firm cannot

[8]The Philippines ranked number 68 in the Logistics Performance Index of 2014, produced by The World Bank. This index is a weighted average of six key dimensions involved in logistical competitiveness, namely: customs (speed, simplicity and predictability of border formalities); infrastructure (quality of ports, roads, railroads, etc.); ease of shipment; ease of tracking and tracing consignments; domestic logistics costs (transport operators, customs brokers) and timeliness. The Philippines ranked below both Thailand (34) and Indonesia (46), its main competitors within ASEAN, and was also ranked lower than Viet Nam. http://lpi.worldbank.org/international/global/2014.

[9] It is estimated that the trucking ban has cost the firm around 92 million pesos and has cost its suppliers around half of this amount (45 million additional pesos) since February 2014. After several months and worsening congestion at the port area, the ban from the local government was rescinded.

carry out just-in-time delivery. The retail price is standardised by the local firm for uniform application and compliance of the dealers. Any higher logistical costs must be absorbed by local firm, putting it at a competitive disadvantage.

Overall, in terms of the government–firm policy interface, there appears to be much potential productivity growth that could be reaped through an improvement in various key policy areas, particularly logistics.

9.4.4. *Additional Policy Considerations*

As reported in the 2005 WTO Trade Policy Review of the Philippines, MFN tariffs on imported passenger cars, trucks and motorcycles range between 20 and 30 per cent. The Government assists assembly of automobiles in-economy by allowing for participants in its Motor Vehicle Development Plan to import completely-knocked-down (CKD) kits at preferential tariff rates of 1–3 per cent. The Philippines has a zero duty policy on the import of capital equipment and spare parts by automotive manufacturers. The Philippine Motor Vehicle Development Program is designed to foster local assembly through low tariffs on components.[10]

These policies have resulted in a very high rate of effective protection for the motor vehicle sector, estimated by the WTO to be around 76 per cent (2004).[11] Presently, imported cars before tariffs are cheaper than domestic cars in the Philippines.[12] In addition, the economy applies non-trade barriers to many automotive inputs, usually in the form of non-automatic import licensing.

The ASEAN Free Trade Agreement for the automotive sector implemented in 2000 has put significant competitive pressure on

---

[10] United States Trade Representative (USTR) (2013). "The Philippines". https://ustr.gov/sites/default/files/2013%20NTE%20Philippines%20Final.pdf.

[11] Trade Policy Review of the Philippines, Report by the WTO Secretariat, WT/TPR/S/149, June 2005, page 89.

[12] "Japan urges Philippines to make strong auto industry policy soonest". Manila Bulletin. June 16, 2014. http://www.mb.com.ph/japan-urges-philippines-to-make-strong-auto-industry-policy-soonest/.

automobile assembly manufacturers in the Philippines over the past decade, as it substantially reduced the effective rate of protection in this sector, requiring adjustment to a new competitive environment. The realisation of the ASEAN Economic Community (AEC) by end 2015, according to government plans, may also bring additional competitive pressures, particularly in the services area, affecting the services that are embedded in the production of the local car. According to The Economist (2014), the AEC signifies the arrival of a more level playing field, less protectionism and greater competition.[13]

Possibly in recognition of this potential future development, the Philippine Government announced its intention in March 2015 to support the automotive industry in order for the economy to become a regional automotive production centre.[14] For this objective, it declared that it would be offering tax breaks to up to three automakers. The Manila Bulletin (2014) reported that the government has considered an annual production of 40,000 vehicles per model as a qualification for the incentive. The incentive is part of a new program entitled the "Comprehensive Automotive Resurgence Strategy" that has been announced for implementation during 2015. The Government has stated that expanding the automobile supply chain would feed into other industries as well, providing a boost to manufacturing beyond the car industry. The extent of investment and growth in the automobile industry in the Philippines will be in large part a function of foreign direct investment, where currently the Philippines lags behind Thailand and Indonesia.[15]

---

[13] "The automotive sector after the AEC". The Economist. 30 July 2014. http://country.eiu.com/article.aspx?articleid=722103856&Country=Malaysia&topic=Economy.

[14] In 2015, the Philippines is projected to produce 90,000 light automotive vehicles, compared to 1.3 million in Indonesia and 2.14 million in Thailand, according to projections by IHS Automotive in Bangkok.

[15] President Benigno Aquino has made investing in infrastructure a priority during his term, which comes to an end in June 2016, in part to boost growth and also attract foreign direct investment. In 2014, the Philippines attracted $6.2 billion in FDI, compared to $12.6 billion in Thailand and $8.9 billion in Viet Nam.

The services and investment policies in the economy seem to be supportive of an expanded automobile manufacturing objective on the part of the Philippines Government, to the degree that there are no restrictions on foreign direct investment in this sector. Additionally, as reported by the WTO Trade Policy Review (2005) latest report on the Philippines, competition in the financial sector seems to be healthy, with no evidence of monopoly or collusive behaviour, and quite a few foreign banks or subsidiaries in the economy.[16] The telecommunications sector is also cited in the WTO Report as showing relatively unfettered market entry and exit, and healthy competition.[17] Market access has been liberalised in nearly all telecom services, including local, domestic long-distance, international, mobile, data, telex, leased line, paging, cable television and satellite. However, the WTO report cited evidence of restricted competition in the transport sector, and particularly between private and public ports, as well as among cargo handling operators. The need for infrastructure improvement and greater efficiency and openness in the transport sector (maritime, ports and domestic shipping) is underlined.

Despite the supportive environment in key services and investment in finance and telecommunications, the Philippines will need to differentiate its automobile products from its successful exporting neighbours in ASEAN by developing niche markets as well as by lowering its internal production costs in order to maintain and expand competitiveness. To achieve this, it will need to address the sources of higher costs for its automobile manufacturers identified in this study and to improve its logistics policies.

Medium-size manufacturing, including automobiles, is at a relative deficit compared to other production areas in which the Philippines is a bigger player on the global market, for example, medical devices, aircraft parts and platforms for oil and gas.

[16]Trade Policy Review of the Philippines, Report by the WTO Secretariat, WT/TPR/S/149, June 2005, page 93.

[17]*Ibid*., page 103.

APPENDIX

Table A.1: Services used in the Pre-Production/Production/Post-Production/Sales and Post-Sales Stages of the Value Chain.

Stage/category	Service	CPC Ver.2 Code	Supplied in-house	Supplied by another company in the group	Outsourced to third-parties	Reasons of outsourcing
		Raw materials, inputs/pre-production stage				
1. Procurement of raw materials	(1) Procurement agent for raw material sourcing	85999 — Other support services n.e.c.			Yes. Outsourced to qualified brokers for authorised that can access the port	Legal obligations to outsource by Philippine law
	(2) Customs-related services for raw materials imported	85999 — Other support services n.e.c.			Yes. Outsourced to qualified brokers for imports. Several brokers are involved	Legal obligations to outsource by Philippine law
	(3) Quality assurance services (of raw materials)	83441 — Composition and purity testing and analysis services	Yes. In-house inspection for raw materials and parts		Yes. Philippines government requires pre-shipment inspection and quality and quantity	

(Continued)

Table A.1: (*Continued*)

Stage/category		Service	CPC Ver.2 Code	Supplied in-house	Supplied by another company in the group	Outsourced to third-parties	Reasons of outsourcing
2. Logistics	(4)	Freight transportation services (of raw materials) by road, rail, sea, or air	Division: 65 — Freight transport services			Yes. Parts are imported from plants in other parts of the Asia	
3. Storage	(5)	Storage of raw materials — General storage	67290 — Other storage and warehousing services			Yes. Outsourced to SMEs	Efficiency
4. Product design	(6)	Conception and design of product	83920 Design originals		Yes. Royalty paid for design in other companies in group for design		
	(7)	Industrial design	83912 Industrial design services		Yes. Royalty paid for design in other companies in the group for design		

	(8)	Patent filing and acquisition	83960 Trademarks and franchises n/a		Yes. In-group patent filing department	Yes. Government services	Handling of patent filing is a government services
Manufacturing Stage							
1. Production administration	(9)	Production administration — Production management	83115 — Operations management consulting services	Yes. In-house management team decide production schedule	Yes. Group in Japan decides on car models for production		
	(10)	Production administration — Quality assurance and compliance with ISO	83441 — Composition and purity testing and analysis services	Yes. In-house team is responsible for quality assurance			
2. Services supporting factory daily operations	(11)	Cleaning services of factory	85330 — General cleaning services			Yes. Outsourced to local firms	
	(12)	Engineering Services	83310 — Engineering advisory services Class: 8332 — Engineering services for specific projects	Yes. In-house engineers	Yes. Technical assistance service for special skills is outsourced to the group		

(*Continued*)

Table A.1: (*Continued*)

Stage/category	Service	CPC Ver.2 Code	Supplied in-house	Supplied by another company in the group	Outsourced to third-parties	Reasons of outsourcing
	(13) Gardening services for factory lawn	85970 — Landscape care and maintenance services			Yes. Outsourced to local firm	
	(14) Security guards for factory and warehouses	85250 — Guard services			Yes. Outsourced to local SMEs	
	(15) Sewage water treatment services	94110 — Sewerage and sewage treatment services	Yes. In-house management of outsourced team		Yes. A team of 4 handles sewage water treatment	Efficiency, cost
	(16) Repair and maintenance services of machines and equipment	87156 — Maintenance and repair services of commercial and industrial machinery	Yes. In-house-team provide everyday maintenance of machines	Yes. Support will be provided by the Group in Japan when necessary	No	Risk aversion
	(17) Waste collection — General	94239 — General waste collection services, other			Yes. To local SMEs	Local employment

	(18) Waste collection — scrap metal disposal	94229 — Collection services of non-hazardous recyclable materials, other			Yes. To a local firm for scrap metal disposals and recycling	
	(19) Waste collection — Sludge hauling services	Class: 9421 — Collection services of hazardous waste			Yes. To local SMEs	Local employment
3. Services from government regulation requirements	(20) Government inspections on fire prevention, health hazards, environmental protection and other aspects	91133 — Public administrative services related to mining and mineral resources, manufacturing and construction 91290 — Public administrative services related to other public order and safety affairs	Yes. The firm has to prepare for inspection		Yes. Government services	Government services
4. Worker- related services	(21) Catering services for workers	63393 — Other contract food services			Yes. Outsourced to local SMEs	

(Continued)

Table A.1: (*Continued*)

Stage/category	Service	CPC Ver.2 Code	Supplied in-house	Supplied by another company in the group	Outsourced to third-parties	Reasons of outsourcing
	(22) Personnel search and referral services — Recruitment of factory workers	85112 — Permanent placement services, other than executive search services	Yes. In-house human resources department hires workers			
	(23) Transportation services for employees	64114 — Local special-purpose scheduled road transport services of passengers			Yes. Outsourced to local SMEs	
5. Others	(24) Accommodations for dealers	63111 — Room or unit accommodation services for visitors, with daily housekeeping services	Yes. Firm manages on-site accommodation for visiting dealers			
	(25) Moving cars from assembly line to storage parking yard	67290 — Other storage and warehousing services			Yes. Outsourced to a large company	

	(26) Repair and maintenance services for car carriers and busses	87141 — Maintenance and repair services of motor vehicles		Yes. Outsourced to SMEs	Local employment
	(27) Prototype testing services	83443 — Testing and analysis services of integrated mechanical and electrical systems	In-house		
Delivery and sales post-manufacturing stage					
1. Delivery to wholesaler/retailer	(28) Land transport of vehicles to dealers	65112 — Road transport services of freight by tank trucks or semi-trailers		Yes. Outsourced to a large company	
2. Sales services	(29) Retail trade services — By store	Group of 621 — Non-specialised store retail trade services Group of 622 — Specialised store retail trade services		Outsourced to authorised dealers	

(Continued)

Table A.1: (*Continued*)

Stage/ category	Service	CPC Ver.2 Code	Supplied in-house	Supplied by another company in the group	Outsourced to third-parties	Reasons of outsourcing
	(30) Retail trade services — By other non-store retail trade services	Group of 624 — Other non-store retail trade services			Outsourced to authorised dealers	
	(31) Retail services on a fee or contract bases	Group: of 625 — Retail trade services on a fee or contract basis			Outsourced to authorised dealers	
	(32) Retail administration — Operation management	83115 — Operations management consulting services 83116 — Supply chain and other management consulting services	In-house management for dealers — appointment, evaluations and supports		Outsourced to authorised dealers for daily operations	
	(33) Retail administration — Site development for new shops	83911 — Interior design services Group of 546 — Installation services Group of 547 — Building completion and finishing services			Outsourced to authorised dealers	

	(34) Storage and warehousing services for vehicles at retail outlets	67290 — Other storage and warehousing services			Outsourced to authorised dealers	
	(35) Installation services for car accessories	87390 — Installation services of other goods n.e.c.			Outsourced to SMEs	Specialised. SMEs can also install in accordance to specific requirements of customers
Back office, utilities and general services						
1. Finance	(36) Auditing on financial accounts	Group of 822 — Accounting, auditing and bookkeeping services			Outsourced	Legal requirement
	(37) Financial services	71121 — Deposit services to corporate and institutional depositors 71313 — Group pension services 71701 — Services of holding equity of subsidiary companies		Yes. A major shareholder of the firm provides financial services		

(Continued)

Table A.1: (*Continued*)

Stage/ category	Service	CPC Ver.2 Code	Supplied in-house	Supplied by another company in the group	Outsourced to third-parties	Reasons of outsourcing
	(38) Internal auditing (including audits of financial accounts and corporate governance)	Group of 822 — Accounting, auditing and bookkeeping services 83118 — Head office services	Yes			
	(39) Insurance services for machinery and inventory	71332 — Marine, aviation and other transport insurance services			Outsourced. Insurance for inventory	
	(40) Insurance for staff	71321 — Accidental death and dismemberment insurance services 71322 — Health insurance services			Outsourced	
2. General management	(41) Business and management consultancy services	Class of 8311 — Management consulting and management services	Yes			

(42) Company secretary services	83990 — All other professional, technical and business services, n.e.c.	Yes			
(43) Corporate communications, marketing and public relationship	83114 — Marketing management consulting services 83121 — Public relations services	Yes			
(44) Courier, postal and local delivery services	Group of 681 — Postal and courier services			Yes	Efficiency and cost
(45) Estate management	72112 — Rental or leasing services involving own or leased non-residential property	Yes			
(46) Human resources management	83113 — Human resources management consulting services	Yes			
(47) Human resources management — Personnel search and referral services for back office staff	Class of 8511 — Personnel search and referral services	Yes			

(*Continued*)

Table A.1: (*Continued*)

Stage/ category	Service	CPC Ver.2 Code	Supplied in-house	Supplied by another company in the group	Outsourced to third-parties	Reasons of outsourcing
	(48) Training of staff	92919 — Other education and training services, n.e.c.	Yes (to maintain personnel and quality standards)			
	(49) IT and information system management, consulting and support, with webpage development	Class of 8313 — IT consulting and support services Class of 8314 — IT design and development services Class of 8316 — IT infrastructure and network management services			Yes. IT agency has full-time staff on-site to offer help	Specialised, efficiency, cost
	(50) Safety and security services	85230 — Security systems services 85250 — Guard services			Yes. Outsourced to local SMEs	

	(51) Telecommunication services	Group: 841 — Telephony and other telecommunications services 84210 — Internet backbone services 84221 — Narrowband Internet access services 84222 — Broadband Internet access services		Yes. To tele-communication companies	
	(52) Uniform — Design and alternation	83919 — Other specialty design services		Possibly outsourced	Specialised
	(53) Uniform — Laundry	97130 — Other textile cleaning services		Possibly outsourced	Specialised
3. Legal	(54) Legal services	82120 — Legal advisory and representation services concerning other fields of law 82130 — Legal documentation and certification services		Yes	Specialised

(*Continued*)

Table A.1: (*Continued*)

Stage/ category	Service	CPC Ver.2 Code	Supplied in-house	Supplied by another company in the group	Outsourced to third-parties	Reasons of outsourcing
4. Research and development	(55) Product development/ R&D	81129 — Research and experimental development services in other engineering and technology 81400 — Research and development originals	Yes. In-house team advice on local specialty on design	Yes. Outsourced to Group and other plants in Japan and Thailand		
5. Advertisement	(56) Retail Administration — Advertising	83611 — Full service advertising	Yes. In-house team to manage outsourced agencies. Also distributes advertisement materials to dealers		Yes. Professional advertisement agency contracted for marketing events for media	Specialised
	(57) Advertisement — Advertisement agencies services	83611 — Full service advertising 83620 — Purchase or sale of advertising space or time, on commission				
	(58) Advertisement — designers' services	83611 — Full service advertising				
	(59) Advertisement — Direct marketing and mail-in advertisements	83612 — Direct marketing and direct mail services				

	(60)	Advertisement — Cameramen, directors, photographers and models	83611 — Full service advertising		
6. Utilities	(61)	Electricity supply	Class of 8631 — Support services to electricity transmission and distribution	Yes. Outsourced to national network	Efficiency
	(62)	Gas supply	86320 — Gas distribution services through mains (on a fee or contract basis)	Yes. Outsourced	
	(63)	Water supply	86330 — Water distribution services through mains (on a fee or contract basis)	Yes. Outsourced	
7. Other related services	(64)	Support for labour relocation	91138 — Public administrative services related to general economic, commercial and labour affairs	Yes	Government services
	(65)	Government relations/labour relations	83121 — Public relations services	Yes. firm contracts one agency to deal with all Philippine government departments	Specialised, requires local know-ledge and interface connections

(Continued)

Table A.1: (*Continued*)

Stage/ category	Service	CPC Ver.2 Code	Supplied in-house	Supplied by another company in the group	Outsourced to third-parties	Reasons of outsourcing
		Post-sales services				
1. Sales/after-sales customer services	(66) Customer services — Complains and compliments handling	85931 — Telephone call centre services	In-house			
	(67) Customer services — Technical support	85931 — Telephone call centre services	In-house			
	(68) Customer service — Warranty and repair services	87141 — Maintenance and repair services of motor vehicles	In-house			
	(69) Sales administration — Secret customers	83700 — Market research and public opinion polling services			Outsourced to market research firms	
	(70) Customer satisfaction surveys	83700 — Market research and public opinion polling services			Outsourced to market research firms	

Source: Author.

REFERENCES

Aldaba, RM (2008). Globalization and the need for strategic government-industry cooperation in the Philippine Automotive Industry. Discussion Paper Series No. 2008–21. Philippine Institute for Development Studies. http://www.eaber.org/sites/default/files/documents/PIDS_Aldaba_2008_3.pdf [20 March 2015].

Aldaba, RM (2007). Assessing the competitiveness of the Philippine Auto Parts Industry. Discussion Paper Series No. 2007–14. Philippine Institute for Development Studies. http://www.eaber.org/sites/default/files/documents/PIDS_Aldaba_2007.pdf [18 March 2015].

Le, TM (2003). *Value Added Taxation: Mechanism, Design, and Policy Issues.* Washington, D.C.: The World Bank Course on Practical Issues of Tax Policy in Developing Countries.

Magkilat, B (2014). Japan Urges Philippines to make strong auto industry policy soonest (June 2014). *Manila Bulletin.* http://www.mb.com.ph/japan-urges-philippines-to-make-strong-auto-industry-policy-soonest/ [15 March 2015].

The Economist (2014). The automotive sector after the AEC (July 2014). http://country.eiu.com/article.aspx?articleid=722103856&Country=Malaysia&topic=Economy [15 March 2015].

Trade Policy Review Report by the Philippines (2012). Geneva: WTO. https://www.wto.org/english/tratop_e/tpr_e/tp361_e.htm [15 March 2015].

—— (2005). Trade Policy Reviews WTO. http://www.wto.org/english/tratop_e/tpr_e/tp249_e.htm [15 March 2015].

United States Trade Representative (2013). The Philippines. https://ustr.gov/sites/default/files/2013%20NTE%20Philippines%20Final.pdf [15 March 2015].

Chapter 10

MANUFACTURING OF THERMAL POWER GENERATION EQUIPMENT

Gloria O. Pasadilla

10.1. INDUSTRY OVERVIEW

The electric power industry has several major activities: power generation, transmission and distribution. After electricity is generated, it passes through high voltage transmission lines. Power then goes to a substation where a transformer adjusts the voltage to a lower level to be distributed for consumer and industrial use. Figure 10.1 describes the electricity industry value chain, where power generation is shown to be an upstream activity. The focus of this case study is on power generation because the firm is a global player in power generation technologies.

The electricity industry has undergone deregulation all over the world, particularly through the privatisation of many previously state-owned electricity companies and dismantled monopolies. In general, industry restructuring has carved out the power generation and retail supply segments (billing, metering and installation) and introduced more market competition in these sub-industries to increase efficiency. In most places, transmission and distribution remain as monopolies, however, because it is unviable to build competing grids.

Power generation, the upstream part of the electricity value chain, involves the transformation of mechanical energy into electrical energy. Central to virtually all power generation is the

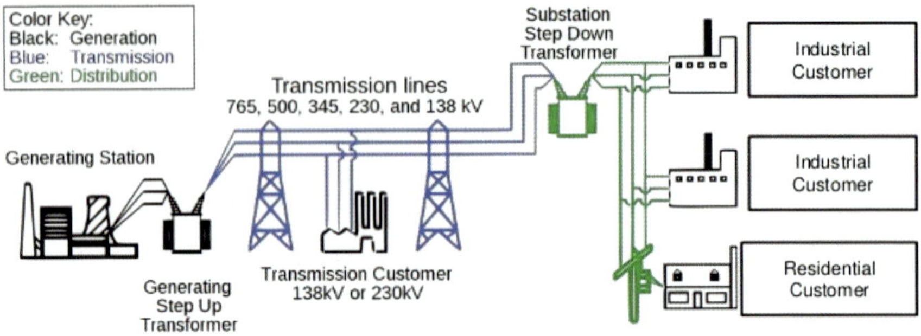

Figure 10.1: Electricity Industry Value Chain.

Source: Courtesy of North American Electric Reliability Corporation.

turbine. When the blades on the shaft of a turbine are rotated, the generator produces electricity through a process called magnetic induction.[1] The sources of energy that help to turn the blades of the turbine to generate electricity vary. Coal is the cheapest energy source but emits the most amount of harmful substances into the environment. Other energy sources are nuclear, natural gas, geothermal, hydro as well as renewable sources like solar energy or wind.

Electricity generation causes more greenhouse gas emission than transportation.[2] Increased global awareness of the need for environmental sustainability has driven technological research to find alternative energy sources such as renewable energy as well as to create improved designs of turbine and other power plant equipment that minimise emissions and reduce waste. The company in this study presents itself as a provider of both thermal power and environmental technologies because its new designs of power plant equipment use technologies that are environmental friendly.

---

[1]'Electricity — a Visual Primer'. Available at: http://www.slideshare.net/MavensManor/electricity-a-visual-primer-5913813.

[2]US Environmental Protection Agency. Available at: http://www.epa.gov/climatechange/ghgemissions/sources.html.

10.2. BACKGROUND INFORMATION ON THE FIRM[3]

The company is headquartered in Japan and is a big player in the production of thermal power generation machines and equipment. It has five production sites in Japan, but some products and/or parts and components are also manufactured in different continents either through a fully owned subsidiary or a joint venture company. For example in Asia, it has manufacturing plants in China, India and the Philippines; In Europe, there is a manufacturing plant in Germany, while UK and Belgium have repair facilities; North America, Canada and the US have manufacturing plants. This is in line with the company strategy of manufacturing at sites close to where products are to be delivered. In total, its company brochure lists 54 different enterprises across the globe, either subsidiaries or joint venture companies that are engaged in either manufacturing, maintenance services and repairs, corporate functions and sales or a combination of several functions. Of these 54 enterprises, it is significant to note that the majority are services companies.

The company's major products are gas turbine combined cycle power plants; integrated gasification combined cycle power plants; boiler and turbine generation plants; geothermal power plants; gas turbines; boilers; steam turbines (Fig. 10.2); generators; equipment peripheral to power generating plants and fuel cells. Its research and development activities have helped to upgrade and redesign products either to minimise waste or limit harmful gas emissions and thus make electricity generation more environmentally friendly.

Beyond producing products for power plants, the company's business also includes design, build and maintenance of power plants as well as a whole suite of after-sales services and support. Example of after-sales services include: (1) preventive maintenance which includes remaining life estimation service, maintenance

---

[3]Information on the firm has been sourced from its corporate website and company brochure.

Figure 10.2: Example of Firm's Product: Steam Engine and Generator Installed in an Energy Centre in the United States.

Source: Courtesy of the firm.

schedule management, as well as maintenance personnel education programs; (2) original spare parts supply; (3) operation support to reduce power consumption or greenhouse gas emissions; (4) performance enhancement through operation control adjustments or equipment rehabilitation and (5) integrity inspections, for example for the high pressure and temperature vessel or piping in boilers.

10.3. DESCRIPTION OF THE VALUE CHAIN

This case study considers an EPCM (engineering, procurement, construction and maintenance) project where the firm may be the lead, or a member, of a consortium bidding for the construction of a power plant. There are, in fact, few EPCM projects in which the company is the prime contractor; in most projects, the firm is only a member of a consortium where its primary responsibility is to supply the key machines and equipment for the power plant. The value chain, in this case, begins with the bidding stage, then moves to the equipment design and pre-construction of power plant stage, followed by the manufacturing and construction stage, and ends with the commissioning and operation of the power plant (Fig. 10.3).

Other types of projects, which do not involve construction from scratch, also exist. Some projects may simply involve retrofitting

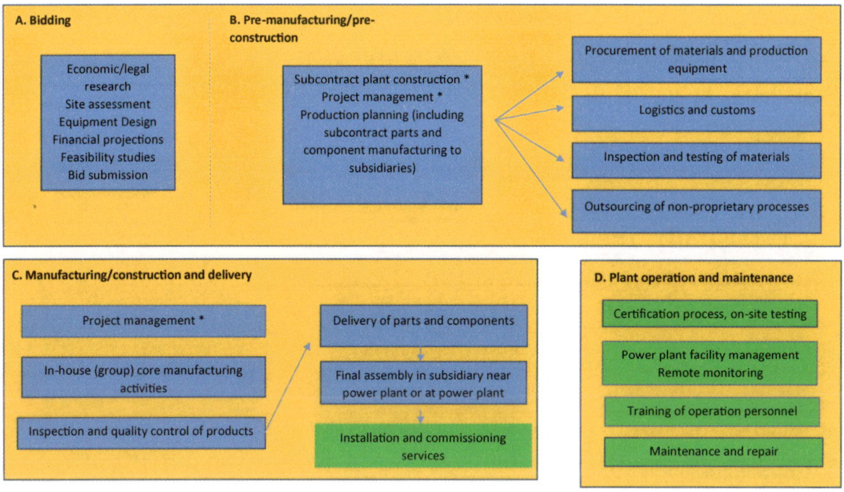

Figure 10.3: Dimensions of the Value Chain Covered by Case Study.

Note: Optional activities in the value chain are indicated by green boxes. (*) means 'if firm is the prime contractor'.
Source: Author.

and upgrading existing power plant facilities, or the construction of additional power plant features for environmental sustainability purposes. These types of projects will have a shorter value chain and will, in some respects, exhibit similar characteristics as an EPCM turnkey project. On the other hand, the value chain described above omits an increasingly important part of the value chain, i.e. the disposal of old machines.

10.3.1. *Bidding Stage*

If the firm is the prime contractor or the leader of the consortium bidding for a project, it will identify potential partners, especially the local construction company that will carry out the plant construction. Then it will carry out research on the economic and legal environment, and the relevant policy framework, especially in respect of environmental policies. It will check the site or location of the power plant to assess if its geographical configuration necessitates major redesign or adjustments in the company's existing equipment should it win the contract. It will carry out feasibility studies and a financial projection before submitting the bid.

In cases where the firm is only part of a consortium led by another prime contractor, the latter is considered the firm's 'customer'. Usually, these 'customers' are either utility companies or independent power providers in the economy where the power plant is to be built. In this case, all the feasibility studies and financial projection of the entire project are the responsibility of the prime contractor, but the firm also does its own research on the economic conditions, the legal environment of the economy, as well as assess the site or location of the proposed plant to be able to configure the right machines according to land type or topography. The firm participates in the preparation of the bid submission. In power plant projects, the cost that the case study firm provides to the consortium will typically constitute a major part of the total project cost and could help the consortium win or lose the bid.

10.3.2. *Design, Pre-manufacturing and Pre-construction of the Power Plant*

If the firm is in-charge of the plant construction, it will typically outsource the construction work to a local subcontractor which is typically a construction or engineering company. The construction company then takes care of securing all the requisite government permits and often the overall project management of the construction process under the overall supervision of the firm.

Once the contract is awarded, the firm proceeds with the equipment design. Sometimes the design has either been done during the bidding phase, or is not necessary since the project's requirement could be served by the company's existing suite of products. If at all, minimal design changes may be needed only to configure the machine to the specific local or geophysical conditions.

This pre-manufacturing phase follows the same pattern as other manufacturing activities. The firm procures materials and services, and transports them to the manufacturing plants. Most outsourced parts and component manufacturing are usually supplied within the company group. These are then assembled either in Japan or in one of its manufacturing plants that is nearest the soon-to-be constructed power plant.

10.3.3. *Manufacturing of Machines/Construction Stage and Delivery*

In this stage, the firm's involvement depends on a per project basis. In some, the firm may be heavily involved in the engineering services for building projects, while in others it only needs to focus on the engineering services related to the machine manufacturing. Once the machine is ready for delivery and the power plant is ready to receive it, the firm takes care of the delivery, storage, and other logistical arrangements to bring the machine to the site of the power plant. Because of the bulkiness of these machines, the firm's strategy has been to manufacture or assemble their products in the nearest manufacturing plant as much as possible to save on delivery cost. The parts and components, however, are procured globally from its other subsidiaries or third-party suppliers. Examples of parts and components procured globally include: rotors for gas turbines, tube materials for boilers, steel plates and cylinders. The firm may or may not install the machines themselves at the power plant site.

10.3.4. *Commissioning and Operation of the Power Plant*

The firm ensures that the machines are working according to specification, tests them on-site, provides the necessary personnel to install them (if asked) and trains local operators. If installation services are not part of the contract, the firm, nevertheless, sends a guide to help with very sophisticated machines. Depending on the project, it may entrust the facility management and the operation of the plant to the firm. In this case, engineering services during operation are usually entrusted to the affiliated company in the relevant economy. This is why the services global network of the firm is extensive. Besides attending to after-sales services such as maintenance and repair, the firm may sometimes be called upon to take responsibility for the engineering services of the entire plant facility.

10.4. SERVICES ALONG THE VALUE CHAIN

What services activities are involved at each stage of the chosen value chain? Figure 10.4 shows a few examples of the services at each stage of the EPCM value chain. It is not an exhaustive list and only covers the major services associated with bidding, pre-manufacturing/construction, construction and operation of the plant.[4] Tables in Appendix identify at least 39 major service categories; 74 if services sub-categories of the central product

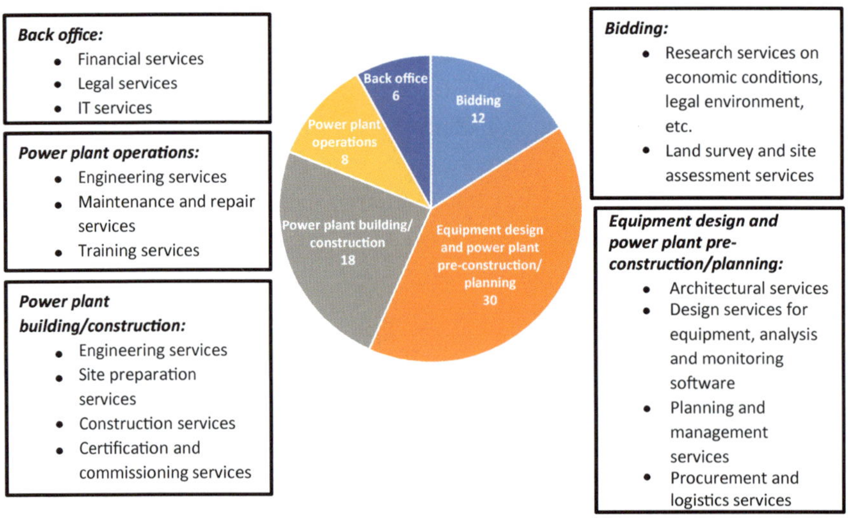

Figure 10.4: Examples of Services in EPCM Value Chain.

Source: Author.

[4] For example, cleaning services and water treatment services could be among the services listed but are not included. Only the major services have been chosen in the figure. More details are available in Tables A.1–A.5.

classification (CPC) are considered. For example site preparation services in the construction phase can be further subdivided into demolition services, site formation and clearance services, and excavating and earthmoving services. The same goes for other major service categories like transportation services which can be either land, rail, water or air transport services.

The firm was unable to give an estimate of the value of these services. But for the firm as a whole, the rough estimate of the current share of services in the value of the company is about one-third, while manufacturing is two-thirds. However, the services share is expected to increase due to aging of existing power plant facilities all over the world and the rising cost of building completely new facilities.

10.4.1. *Importance of After-sales Service to the Firm*

In particular, the firm expects to generate more revenues from after-sales services although it acknowledges the tough competition in this market segment from other global players. Based on the after-sales services offerings on its corporate website, it appears that this segment of its business is given a high priority. The company provides detailed descriptions of various after-sales services. For example it offers advanced inspection techniques or robotic coating of damaged tubes or parts, both for their own products as well as those of third-party supplied equipment. Just to give a flavour of the services offered in this industry, the maintenance and repair of a steam engine consists of several services — mostly engineering — such as: rotor welding and machining, blading replacement, valve inspection, repair and calibration, vibration testing and analysis, dynamic rotor balancing, and technical advisory services.

After-sales service business is defined by the firm as composed of the following: supply of parts; repair of existing parts or machines; installation services; removal of some machine parts; monitoring of operational status of customer power plants through which they can anticipate and propose repairs and maintenance needs. In some cases, the firm signs a long-term service agreement for equipment which lasts between 6 and 12 years, where the company provides the necessary manpower to carry out inspection, provides replacement parts and maintenance engineering support, trouble shoots, and monitors remotely around the clock. In addition, the firm also offers a parts supply agreement in which the firm guarantees the supply of parts for the duration of the contract at an agreed price.

Besides repair and maintenance, other services that the firm offers are: repowering of existing facilities to raise its output and efficiency, and conversion or rehabilitation of existing machines like boilers to limit gas emissions and enhance performance. It has advanced integrity inspection technology to evaluate the deterioration of the machine from creep,[5] fatigue, corrosion, etc. and assess the remaining life of the machines, thereby be in the position to propose modernisation and upgrading program.

10.4.2. *Outsourcing, Bundling and other Aspects of Services Supply*

The large scope of an EPCM value chain necessitates the outsourcing of many activities. If the firm is the prime contractor, it usually subcontracts the construction of the facility to a local subcontractor who knows better the legal and regulatory environment for everything related to construction. In the subcontracting agreement, many services are typically bundled under construction services, such as staffing services, staff housing services, the acquisition of various government permits, transportation and logistics, as well as the testing of construction materials. This, in itself, is a large value chain that would merit a separate study.

In the more frequent case of the firm being only part of a consortium, its task is limited to manufacturing and delivering the machines, including installation in some cases. Once the firm has

[5]The tendency of a solid material to move slowly or deform permanently under the influence of mechanical stresses. It can occur as a result of long-term exposure to high levels of stress or temperature (Wikipedia.com).

delivered, tested on-site and certified the machines, the rest depends on the plant operator, except in cases when the machines are still under warranty. Yet, even in the case where the firm is not the prime contractor, it is closely involved throughout the EPCM value chain. In the bidding phase, for example a project in-house assessment is made in terms of financial viability. Independent from the site assessment services undertaken by the prime contractor or local construction company, the firm also outsources either to a subsidiary or third party the service of assessing the topography of the plant site. The firm is sure, however, to do its own research (with help from a subsidiary or third party) on environmental and other relevant regulations in order to adjust its machines if necessary.

In the pre-manufacturing phase, the design and engineering front-end services[6] for the machine, the procurement services, testing and inspection, and other engineering services are all done in-house or by a subsidiary/joint venture (JV) company. For information technology (IT) design and software for control systems, the services may either be provided in-house or outsourced to a third party, usually an instrument and control (I&C) company. Manufacturing may take place in the different subsidiary manufacturing plants, exploiting each plant's comparative expertise, say, in manufacturing boiler tubes, blades, etc. These are then delivered for final assembly to the strategically chosen subsidiary that is closest to the power plant site. For example if the power plant is being constructed in an Asian economy, the final assembly takes place in the same economy if one exists, or else in the nearest subsidiary.

Delivery and other logistical arrangements are usually outsourced to a third-party logistics provider. Installation may be undertaken by in-house personnel or by the contractor, but the certification and commissioning services as well as on-site testing

---

[6]Defined as the work required to produce process and engineering documentation of sufficient quality and depth to adequately define the project requirements for detailed engineering, procurement and construction or manufacturing. See http://www.fluor.com/epc-services/engineering/Pages/front-end-engineering-design.aspx.

are usually provided by the firm's staff. Repairs and maintenance are done by the firm or its subsidiary, and this depends on the contract. Their machines may also be serviced by third-party service providers.

In general, most of the services are carried out in-house or by the firm's subsidiaries. Because of its extensive subsidiary network, most of the required services and materials are available within the corporate family, except for services in which the group may not have specific expertise, such as in logistics or I&C software design. Likewise, because most of the inputs are from its own subsidiaries, the trust factor on the quality of inputs and services is high, resulting in less need for strict testing and inspection, except on a random basis.

10.5. POLICIES AFFECTING THE VALUE CHAIN

Policy discussions with the firm covered several fronts. Interestingly, on whether the policy environment determines the firm's decision to establish in any economy, the firm said that it is not a principal determinant in the decision. Rather, more important is whether they can find a good and reliable partner in the place. In the past, establishment in various economies was part of a tariff jumping strategy because tariffs for their products were then very high.

Currently, investments in new manufacturing plant are no longer a priority because the firm thinks it has enough capacity to meet growth in the energy market in Asia and the world. Rather, more recent investments are in support or services centres, or subsidiaries with a service focus. The decision on where to locate new services subsidiaries usually depends on the number of products sold in the market, as this drives the need for more services support.

10.5.1. *Trade Policy*

The firm usually faces no major problems with customs because their products are specialised machines. But uncertainty about whether or not its imports will be accorded a preferential tariff obliges them to assume in their bid that the highest tariff will be payable. This makes the cost of the project (and their bid price) higher than if they knew for certain beforehand what tariff rate will be used. Ultimately, the higher cost occasioned by this uncertainty is borne by the customer (the utility) and eventually by the end-users. Clear guidelines and agreement on tariffs and other relevant policies at the outset would be helpful in lessening the cost of the project.

10.5.2. *Human Capital Needs*

A major concern is the large employee turnover, especially of engineers, in their subsidiaries in developing economies. This is related to the small pool of quality engineers relative to those in developed economies. In turn, this is connected to the quality of education in the respective host markets. The firm's strategy is to train local talent in order to minimise the need to use many Japanese expatriates which costs the company more. But the firm also sees the need for different Asian economies to upgrade their education and skills development. Human capital development and skills matching is important to support large manufacturing and service facilities that are set up in the economy.

10.5.3. *Labour Mobility*

To support their customers, sometimes it is necessary to dispatch engineers from Japan or elsewhere in less than 24 hours to prevent a plant shutdown.[7] But in some economies with visa restrictions, it takes time to send the necessary personnel. Though the application process in itself is not difficult for the firm, it is still time consuming, and delays can become costly. The firm thinks that minimising

[7]The firm already uses advance technology to carry out remote monitoring, obviating the need for frequent plant visits by their employees. Still, a regular physical visit, usually by employees in the subsidiary located nearest the plant, is undertaken as a preventative measure.

the visa processing period would greatly help, particularly for very short-term and temporary stays of intra-corporate transferees.

Other labour-related regulations that pose a threat are indirect regulations on corporate officers' qualifications. For example in one Asian economy, a regulation being discussed in its legislative assembly is the language test requirement for managers and directors. If approved, this requirement effectively puts restrictions on foreign directors, putting unnecessary burden on them in carrying out their responsibilities.

Services provided by foreign personnel, such as the installation and commissioning of machinery or to act as EPCM project technical advisor, are usually subject to a withholding tax if the personnel stays beyond a certain threshold number of days, which in some economies is up to 60 days. The additional cost incurred is something that the firm takes as a given. But in some economies, particularly in those where Japan has no bilateral agreement on double taxation, the withholding tax can go beyond the fees for technical services and also include a tax on royalties.

10.5.4. *Intellectual Property (IP)*

Some economies require the local transfer of IP, whether for imported components or for manufacturing. The firm has resolved the potential difficulty arising from such a regulation by forming a joint venture to which it licenses the IP with restricted conditions, i.e. only for the purpose of manufacturing for the specific market (exclusive market).[8] The joint venture receives all the drawings and technical designs but must ensure no leakage of the IP. Any improvement work on the licensed IP, however, will belong to the joint venture.

So far, the arrangement appears satisfactory. The firm is, however, aware that other companies have experienced either

[8]In general, as long as the local partner is reliable and qualified, both in the technical and business sense, a simple licensing agreement is sufficient to protect intellectual property, without the need to form a joint venture.

leakage in the use of their IP or non-observance of the restrictions on IP use by their local partner. In particular, these companies found machines being sold in other Asian economies that made use of their licensed technology when its use is supposed to be restricted to one particular economy in Asia. Improved IP protection and the implementation of rules in different economies, as well as the careful choice of local partners, are important to safeguard corporate IP.

10.5.5. *Equity Limitation*

As a rule, the firm prefers to be free to decide whether to acquire local partners, i.e. have a fully owned subsidiary, or form a joint venture. This decision is made on a case-by-case basis. In some places, acquisition is the best way to enter the market. In certain economies, however, there is preference for joint ventures or even an expectation that these will be established. The vehicle for ensuring that this occurs is rules on foreign equity shares. In some cases where the firm finds a good local partner, it does not consider foreign equity limits an operational constraint. This view is clearest with respect to manufacturing, but less so with after-sales services. The reason for this is that the firm considers that its reputation, and possibly its IP, is at greater risk in this market segment.[9]

Though there has been a relaxation of equity restrictions in some economies in the case of the energy industry, restrictions still abound in construction services. Some of these restrictions include permits or licenses to become a contractor, with specified conditions on how to obtain such licenses, for example years of engineering

[9]For example, the sales partner or maintenance service partner in the host economy must satisfy a certain technical level required to supply the necessary customer service or be able to detect problems and inform the company in time. For this to happen, the case study firm has to train local staff, including possibly making them undergo short stints at the factories in Japan. Likewise, the case study firm can dispatch its engineers to the host economy to train local engineers. Such intensive investments in human capital training call for a long-term, stable relationship based on shareholding.

experience, education, etc. If licensing for the engineering profession is restrictive, establishing a foreign-owned construction company in the economy concerned becomes virtually impossible. Restrictions like this in construction services affect the firm's capacity to lead consortia as the prime contractor in particular economies.

10.5.6. *Local Content Requirements*

The firm feels that there is need for clear guidelines on local content requirements. It noted that in some projects in the region, the local content requirement is too high and detailed rules on its computations are not very clear. It is particularly problematic especially if there are not enough companies in the host economy that can satisfy the stringent technical requirement by the firm for power plant construction projects. In some cases, it is virtually impossible to meet the local content requirement.

10.5.7. *Health, Safety and Environment Regulations*

To ensure compliance with government regulations on health, safety and environment, there is a need for government to visit and inspect facilities. However, overly frequent visits add unnecessarily to firm costs, particularly because the inspectors' costs of transportation, accommodation and other expenses are usually borne by the power plant management. For large enterprises, these costs may be negligible. But if an independent power provider — the firm's customer — is small, it will try to recoup these additional costs either through hard bargaining with the equipment provider and other suppliers on price, or pass the cost on to final end-users of electricity.

10.5.8. *Transparency*

The energy generation business has to deal with various government regulations, for example on environmental issues, land ownership, permits and various taxes. But in many economies, information is difficult to obtain and the firm has to rely on local companies to get the information.

Likewise, when there are delays in plant construction, the firm has often felt that it is left unaware of the project status and of the difficulties that arise in the implementation process.[10] Since the firm bases its manufacturing schedule on the plant construction schedule, delays in project implementation have a huge impact on cost. For example the firm might have finished manufacturing the bespoke equipment for a project and if the plant is not yet ready, it incurs storage and maintenance costs. If the firm delay the manufacture of particular types of equipment for a specific power plant, the delay cascades down on all their suppliers as well as on the schedules of the different subsidiaries that may be involved in supplying parts for the equipment. Most of the factors causing delay are difficult for foreign firms to observe or anticipate, particularly if the government is not transparent.

The risk of delays and other similar risks that are frequently prevalent in developing economies are typically factored into project costs, but only to a limited degree because of competition from other bidding consortia. In sum, for big infrastructure projects like power plants, a government's capacity to implement the project according to agreed timelines can minimise overall cost and thus ultimately improve the provision of electricity at a cheaper cost.

[10] Delays may be caused by factors beyond the control of the utility or customer. For example, a large tract of land where the plant is to be constructed may not be ready as planned because the government has failed to evict or transfer dwellers in the area. This type of situation is usually hard for foreigners to observe and anticipate.

APPENDIX

List of Potential Services for Designing, Building and Operating a Power Plant.

Table A.1: Services during Bidding Stage.

Service	Corresponding CPC Ver. 2 Code	Supplied in-house	Outsourced to affiliated companies and reasons[1]	Outsourced to third-party suppliers and reasons	Remarks
1 Research on economic conditions, legal environment, etc. on the economy where power plant will be built	81212 — Research and development services in economics	Yes	Yes		
	81213 — Research and development services in law	Yes if prime contractor; or else No	Yes if prime contractor; or else No		
2 Land survey; site assessment services	83411 — Geological and geophysical consulting services	No	Yes, proximity	Yes	
	83412 — Geophysical services	No	Yes, proximity	Yes	
	83421 — Surface surveying services	No	Yes, proximity	Yes	

[11] Examples of reasons: (i) government services; (ii) required by laws and regulations; (iii) lack of feasibility to supply service in-house; (iv) external economies of scale; (v) lack of expertise; (vi) need to ensure access to the best services; (vii) need for strong relationship with related stakeholders (government agencies in many instances) to supply the services and (viii) outsourcing during peak seasons.

	83442 — Testing and analysis services of physical properties	No	Yes, proximity	Yes	
	Preparation of bidding document	Yes; with prime contractor[12]	Yes		
	Headquarter services	Yes			
4 Design services	83912 — Industrial design services	Yes			If necessary to adjust existing design to local conditions
5 Telecommunication services	841 — Telephony and other telecommunication services			Yes, needs economies of scale	
	842 — Internet telecommunication services			Yes, needs economies of scale	
6 Planning and management services	Selection of contractors for facility construction	Yes if prime contractor			

[12] With respect to the equipment firm is supplying.

Source: Author.

Table A.2: Services during design Stage of Equipment and Pre-Construction/Planning Stage of Power Plant.

Service	Corresponding CPC Ver. 2 Code	Supplied in-house	Outsourced to affiliated companies and reasons[13]	Outsourced to third-party suppliers and reasons	Bundled
7 Site assessment services	83411 — Geological and geophysical consulting services	Yes, if prime contractor		Yes[14]; use of outside expertise	
	83412 — Geophysical services	Yes, if prime contractor		Yes; use of outside expertise	
	83421 — Surface surveying services	Yes, if prime contractor		Yes; use of outside expertise	
	83442 — Testing and analysis services of physical properties	Yes, if prime contractor		Yes; use of outside expertise	
8 Architectural services for power plant building	8321 — Architectural services and advisory services	Depends on project		Yes to construction services company; engineering company; lack of in-house expertise	

[13] Examples of reasons: (i) government services; (ii) required by laws and regulations; (iii) lack of feasibility to supply service in-house; (iv) external economies of scale; (v) lack of expertise; (vi) need to ensure access to the best services; (vii) need for strong relationship with related stakeholders (government agencies in many instances) to supply the services and (viii) outsourcing during the peak season.

[14] Normally works with prime contractor to check if there is need to adjust the equipment to the specific land topography, type, proneness to calamities, etc.

		8323 — Landscape architectural services and advisory services			Yes to construction services company; engineering company; lack of in-house expertise	
9	Design services for equipment as well as analysis and monitoring software	83912 — Industrial design services	Yes			If necessary to design a new one; or else adjust existing ones
		8392 — Design originals; original design concepts, produced on own account: • industrial product designs • aesthetic designs • graphic designs	Yes			If necessary to design a new one; or else adjust existing ones.
		83141 — IT design and development services for applications	Depends on project; Yes (depends on project and economy)		Yes, to Instrumental and Control (I&C) Company; lack of expertise	Simple control design requirement from economies may be done in-house

(*Continued*)

Table A.2: (*Continued*)

Service	Corresponding CPC Ver. 2 Code	Supplied in-house	Outsourced to affiliated companies and reasons[13]	Outsourced to third-party suppliers and reasons	Bundled
	83142 — IT design and development services for networks and systems	Depends on project; Yes (depends on project and economy		Yes, to Instrumental and Control (I&C) Company; lack of expertise	Simple control design requirement from economies may be done in-house
	83143 — Software originals — copyrighted IP produced without contract for outright sale	Project basis		Yes; lack of expertise	
10 Government-related pre-building services (e.g. securing government permits for construction)	91132 — Public administrative services related to fuel and energy	Prime contractor		Yes, for efficiency	Usually bundled with construction services
	91133 — Public administrative services related to mining and mineral resources, manufacturing and construction	Prime contractor; Project basis		Yes, for efficiency	Usually bundled with construction services

11	Planning and management services	83223 — Project site master planning services	Project basis		
		83330 — Project management services for construction projects	Project basis		Yes, construction company/engineering company; efficiency considerations
		Selection of contractors for facility construction	Yes if prime contractor	Yes if prime contractor	
		83190 — Other management services, except construction project management services (selection of contractors for architectural design)	Yes if prime contractor	Yes if prime contractor	
		83190 — Other management services, except construction project management services (Selection of contractors for engineering design)	Project basis		

(*Continued*)

Table A.2: (*Continued*)

Service	Corresponding CPC Ver. 2 Code	Supplied in-house	Outsourced to affiliated companies and reasons[13]	Outsourced to third-party suppliers and reasons	Bundled
12 Construction services	54 — Construction services		Yes if prime contractor	Yes;[14] outside expertise	*Bundled with construction materials, leasing of equipment, labour supply as well as accompanying components such as housing, medical and insurance
13 Research and development services	8112 — Research and experimental development services in engineering and technology	Yes	Yes; simple engineering or adjustments to machines		
14 Procurement services	83116 — Supply chain and other management consulting services	Yes	Yes		

		85999 — Other support services n.e.c.	Yes	Yes	
15	Customs clearance services and logistics of raw materials	67110 — Container handling services	No		Yes; outside expertise
		85999 — Other support services n.e.c. (business brokerage)	Yes		Yes
16	Technical testing of raw materials	83441 — Composition and purity testing and analysis services	Yes	Yes	No
17	Transport services of raw materials	651 — Land transport services of freight	No	No	Yes; efficiency
		652 — Water transport services of freight	No	No	Yes; efficiency
		6531 — Air transport services of freight	No	No	Yes; efficiency
		67910 — Freight transport agency services and other freight transport services	No	No	Yes; efficiency
18	Environmental consulting services	83931 — Environmental consulting services			Yes; lack of expertise

Source: Author.

Table A.3: Services during Building/Construction Stage of Power Plant.

Service	Corresponding CPC Ver. 2 Code	Supplied in-house	Outsourced to affiliated companies and reasons	Outsourced to third-party suppliers and reasons	Bundled
19 Engineering services during building/ construction	83310 — Engineering advisory services	Yes			
	83321 — Engineering services for building projects	Project basis		Construction/ engineering company; due to economies of scale	
	83324 — Engineering services for power projects	Yes			
20 Site preparation services	54310 — Demolition services	Project basis			
	54320 — Site formation and clearance services	Project basis			
	54330 — Excavating and earthmoving services	Project basis			

21 Importation of power plant equipment: Customs clearance and logistics	85999 — Other support services n.e.c.		Yes	Yes
22 Importation of power plant equipment: Freight transportation services	6511 — Road transport services of freight		Yes	Yes
	6512 — Rail transport services of freight		Yes	Yes
	652 — Water transport services of freight		Yes	Yes
	653 — Air and space transport services of freight		Yes	Yes
23 Importation of power plant equipment: Storage and warehousing services	67220 — Bulk liquid or gas storage services	Prime contractor		
	67290 — Other storage and warehousing services		Yes; proximity	

(Continued)

Table A.3: (*Continued*)

Service	Corresponding CPC Ver. 2 Code	Supplied in-house	Outsourced to affiliated companies and reasons	Outsourced to third-party suppliers and reasons	Bundled
24 Construction services	54 — Construction services	Project basis		Construction/ engineering company	*Bundled with construction materials, leasing of equipment, labour supply as well as accompanying components such as housing, medical and insurance
25 Installation services for equipment and related components such as wiring	546 — Installation services	Project basis	Yes	Construction/ engineering company	

26	Certification and commissioning services of power plant building and equipment by firm	8344 — Technical testing and analysis services	Yes	Yes	
27	Government inspections on fire prevention, health hazards, environmental protection and other aspects	91132 — Public administrative services related to fuel and energy	Project basis		Yes; required by law
		9129 — Public administrative services related to other public order and safety affairs	Project basis		Yes; required by law

Source: Author.

Table A.4: Services during Operations Stage of Power Plant.

Service	Corresponding CPC Ver. 2 Code	Supplied in-house	Outsourced to affiliated companies and reasons	Outsourced to third-party suppliers and reasons	Bundled
28 Engineering services during operations	83310 — Engineering advisory services	Project basis			
	83324 — Engineering services for power projects	Yes (if part of contract)	Yes (if part of contract); to affiliated companies in relevant economy		
29 IT services for on-site and remote monitoring of power plant	8315 — Hosting and IT infrastructure provisioning services	Yes (if part of contract)			
	8316 — IT infrastructure and network management services	Yes (if part of contract)			
30 Telephone-based support services	85931 — Telephone call centre services	Yes (if part of contract)	Yes (if part of contract); to affiliated companies in relevant economy		

31 Diagnostic, inspection, maintenance and repair of equipment	87156 — Maintenance and repair services of commercial and industrial machinery	Yes (if part of contract)	Yes (if part of contract); to affiliated companies in relevant economy
32 Installation services for replacement parts and equipment as well as related components such as wiring	546 — Installation services	Yes (if ordered)	
33 Training services for workers	9291 — Other education and training services	Project basis	

Source: Author.

Table A.5: Back Office Services (Before, during and After Plant Construction).

	Service	Corresponding CPC Ver. 2 Code	Supplied in-house	Outsourced to affiliated companies and reasons	Outsourced to third-party suppliers and reasons	Bundled
34	Financial services	71 — Financial and related services	Yes if prime contractor		Yes; economies of scale	
35	Insurance services	713 — Insurance and pension services			Yes; economies of scale	
36	Accounting, auditing and bookkeeping services	822 — Accounting, auditing and bookkeeping services			Yes; required by law	
37	Legal services	821 — Legal services			Yes; lack of in-house expertise	
38	IT services	8313 — IT consulting and support services			Yes; economies of scale	
39	Visa and immigration services for foreign employees	91290 — Public administrative services related to other public order and safety affairs			Yes; efficiency	

Source: Author.

Chapter 11

PRODUCTION OF PRECISION DIE AND MACHINE PARTS IN THAILAND

Denise Cheung and Andre Wirjo

11.1. BACKGROUND INFORMATION

The firm represented in this case study is a manufacturer of dies[1] and machine parts in Thailand. Its parent company is a listed Japanese company engaging in contract manufacturing and sales of dies and machine parts. The firm is located in an industrial estate in the Pathumthani Province, a location chosen by the parent company mainly due to its strong ground and water supply.

Annual sales revenue of the firm in financial year (FY) 2012 was approximately THB 500 million. The firm has approximately 240 employees, including six Japanese. Roughly 140 employees engage in production directly, while the remainder are in service capacities such as management and engineering.

The firm considers itself in the supporting industry for makers of machinery. It does not market its products under its own name but sells them directly to manufacturers who use them as inputs for production. Its main customers are Japanese corporations or the Thai subsidiaries thereof. The biggest customer, a metal product manufacturer which accounts for 70 per cent of the firm's revenue, is located near the firm in the same industrial estate.[2]

---

[1]A die is a device for cutting or moulding metal into a particular shape.

[2]The above mentioned customer together with its parent company also owns 50% stake of the firm being studied.

The firm has four factories on site, each manufacturing a different type of die and machine parts. Dies are highly customised in that the type, shape, dimensions and material of a die needs to be designed with precision according to the specifications of the product that it is supposed to form.

The product selected for this case study is a precision die. A die is a mould used to shape metals. Simply put, the die casting process involves injecting molten metal into the die opening under high pressure. Dies 'can be designed to produce complex shapes with a high degree of accuracy and repeatability'.[3]

11.2. DESCRIPTION OF THE VALUE CHAIN

This case study follows a precision die from the design stage until it is transferred to the customer (Fig. 11.1). The die will subsequently be used by its customer in producing metal parts for home equipment.

The value chain begins with the customer approaching the firm with drawings of the desired metal product that the precision die would help produce. The customer for this value chain is the firm's major customer, as mentioned in Section 11.1, whose production site is near the firm's. The firm then proceeds with industrial design, which includes computer-aided design of the die and determination of material, and production planning. The firm then orders raw materials through an importation agent, who handles procurement, customs clearance and transportation of materials all the way to the firm's factory gate. Once the design specifications are programmed into the machine, production is rather straightforward, requiring operation and monitoring by semi-skilled technicians. The die is then manufactured and tested before it is packaged for transfer to the customer. As the customer is located nearby, the transfer of goods only involves minimal labour.

---

[3]Source: North American Die Casting Association, http://www.diecasting.org/.

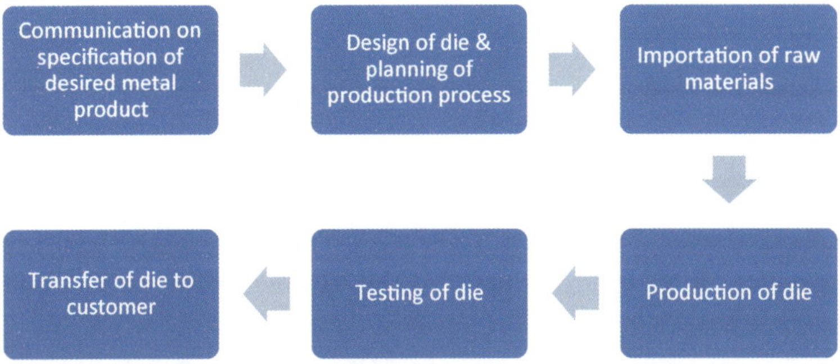

Figure 11.1: Value Chain of Precision Die.

Source: Authors.

11.3. SERVICES ALONG THE VALUE CHAIN

A total of 38 services has been identified in the value chain. These services have been grouped together according to the various stages in which they enter the value chain: (i) pre-production, (ii) in-factory and factory-related services, (iii) delivery and sales stage, (iv) post-sales services and (v) operation and management-related services (Table 11.1).

11.4. ANALYSIS OF THE SERVICES INPUTS IN THE VALUE CHAIN

11.4.1. *Value Contribution*

The firm estimated that services contribute over half of value in the value chains of die and machine parts. Due to the high level of customisation and technology content of precision die, high-value services such as industrial engineering and product testing are necessary to make the product. On the other hand, because the product is produced for a sole on-site customer, many business services that are typical in manufacturing GVCs — downstream as well as upstream — such as logistics and warehousing services are not involved. On balance, the overall value contribution of services

Table 11.1: Services Entering the Value Chain.

Services	Corresponding central product classification (CPC) Rev. 2 code
Raw materials, input/pre-production stage	
1. Procurement agent for raw material sourcing	85999 — Other support services n.e.c.
2. Customs-related services for raw materials imported	85999 — Other support services n.e.c.
3. Quality assurance services (of raw materials)	83441 — Composition and purity testing and analysis services
4. Freight transportation services (of raw materials) by road, rail, sea or air	65 — Freight transport services
5. Storage of raw materials — General storage	67290 — Other storage and warehousing services
6. Industrial design	83912 — Industrial design services
In-factory stage	
7. Production administration — Production management	83115 — Operations management consulting services
8. Production administration — Quality assurance and compliance with ISO	83441 — Composition and purity testing and analysis services
9. Warehousing services for intermediate goods	67290 — Other storage and warehousing services
10. Cleaning services of factory	85330 — General cleaning services
11. Security guards for factory and warehouses	85250 — Guard services
12. Sewage water treatment services	94110 — Sewerage and sewage treatment services
13. Industrial waste disposal	94212 — Collection services of industrial hazardous waste (except medical and other bio-hazardous waste) 94339 — Other non-hazardous waste treatment and disposal services

(Continued)

Table 11.1: (*Continued*)

Services	Corresponding central product classification (CPC) Rev. 2 code
14. Repair and maintenance services of machines and equipment	87156 — Maintenance and repair services of commercial and industrial machinery
15. Financial leasing of machinery	71140 — Financial leasing services
16. Government inspections on fire prevention, health hazards, environmental protection and other aspects	91133 — Public administrative services related to mining and mineral resources, manufacturing and construction 91290 — Public administrative services related to other public order and safety affairs
17. Catering services for workers	63393 — Other contract food services
18. Medical services	93121 — General medical services
19. Personnel search and referral services — Recruitment of factory workers	85112 — Permanent placement services, other than executive search services
20. Transportation services for staff	64114 — Local special-purpose scheduled road transport services of passengers
21. Social insurance for factory workers	91320 — Administrative services related to government employee pension schemes; old-age disability or survivors' benefit schemes, other than for government employees 91330 — Administrative services related to unemployment compensation benefit schemes
Delivery and sales stage	
22. Design of packages	83919 — Other specialty design services
23. Packaging services	85400 — Packaging services
24. Transport of goods by manual labour	65119 — Other road transport services of freight
25. Storage and warehousing services for finished goods	67290 — Other storage and warehousing services

(*Continued*)

Table 11.1: (*Continued*)

Services	Corresponding central product classification (CPC) Rev. 2 code
Post-sales stage	
26. Customer services — Technical support and complains handling	85931 — Telephone call centre services
Operation and management-related services	
27. Auditing on financial accounts	822 — Accounting, auditing and bookkeeping services
28. Financial services	71121 — Deposit services to corporate and institutional depositors
29. Internal auditing (including audits of financial accounts and corporate governance)	822 — Accounting, auditing and bookkeeping services 83118 — Head office services
30. Insurance services for machinery	71334 — Other property insurance services
31. Corporate communications, marketing and public relationship	83114 — Marketing management consulting services 83121 — Public relations services
32. Real estate services	72112 — Rental or leasing services involving own or leased non-residential property
33. Human resources management	83113 — Human resources management consulting services
34. Human resources management • Personnel search and referral services for back office staff	8511 —Personnel search and referral services
35. Information technology (IT) and information system management	8314 — IT design and development services 8316 — IT infrastructure and network management services
36. Legal services	82120 — Legal advisory and representation services concerning other fields of law 82130 — Legal documentation and certification services

(*Continued*)

Table 11.1: (*Continued*)

Services	Corresponding central product classification (CPC) Rev. 2 code
37. Product development/R&D	81129 — Research and experimental development services in other engineering and technology
38. Visa and immigration services for foreign investors/employees	91290 — Public administrative services related to other public order and safety affairs

Source: Authors.

in this value chain is not far off that of other value chains of goods with lower technological content.

Among the services identified, back office services account for the biggest share of value. This may be due to the firm's customer profile. For die and machine parts, the firm relies on a handful of major customers that are also subsidiaries of Japanese corporations. Because of this profile, emphasis needs to be placed on maintaining a strong management team that is capable of communicating and building long-term relationships with the Japanese customers. The reputation of the firm's parent company likely plays a part in winning these contracts. Management that is capable of working closely with the Japanese parent company is therefore also important.

11.4.2. *Outsourcing*

Out of the 38 services identified in the value chain, 17 are fully outsourced, 5 are partially outsourced, and 16 are supplied in-house. Table 11.2 shows the full list of services, with information on whether each service is supplied in-house or outsourced and the respective reason.

Decisions on whether or not to outsource in this case study are generally prompted by three factors. The decision to outsource turns on two considerations — cost and government requirements. The decision not to outsource is usually driven by perceptions of

Table 11.2: Factors Affecting Outsourcing Decisions.

Stage/category	Service	CPC Ver.2 code	In-firm supply	Outsourced and to whom?	Reason(s) for In-house supply/outsourcing	Remarks
(a) Raw materials, input/pre-production stage						
1. Procurement of raw materials	Procurement agent for raw material sourcing	85999 — Other support services n.e.c.		Yes. Import agent handles all tasks up to delivery of materials to the factory	Efficiency and stock control	
	Customs-related services for raw materials imported	85999 — Other support services n.e.c.				
	Quality assurance services (of raw materials)	83441 — Composition and purity testing and analysis services	Yes		Core activity related to production	
2. Logistics	Freight transportation services (of raw materials) by road, rail, sea or air	65 — Freight transport services		Yes. Handled by import agent	Efficiency and stock control	

3. Storage	Storage of raw materials — general storage	67290 — Other storage and warehousing services		Yes. Raw materials are delivered to factory by the import agent just-in-time	Efficiency and stock control
4. Product design	Industrial design	83912 — Industrial design services	Yes. In-house computer-aided design (CAD) engineers		Core activity related to production
(b) In-factory stage					
1. Production administration	Production Administration — Production management	83115 — Operations management consulting services	Yes		Core activity related to production
	Production Administration — Quality assurance and compliance with ISO	83441 — Composition and purity testing and analysis services	Yes		Core activity related to production

(Continued)

Table 11.2: (*Continued*)

Stage/category	Service	CPC Ver.2 code	In-firm supply	Outsourced and to whom?	Reason(s) for In-house supply/outsourcing	Remarks
	Warehousing services for intermediate goods	67290 — Other storage and warehousing services	Yes		Efficiency	
2. Services supporting factory daily operations	Cleaning services of factory	85330 — General cleaning services	Yes. Cleaning of manufacturing production area	Yes. Cleaning of office and general areas	Efficiency/economy of scale	
	Security guards for factory and warehouses	85250 — Guard services		Yes	Efficiency/economy of scale	
	Sewage water treatment services	94110 — Sewerage and sewage treatment services		Yes. Provided by the industrial estate	Efficiency/economy of scale	

Industrial waste disposal	94212 — Collection services of industrial hazardous waste (except medical and other bio-hazardous waste) 94339 — Other non-hazardous waste treatment and disposal services		Yes. Domestic firm licensed to conduct industrial waste treatment	Regulatory requirement	Regulation requires that industrial waste treatment be carried out by licensed firms. Currently only three to four firms in the economy are licensed to conduct such business
Repair and maintenance services of machines and equipment	87156 — Maintenance and repair services of commercial and industrial machinery	Yes, general maintenance	Yes. Maintenance and repair of specialised equipments is undertaken by equipment suppliers	Insufficient in-house expertise	

(*Continued*)

Table 11.2: (*Continued*)

Stage/ category	Service	CPC Ver.2 code	In-firm supply	Outsourced and to whom?	Reason(s) for In-house supply/ outsourcing	Remarks
3. Leasing/hire purchase services	Capital leasing of machinery	71140 — Financial leasing services		Yes	Efficiency/cost	
4. Services from government regulation requirements	Government inspections on fire prevention, health hazards, environmental protection and other aspects	91133 — Public administrative services related to mining and mineral resources, manufacturing and construction 91290 — Public administrative services related to other public order and safety affairs		Yes	Government services	At factory establishment and once every few years thereafter. Paper reports are required periodically and in case of any change of factory structure

5. Worker-related services	Catering services for workers	63393 — Other contract food services		Yes	Efficiency/economy of scale	
	Medical services	93121 — General medical services	Yes, in-house nurse to treat minor illness/injury		Efficiency	
	Personnel search and referral services — Recruitment of factory workers	85112 — Permanent placement services, other than executive search services	Yes		Efficiency	
	Transportation services for staff	64114 — Local special-purpose scheduled road transport services of passengers		Yes. Coach service between Bangkok and the factory	Regulatory requirement and efficiency	License is required to provide transportation services

(*Continued*)

Table 11.2: (*Continued*)

Stage/ category	Service	CPC Ver.2 code	In-firm supply	Outsourced and to whom?	Reason(s) for In-house supply/ outsourcing	Remarks
	Social insurance for factory workers	91320 — Administrative services related to government employee pension schemes; old-age disability or survivors' benefit schemes, other than for government employees 91330 — Administrative services related to unemployment compensation benefit schemes		Yes	Government services	
(c) Delivery and sales stage						
1. Packaging and labelling	Design of packages	83919 — Other specialty design services	Yes		Efficiency	

	Packaging services	85400 — Packaging services	Yes for most products	Yes. Certain exported products are packaged by logistics service provider	Efficiency
2. Delivery to customer	Transport of goods by manual labour	65119 — Other road transport services of freight	Yes		Efficiency
3. Storage and warehousing	Storage and warehousing services for finished goods	67290 — Other storage and warehousing services	Yes		Efficiency
(d) Post-sales stage					
1. Sales/after-sales customer services	Customer services — Technical support and complains handling	85931 — Telephone call centre services	Yes		Core activity related to production

(*Continued*)

Table 11.2: (*Continued*)

Stage/ category	Service	CPC Ver.2 code	In-firm supply	Outsourced and to whom?	Reason(s) for In-house supply/ outsourcing	Remarks
(e) Back office services						
1. Finance	Auditing on financial accounts	822 — Accounting, auditing and bookkeeping services		Yes, outsourced to accounting firm	Regulatory requirement	
	Financial services	71121 — Deposit services to corporate and institutional depositors		Yes. A Thai bank and Bangkok branches of two Japanese banks	Insufficient capacity (impossible to provide in-house)	
	Internal auditing (including audits of financial accounts and corporate governance)	822 — Accounting, auditing and bookkeeping services 83118 — Head office services	Yes			

	Insurance services for machinery	71334 — Other property insurance services			Insufficient capacity (impossible to provide in-house)
2. General management	Corporate communications, marketing and public relationship	83114 — Marketing management consulting services 83121 — Public relations services	Yes		Core activity
	Real estate services	72112 — Rental or leasing services involving own or leased non-residential property		Yes	Economy of scale
	Human resources management	83113 — Human resources management consulting services	Yes	Yes. Renovation of human resource (HR) information system	Insufficient in-house expertise

(Continued)

Table 11.2: (*Continued*)

Stage/ category	Service	CPC Ver.2 code	In-firm supply	Outsourced and to whom?	Reason(s) for In-house supply/ outsourcing	Remarks
	Human resources management — Personnel search and referral services for back office staff	8511 — Personnel search and referral services	Yes. The firm posts job advertisements on the internet and directly deals with candidates.		Efficiency	
	IT and information system management	8314 —IT design and development services 8316 — IT infrastructure and network management services	Yes. Two in-house staff are responsible for routine maintenance of server	Yes. More complicated IT tasks are outsourced	Insufficient in-house expertise	

3. Legal	Legal services	82120 — Legal advisory and representation services concerning other fields of law 82130 — Legal documentation and certification services		Yes	Insufficient in-house expertise	
4. Research and development	Product development/ R&D	81129 — Research and experimental development services in other engineering and technology	Yes		Core activity related to production	R&D activities are mainly done by the parent company. The firm contributes by providing feedback
5. Government services (licensing etc.)	Visa and immigration services for foreign investors/ employees	91290 — Public administrative services related to other public order and safety affairs			Government services	

Source: Authors.

risk associated with a lower level of control over particular functions.

Most of the decisions to outsource are cost related. In one form or another, they relate to external economies of scale. For instance, all services related to the purchase of raw materials are outsourced to an import agent in order to save cost and minimise inventory. The firm's location in an industrial park means that it can benefit from relatively efficient provision of estate management and sewage treatment services. Other illustrations of external economies of scale at work are coach services for staff, repair of specialised machinery, complicated IT tasks, legal services, security guard services, general cleaning services and catering. In many of these cases, the cost advantages of outsourcing relate to the underutilisation of personnel or of physical assets that would result from in-house supply. In other instances it is a matter of the level of investment required for the acquisition of particular skills that would not be fully utilised.

Some services are outsourced simply because the firm is unable to provide them. One could conceive of the rationale as being a matter of scale, but the reality is that the intrinsic purpose of the services means that they must be purchased elsewhere. These kinds of services include banking, financial leasing and insurance.

Finally, outsourcing that results from government regulation includes financial, social insurance and inspections with respect to environmental, health and safety regulations. Governments may also mandate the outsourcing of certain services that in theory could perhaps be managed in-house. In the value chain under study, the Thai government requires that industrial waste treatment be conducted only by licensed industrial waste disposal companies. Of the licensed waste disposal companies in the whole economy, only three to four are qualified to treat the type of waste that the firm produces.

The 16 services that are provided in-house are often core activities relating to production, where greater direct control is commercially desirable. These include industrial design, production management and quality assurance. Other services supplied in-house may be required at very short notice. Examples of these are first aid supplied by an in-house nurse and day-to-day IT maintenance. Finally, some service tasks may be supplied in-house because they are so simple and arise sufficiently infrequently to be taken on by other staff on a part-time basis. Examples of these include the recruitment of factory workers, packaging of finished dies and delivery of finished goods (to the customer next-door).

11.4.3. *Bundling*

In this value chain, the product that the firm sells to its customer is essentially a bundle of services and goods. A number of services including industrial design and quality assurance are bundled with the precision die. The firm bills the customer based on the physical product, but the product would not have been made without the services that enter the value chain in various stages.

Another example of bundled services in this value chain is raw material sourcing undertaken by the import agent. The agent takes care of all the steps needed to deliver the raw material to the firm's factory gate. A range of services including procurement, transportation, customs-related services and storage are all bundled with the raw materials and enter the value chain as physical inputs.

11.5. POLICIES AFFECTING SERVICES IN THE VALUE CHAIN

The firm faces unique opportunities and challenges because of its dual role as manufacturer and service provider (see Appendix). The Thai government's preferential policies in the manufacturing sector and more restrictive policies in certain services put the firm in an interesting situation.

11.5.1. *Investment Promotion*

The Thailand Board of Investment (BOI) offers incentives for investments in seven industries. Classified under A3 in the 'activity-based

Table 11.3: Activity-Based Investment Incentives.

Group	Corporate income tax exemption	Exemption of import duty on machinery	Exemption of import duty on raw or essential materials used in manufacturing export products	Non-tax
Group A: shall receive corporate income tax incentives, machinery and raw materials import duty incentives and other non-tax incentives				
A1	8 year (without cap) + merit	✓	✓	✓
A2	8 year + merit	✓	✓	✓
A3	5 year + merit	✓	✓	✓
A4	3 year + merit	✓	✓	✓
Group B: shall receive only machinery and raw materials import duty incentives and other non-tax incentives				
B1	Merit (some activities)	✓	✓	✓
B2	Merit (some activities)	–	✓	✓

Source: Thailand Board of Investment.

incentives' categories by the BOI, the die and mould industry enjoys exemption of import duty on machinery and raw materials as well as a 5-year corporate income tax exemption upon establishment (refer to Table 11.3).

Thailand's new investment promotion policy, which came into effect on 1 January 2015, introduced 'merit-based' incentives granting additional corporate income tax exemptions to certain projects that firms undertake such as: (i) R&D in technology and innovation including in-house R&D, outsourced research in Thailand or joint R&D with overseas institutes; (ii) donations to Technology and Human Resources Development Funds, educational institutes, specialised training centres, research institutes or governmental agencies in the science and technology field in Thailand, as approved by

BOI; (iii) intellectual property (IP) acquisition/licensing fees for commercialising technology developed in Thailand; (iv) advanced technology training; (v) development of local suppliers with at least 51 per cent Thai shareholding in advanced technology training and technical assistance and (vi) product and packaging design, either in-house or outsourced in Thailand, as approved by BOI.[4]

11.5.2. *Value-added Tax*

While the BOI considers the firm a manufacturing company, the Revenue Department defines the firm's business as 'contract manufacturing' and hence a service subject to value-added tax (VAT) . Currently, Thailand's rate for VAT is 7 per cent. The firm's customers are required to withhold 3 per cent tax on payments to the firm.[5]

In fact, the latest VAT rate as set by the Revenue Code was 10 per cent but the Thai government had extended the previous 7 per cent rate for 2 years in order to boost the economy. Unless another extension is granted, the VAT rate is scheduled to increase to 10 per cent on 1 October 2015.[6]

11.5.3. *Assessment of Customs Duties*

Because the company imports many of the machine parts from its Japanese parent company, intellectual property content embedded in the imports became an issue. The Thai Customs sometimes demands that royalties be added to the invoiced amount in calculating the dutiable value of the imported inputs from Japan, even if no royalty payment has been made in reality.

---

[4]Source: Board of Investment, http://www.boi.go.th/tir/issue_content.php?issueid=118;page=0.

[5]Source: http://www.ey.com/GL/en/Services/Tax/Worldwide-VAT--GST-and-Sales-Tax-Guide---XMLQS?preview&XmlUrl=/ec1mages/taxguides/VAT-2014/VAT-TH.xml.

[6]Source:https://tax.thomsonreuters.com/blog/onesource/indirect-tax/thailand-maintains-vat-7-vat-rate-october-2015/.

Royalties are indeed a common source of problem in international trade. According to the World Trade Organization (WTO) Customs Valuation Agreement, royalties paid, either directly or indirectly, by the buyer must be included in the value for customs duty purposes if the payments meet the following criteria: (i) they are related to the imported goods and (ii) the royalties are paid as a condition of sale. Assessment of the latter criterion is not always straightforward and disputes sometimes arise between companies and customs, as in the case of Nike Thailand, which was taken to the Thai Supreme Court in 2011.[7]

11.5.4. *Challenges with Movement of Natural Persons*

In order to continuously improve its facility, the firm has to regularly bring in highly skilled staff such as engineers from Japan for a short period of time for activities such as training and equipment installation. The firm estimated that such improvements would have increased its revenue by 15–20 per cent and hence should be encouraged by the government.

Instead, there is a 15 per cent international withholding tax on payments to non-residents for services that are defined by the Thai government as involving royalties.[8] For example, when a technician from the parent company came to the firm to provide training to local staff on the operation of a new machine, a 15 per cent withholding tax was charged on the service bill of the Japanese technician.

Visa requirements for work-related trips also discourage transfer of know-how. The firm's management indicated that it was not

[7]Source: http://www.ey.com/Publication/vwLUAssets/EY-TradeWatch-Volume-11/$FILE/ey-TradeWatchVol 11.pdf. In this case, Nike got a favourable ruling that 'royalty payments made by Nike Thailand (importer/licensee) to Nike International (licensor) for the use of the Nike trademark and trade name are not required to be added to the customs value of the imported goods'.

[8]Source: https://www.kpmg.com/Global/en/services/Tax/regional-tax-centers/asia-pacific-tax-centre/Documents/CountryProfiles/Thailand.pdf.

easy to obtain short-term visas although they are not explicit about what the problem is.

11.5.5. *Environmental Regulations*

The firm is required by law to dispose its industrial waste properly[9] and the government has been tightening enforcement of the regulations lately. As stated in Section 11.4, the Thai government requires that industrial waste treatment be conducted only by licensed industrial waste disposal companies. As of April 2014, the Industry Ministry had issued licenses to 1,843 industrial waste disposal factories, including 141 waste processing factories, 1,256 landfill businesses that separate trash and 446 recycling sites.[10] The firm explained that among these licensed waste disposal companies, only three to four are qualified to treat the type of waste that the firm produces.

While the firm has always adhered to the law and utilises the service provided by one of these firms up to this point, management expressed concern on whether they have the capacity to treat all industrial waste in the near future, considering likely increase in demand for their services.

One possibility is for the firm to engage the services of firms not licensed by the government. This is a measure currently taken by some companies. Acknowledging the capacity constraint of licensed waste disposal companies, local governments would unofficially permit companies to hire non-licensed firms for the task. However, the management would strive to adhere to the law for fear that it might face penalties if there were any issues with the waste.

---

[9] Various sections in the Enhancement and Conservation of National Environmental Quality Act B.E. 2535 stated that the owner or possessor of the point of source of pollution has the duty to control pollution through various means including constructing an on-site facility to treat waste and sending waste to a centralised waste treatment facility. For example, sections 70-77 focus on control of water pollution while sections 78–79 focus on other pollution and hazardous waste. More details can be obtained at: http://www.pcd.go.th/info_serv/en_reg_envi.html.

[10] Source: http://thainews.prd.go.th/centerweb/newsen/NewsDetail?NT01_NewsID=WNECO5704250010014#sthash.1vet2iHr.dpuf.

APPENDIX

Policies Affecting Services in the Value Chain.

Government policies/ services	Authority(ies) in charge	Details	How the policy affects services in the value chain
Investment incentives (Investment Promotion Act B.E. 2520)	Board of Investment	The mould and die industry is eligible for investment incentives offered by the BOI	The firm can enjoy tax exemptions on raw materials and machinery
Value-added tax on services	Revenue Department	Services are subject to VAT	The core activity of the value chain is defined as contract manufacturing (a service), by the Revenue Department. Customers' payments to the firm are subject to a 3 per cent withholding tax
Assessment of custom duties	Customs Department	Direct or indirect royalty payments are to be included in customs valuation of imported goods	The Customs Department sometimes demands that royalties be added to dutiable values of goods that the firm imports from its Japanese parent company

International withholding tax	Revenue Department	15 per cent tax is charged on payments to non-residents for services that are defined by the Thai government as royalty moves	Staff from the parent company providing technical support in Thailand are subject to 15 per cent withholding tax. This has discouraged technological transfer
Visa requirement	Immigration Bureau	Visas are required for all work-related travel into Thailand regardless of length of stay	The administrative burden has discouraged technological transfer
The Factory Act of Thailand — Industrial Waste Treatment	Ministry of Industry	Companies are required to use licensed service providers for industrial waste disposal/treatment	The firm has been adhering to the law so far, but foresees that it will become more difficult as the licensed waste disposal companies may not keep up with demand
The Factory Act of Thailand — Health and Safety Standards	Ministry of Industry	The Factory Act sets out inspection and reporting requirements with respect to occupational safety	No major impact on value chain. Quarterly safety reports and annual fire drill reports are submitted to respective agencies. Government inspections are conducted at factory establishment and when any structural change is made to the building. Other than that, inspections only take place every few years

Source: Authors.

Chapter 12

MANUFACTURING OF REFRIGERATORS

David Sit

12.1. INTRODUCTION

This series of case studies seeks to understand better the role of services in global value chains (GVCs) and to examine how policies may impact the configuration, location and operation of value chains. The company in this case study is a Japan-based multi-product conglomerate that specialises in infrastructure systems, energy solutions and information and telecommunication systems. It also produces a range of capital goods, consumer goods, and services, including logistics, distribution and financial services. This case study explores a value chain that manufactures, sells, and services home appliances.

Total revenue of the group was over USD 4 billion in FY2013. Home appliances account for approximately 10 per cent of the group's revenue. Key products for the consumer market include refrigerators, air conditioners, vacuum cleaners, rice cookers, and other home appliances.

The company has a global production and distribution network. In addition to manufacturing plants in Japan, within the Asian region the company has established important manufacturing bases in Thailand, Malaysia, the Philippines and China. Each of the facilities is responsible for the production of one or more product categories for its domestic market and for exports. The company's biggest market is its home market, where products are sold in the company's own 4,500 sales outlets as well as through other retailers. The company also owns a number of subsidiaries in South East Asia, India, Middle East, the United States and the United Kingdom for sales, marketing and after-sales services.

This case study focuses specifically on the manufacture of refrigerators in Thailand. The company's Thai factory is located in an industrial estate in the neighbourhood of Bangkok, and it produces a wide range of products in the company's portfolio. The factory has approximately 5000 staff, with 30 Japanese engineers. Refrigerators are manufactured in this factory and shipped to markets in Southeast Asia and East Asia. Issues arising in respect of the value chains of other products manufactured at other production sites will also be discussed to provide a comprehensive view of policies affecting the home appliance sector.

The case study is structured in the following way. Section 12.2 will map out the value chain. Section 12.3 will analyse how services enter the value chain. Section 12.4 will discuss how policies relating to foreign investment restrictions, foreign labour restrictions and other areas affect the value chain.

12.2. MAPPING THE VALUE CHAIN

The company, together with its affiliated companies, is involved in every stage of the value chain from pre-production procurement of raw materials and parts to after-sales customer services.

The value chain begins at the procurement of raw materials and parts. Modern refrigerators require a long list of inputs. The exterior shell is usually made with metal and coated with paint. The interior cabin is usually made from plastics. Metal sheets, chemicals for the paint, and plastics are procured locally from third-party suppliers. Key components such as compressors and motors involving technologically advanced manufacturing are procured within the firm's family of companies. To guarantee quality, most of them are manufactured in either factories of the group in Thailand or imported from the group's factories in Japan.

Modern refrigerators such as the company's product involve the use of computer chips and programming software, which are also provided from within the company group. All the materials and components are shipped from suppliers to the manufacturing plant by an affiliated logistics company. This logistics company supports the group's international operations by providing international forwarding, warehousing and distribution services. The Thai subsidiary of this logistics company is an affiliate in which the group owns minority shares. Raw materials and components are shipped to the factory by sea, road or rail.

After the raw materials arrive in the factory, they are stored in on-site warehouses before they are used on the manufacturing line. A typical refrigerator manufacturing process requires the assembly of three parts, namely the exterior shell, the interior cabin and the cooling mechanism. Standard procedures for manufacturing the outer shell involve cutting and welding sheets of metal into appropriate sizes and spraying them with a coat of paint. The interiors of the refrigerators are usually made of plastic that require moulding and shaping before being placed on the assembly line. Tubes for refrigerant, water and electrical wires must be installed within the inner cabinet and connected to the mechanical parts. The back panel is then assembled with the compressor and storage tank for refrigerant. Finally, the doors are installed and fitted with seals. The assembly line will put together all these components.[1] After the refrigerators pass their post-production quality inspections, they will be packed with foam and boxes and shipped to distributors in Thailand as well as overseas markets.

The company has subsidiaries in many economies who act as distributors. The company does not own retail outlets in Thailand, but they work with local retail stores and distribution channels that eventually become their long-term business partners. The company works with these partners on marketing, sales and post-sales services. The company prepares advertising materials and plans advertisement campaigns for its products. In addition, the company hires agents to promote the company's products in retail stores. Home electrical appliances of the company come with warranty services for repair and maintenance for up to 5 years. Many of the repair

[1]Manufacturing procedures for refrigerators are described with reference to http://www.madehow.com/Volume-1/Refrigerator.html.

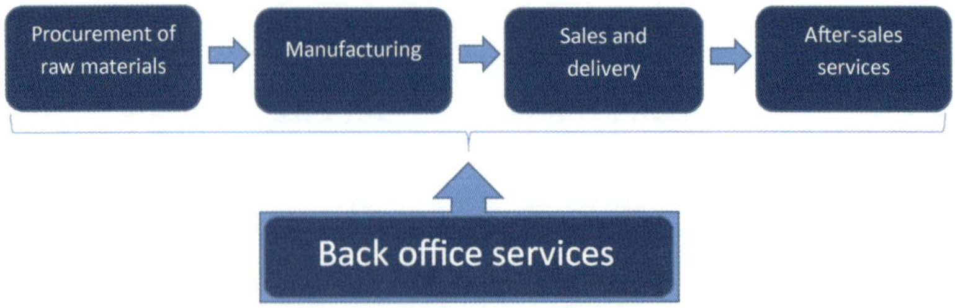

Figure 12.1: Overview of Value Chain.

Source: Courtesy of the company.

service stations are outsourced to retailers, while some are maintained in-house by the company. Complicated maintenance operations requiring specialised engineering services will be brought back to the company. The company maintains call centres for technical supports to customers. In addition to typical after-sales services, the company also provides recycling services in Japan. Business customers who purchase the company's products may also negotiate hire purchase plans supported by the financial services arm of the group.

To support a GVC, a comprehensive set of back office services is required. Among them are various management services, utilities and research and development for new products, which help the company to maintain its cutting edge. Service companies owned by the group contribute to product designs by collecting feedback from customers and user experiences. Figure 12.1 shows a brief overview on the value chain.

12.3. SERVICES IN THE VALUE CHAIN, OUTSOURCING AND BUNDLING

For ease of analysis, the value chain is disaggregat ed into five stages, namely pre-manufacturing, manufacturing, sales, back office services and after-sales services. A list of services inputs to the value chain can be found in Appendix.

The number of services that are used in this value chain is estimated to be 55. Approximately half (25) of the services are back office services, followed by 12 services in the manufacturing stage, and 9 services in the procurement process for raw materials procurement.

The large number of service inputs at the back office stage is not surprising for a complicated GVC that requires support services at every stage of the chain.

12.3.1. *Outsourcing*

Among the 55 services in the value chain, 31 services (56 per cent of all services) are outsourced or partially outsourced to third-party services providers who are not members or affiliates of the group.

There are three main reasons why the company outsources service inputs in the value chain. The most common reason for outsourcing is for efficiency and cost saving. With modularisation of tasks in GVCs, third parties can compete to be part of the value chain of multinational corporations (MNCs), offering quality services at lower costs. The company, as the lead firm in the value chain, can outsource services for cost minimisation. In some industries, outsourced suppliers may be able to attain economies of scale therefore they are capable of providing services to the company at a lower cost than the company would be able to provide itself. An example is the MNC that provides information technology (IT) services to the company. The IT service provider is able to take advantage of its existing network, allowing it to operate IT functions at a lower cost than if the company operated the functions itself.

Lack of expertise, local knowledge or local connections are other reasons for outsourcing. Local connections in some economies are essential for the company to expand to the market more efficiently. That is one of the reasons why the company cooperates with local retailers with long-term business relationships rather than operating in-house retail stores.

Government regimes may sometimes necessitate outsourcing. When these services are efficiently supplied, they add value to the company and to society as a whole. For example safety inspections in factories help to minimise the likelihood of industrial accidents; standard conformity requirements ensure products are safe for public use, also adding to the goodwill enjoyed by the company. On the other hand, if government lacks the capability to carry out policy efficiently, deadweight social and private costs may be incurred.

12.3.2. *Bundling*

Bundling is a way of offering goods together with services in a single offering. This can enhance efficiency, support innovation, and help the company generate higher revenues from their products. In this value chain, the company offers warranties for their products. This is essentially bundling insurance, and repair and maintenance services with the physical product. The company expands its product portfolio in this way. It also creates a positive signalling device to customers on the quality of the products, enabling company to sell more products and at a price premium. The same happens on the input side when the company purchases machinery that is bundled with maintenance.

12.4. POLICY INTERFACES

12.4.1. *Foreign Investment Restrictions*

Restrictions on foreign equity ownership are not uncommon in many economies, particularly developing economies. While foreign equity restrictions in respect of manufacturing activities tend to be more relaxed, restrictions in many services sectors are still prominent. These restrictions can hinder value chain operations and reduce efficiency.

A 2014 OECD report on foreign direct investment (FDI) restrictions in ASEAN economies identifies three major types of restrictions — namely, limitations on foreign equity ownership, screening or approval for foreign investments, and restrictions on land ownership. The OECD's FDI Regulatory Index suggests that the restrictions are generally more severe in ASEAN economies than the OECD average, particularly in services sectors such as distribution, construction, media, communications, financial services, business services and real estate investment.

The wholesale and retail sector, which is one of the most relevant services to the company in this case study, faces a range of different restrictions in several ASEAN economies. The restrictions include foreign equity ownership limitations, scale of operation requirements, and limitations on the type of goods sold (Aldaba and Pasadilla, 2010; OECD, 2014).

The company in this case study, for example has been affected by such restrictions in one ASEAN economy. A newly enacted government order restricts foreign equity ownership for wholesale, distribution and after sales services to 33 per cent. The company had established a local distribution subsidiary years before the new rule was promulgated. The equity was shared between the group (67.5 per cent) and local partners (32.5 per cent). A grandfather clause in the new arrangement allowed the company to maintain the existing ownership structure at the same scale of equity capital. However, the new rule mandates any subsequent changes in the size of equity capital must observe the 33 per cent rule.[2]

This new rule hinders the company's ability to inject new capital into the subsidiary in question. This is because the local partner is unwilling to invest new capital. The company does not want to incur the costs and uncertainty of seeking equity investments from other local partners, leaving the company no room to further expand business in the economy. The company also reported frequent changes to requirements of this nature, which increases uncertainty in the business environment.

In other economies, the company works well with local partners in distribution activities despite holding minority shares in joint venture arrangements. Although joint venture relationships can be difficult if government regulations are constraining, local partners contribute to the value chain with local expertise, market experience and local connections. The company treasures trustworthy local partners for long-term business relationships that also help them guarantee quality services to customers.

12.4.2. *Foreign Labour Restrictions*

Foreign labour regulations restrict the conditions under which foreign employees can work in an economy. Work permits for

---

[2]Unless the change in equity share is for expansion purposes. In the case study firm's case, the intended equity infusion is to pay off loans to the company group's finance affiliate, a purpose which is not covered by the grandfather concession.

foreign personnel may be subject to quota or outright prohibition. The latter is the case in some ASEAN economies for professional services in particular (e.g. medical and legal).

While protection for local employment may make sense in some situations, restrictions on professionals and technicians can hinder value chain operations if there is a lack of available skills locally. In addition, excessive restrictions on foreign skilled labour are likely to impede knowledge transfer to local workers — a vital issue from a developmental perspective.

An example of such a restriction is Thailand's Alien Working Act, B.E. 2551 (2008), which governs approval of work permits for employment of foreign workers. Based on this act, the government outlines clear requirements on paid-up capital associated with foreign labour employment. For every THB 2 million paid-up capital, a company can employ one additional foreign employee up to a maximum of ten. If a company wishes to employ additional foreign workers, it has to employ more than 100 local personnel, pay at least THB 3 million in corporate income tax during previous year, export at least THB 30 million worth of goods or bring in more than 5,000 foreign tourists in the previous year.

The company's manufacturing plants in Thailand employ more than 5,000 workers, so the company generally has no difficulty in acquiring work permits. However, some services businesses, particularly in distribution, may have been affected. Generally speaking, work permits for foreign staff in the Asia-Pacific region are not very challenging to the company.

12.4.3. *Issues Relating to Standards Conformity*

Standards are becoming more pervasive and challenging as production and consumption activities increasingly cross frontiers, and awareness grows of such matters as environmental protection and consumer health and safety. Standards promulgated and enforced by governments are largely predicated on public policy objectives. In addition, standards may help to enhance GVC participation by addressing information asymmetries among different parties involved in GVCs, act as a signalling tool on quality issues, and enhance competition through standardisation and modularisation (Tijaja, 2013).

An example of a cost-augmenting standards regime faced by the case study firm comes from uncertainty in implementation. In 2008, an economy in East Asia implemented new standards on 13 categories of IT-related products. The new standards regime appears to require that certain information relating to source codes be submitted for the products that use these IT components. The guidelines are not very clear on the extent source codes shall be disclosed. In order to minimise risk and protect its most up-to-date source codes from disclosure, the company decided not to sell its most modern product models to the market concerned. Other common issues that add to cost are overlapping standards conformity assessment requirements and complicated paperwork.

12.4.4. *Customs Regimes and Efficiency*

As the law enforcement entity at the border, the customs authorities provide important regulatory services. Customs collect tax revenue, prevent smuggling of dangerous goods and protect consumers by checking conformity of standards and guarding against counterfeit goods (Grainger, 2008). However, customs procedures can add extra costs to value chain operations if not carried out efficiently.

The company featured in this case study often experiences delays in customs procedures in a number of South East Asian economies. Delays in imports of key components can cause delay in production schedules, which adds cost in storage, transportation and manufacturing.

To get around this issue, the logistics company in the group often hires local agents to handle customs procedures. In some economies, customs agents have to be licensed third parties. While firms may choose to outsource these functions to a third party even if they were not obliged to, the costs arise if the service is unduly expensive or inefficient. Table 12.1 summarises how policies affect the value chain.

Table 12.1: Policies Affecting the Value Chain.

Government policies/services	Authority(ies) in charge	Details	How the policy affects services in the value chain
Customs procedures • Inefficiency in customs clearance • Compulsory customs agents	Customs	The company is required to clear imports with customs	Inefficient customs procedures dictate the need for a local agent, imposing extra costs on the company
Foreign equity restrictions • limitation on share of foreign equity • limitation on sector of business • restriction on ownership of key resources such as land	Ministry of Commerce/ Ministry of Investment	The company is prohibited from ownership of equity beyond a certain percentage	May necessitate involuntary partnership with local agents or other service suppliers, adding extra costs for the company
Foreign labour restrictions • Quota on work permits • Forbidden sectors for foreign workers	Ministry of Labour	Work permits for foreign labour may be restricted by quantity or other terms. Some sectors may only be open to local employees	May cause a lack of skills for operations
Standards on products	Standards conformity regulatory agencies	Standards requirements may sometimes not well defined. Duplicating conformity and certification requirements exist	Add uncertainty and cost to the value chain for testing and certification

Source: Author.

APPENDIX

Table A.1: Raw materials, Input/Pre-production Stage.

Stage/category	Service	CPC Ver. 2 code	Outsourced or in-house/in-group provision
1. Procurement of raw materials	1. Procurement agent for raw material sourcing	85999 — Other support services n.e.c.	In-house or in-group. The company procures raw materials directly from third parties
	2. Customs-related services for raw materials imported	85999 — Other support services n.e.c.	Outsourced. The company employed local agents to handle imports and exports
	3. Quality assurance services (of raw materials)	83441 — Composition and purity testing and analysis services	In-house
2. Logistics	4. Freight transportation services (of raw materials) by road, rail, sea or air	Division: 65 — Freight transport services	Possibly in-group and outsourced. Logistics services are provided by in-group logistics company. Actual delivery may be outsourced to third parties
	5. Repair and maintenance for fleets	87143 — Maintenance and repair services of trailers, semitrailers and other motor vehicles n.e.c.	In-house and outsourced

(Continued)

Table A.1: (*Continued*)

Stage/category	Service	CPC Ver. 2 code	Outsourced or in-house/in-group provision
3. Storage	6. Storage of raw materials — general storage	67290 — Other storage and warehousing services	Possibly in-house. Must be located near factories
4. Research and Development	7. Conception and design of product	83920 — Design originals	In-group. The company houses industrial design and engineering centres
	8. Patent acquisition	83960 — Trademarks and franchises n/a	In-house
	9. Product development/R&D for new technology	81129 — Research and experimental development services in other engineering and technology 81400 — Research and development originals	In-group research and development centres

Source: Author.

Table A.2: Manufacturing Stage.

Stage/category	Service	CPC Ver. 2 code	Outsourced or in-house/ in-group provision
1. Production administration	10. Production administration — Production management	83115 — Operations management consulting services	In-house. In-house team manages production
	11. Production administration — Quality assurance and compliance with ISO	83441 — Composition and purity testing and analysis services	Possibly in-house
2. Services supporting factory daily operations	12. Cleaning services of factory	85330 — General cleaning services	Possibly in-house
	13. Engineering services	83310 — Engineering advisory services Class: 3332 — Engineering services for specific projects	In-house. The company's engineer team provides engineering services

(*Continued*)

Table A.2: (*Continued*)

Stage/category	Service	CPC Ver. 2 code	Outsourced or in-house/in-group provision
	14. Security guards for factory and warehouses	85250 — Guard services	Possibly outsourced. Specialised security firms enjoy benefits from network effects
	15. Sewage water treatment services	94110 — Sewerage and sewage treatment services	Possibly outsourced. Infrastructure is generally provided by host economies
	16. Repair and maintenance services of machines and equipment in the factory	87156 — Maintenance and repair services of commercial and industrial machinery	Outsourced. Machinery and equipment providers offer maintenance for machines
	17. Waste collection and recycling services	Class: 9421 — Collection services of hazardous waste 94229 — Collection services of non-hazardous recyclable materials, other 94239 — General waste collection services, other	Possibly outsourced

3. Services from government regulation requirements	18. Government inspections on fire prevention, health hazards, environmental protection and other aspects	91133 — Public administrative services related to mining and mineral resources, manufacturing and construction 91290 — Public administrative services related to other public order and safety affairs	Possibly in-house and outsourced. Government provides inspection services. Company may have to prepare documentation for inspection from time to time
4. Worker-related services	19. Catering services for workers	63393 — Other contract food services	Possibly outsourced to local suppliers
	20. Dormitory for factory workers	63220 — Room or unit accommodation services for workers in workers hostels or camps	Possibly in-group. Japanese large companies usually offer dormitory for factory staff
	21. Transportation services for employees	64114 — Local special-purpose scheduled road transport services of passengers	Possibly outsourced to local transportation companies

Source: Author.

Table A.3: Delivery and Sales/Post-manufacturing Stage.

Stage/category	Service	CPC Ver. 2 code	Outsourced or in-house/in-group provision
1. Delivery to wholesaler/retailer	22. Land transport of products to distributors	65112 — Road transport services of freight by tank trucks or semi-trailers	Partially in-group and partially outsourced. In-group logistics company arranges deliveries
	23. Sea transport of products to distributors	65219 — Other coastal and transoceanic water transport services of other freight	Partially in-group and partially outsourced. In-group logistics company arranges deliveries
2. Sales services	24. Retail trade services — distributors	Group of 621 — Non-specialised store retail trade services Group of 622 — Specialised store retail trade services	In-house. Subsidiaries of the firm act as distributors in different economies
	25. Retail administration operation management	83115 — Operations management consulting services 83116 — Supply chain and other management consulting services	Outsourced to retailers
	26. Storage and warehousing services for machineries at distributors	67290 — Other storage and warehousing services	In-group. In-group logistics firm and subsidiaries provide warehousing for products to be sold
	27. Financing services for customers	71140 — Financial leasing services	In-group. In-group financial company offers hire purchase schemes to customers

Source: Author.

Table A.4: Back office, Utilities and General Services.

Stage/category	Service	CPC Ver. 2 code	Outsourced or in-house/in-group provision
1. Finance	28. Auditing on financial accounts	Group of 822 — Accounting, auditing and bookkeeping services	Outsourced as required by legislation. In-company accounting department assists in preparing documents for auditing
	29. Financial services	71121 — Deposit services to corporate and institutional depositors 71313 — Group pension services 71701 — Services of holding equity of subsidiary companies	Possibly outsourced to large multinational banks
	30. Internal auditing (including audits of financial accounts and corporate governance)	Group of 822 — Accounting, auditing and bookkeeping services 83118 — Head office services	In-group
	31. Insurance services for machinery used in factory	71334 — Other property insurance services	Possibly outsourced to large multinational insurance companies
	32. Insurance for staff	71321 — Accidental death and dismemberment insurance services 71322 — Health insurance services	Possibly outsourced to large multinational insurance companies

(*Continued*)

Table A.4: (*Continued*)

Stage/category	Service	CPC Ver. 2 code	Outsourced or in-house/in-group provision
2. General management	33. Business and management consultancy services	Class of 8311 — Management consulting and management services	In-house
	34. Company secretary services	83990 — All other professional, technical and business services, n.e.c.	In-house. As listed company, it employs its own company secretary
	35. Corporate communications, marketing and public relationship	83114 — Marketing management consulting services 83121 — Public relations services	In-group corporate communications department responsible for corporate communications and marketing
	36. Courier, postal and local delivery services	Group of 681 — Postal and courier services	Possibly outsourced
	37. Estate management	72112 — Rental or leasing services involving own or leased non-residential property	Possibly in-group
	38. Human resources management	83113 — Human resources management consulting services	In-group human resources department.
	39. Human resources management — Personnel search and referral services	Class of 8511 — Personnel search and referral services	In-group human resources department. Agents might be needed to recruit locally

40. IT and information system management, consulting and support, with webpage development	Class of 8313 — IT consulting and support services Class of 8314 — IT design and development services Class of 8316 — IT infrastructure and network management services	In-group and outsource. In-group IT division and outsourced MNC IT service provider provide comprehensive IT services
41. Safety and security services	85230 — Security systems services 85250 — Guard services	Possibly outsourced. Local security firms enjoy network effects and connection with legal enforcement
42. Telecommunication services	Group: 841 — Telephony and other telecommunications services 84210 — Internet backbone services	Outsourced. Host economies provide telecommunication services
	84221 — Narrowband Internet access services 84222 — Broadband Internet access services	
43. Uniform — Design and alternation	83919 — Other specialty design services	Possibly outsourced
44. Uniform — Laundry	97130 — Other textile cleaning services	Possibly outsourced

(*Continued*)

Table A.4: (*Continued*)

Stage/category	Service	CPC Ver. 2 code	Outsourced or in-house/in-group provision
3. Legal	45. Legal services	82120 — Legal advisory and representation services concerning other fields of law 82130 — Legal documentation and certification services	In-group and possibly outsourced. Legal department exists within corporate structure. Possibly outsourced to local legal experts in some economies
4. Advertisement	46. Retail Administration — Advertising	83611 — Full service advertising	Partially in-house and partially outsourced. Subsidiaries in different economies plan advertisement campaigns. Retailers also promote products on their own
	47. Advertisement — advertisement agencies services	83611 — Full service advertising 83620 — Purchase or sale of advertising space or time, on commission	
	48. Advertisement — designers' services	83611 — Full service advertising	
	49. Advertisement — cameramen, directors, photographers and models	83611 — Full service advertising	

5. Utilities	50. Electricity supply	Class of 8631 — Support services to electricity transmission and distribution	Possibly outsourced to local utilities suppliers
	51. Gas supply	86320 — Gas distribution services through mains (on a fee or contract basis)	Possibly outsourced to local utilities suppliers
	52. Water supply	86330 — Water distribution services through mains (on a fee or contract basis)	Possibly outsourced to local utilities suppliers

Source: Author.

Table A.5: After-sales Services.

Stage/category	Service	CPC Ver. 2 code	Outsourced to third parties
1. Sales/after-sales customer services	53. Customer services — complains and compliments handling	85931 — Telephone call centre services	In-house. Provided by subsidiaries
	54. Customer services — Technical support	85931 — Telephone call centre services	In-house. Provided by subsidiaries
2. Repair and maintenance	55. Repair and maintenance of machines for customers	87156 — Maintenance and repair services of commercial and industrial machinery	Partly in-house and party outsourced. Service stations are set up mostly by retailers for maintenance. Complicated repairs will be brought to the company

Source: Author.

REFERENCES

Aldaba, RM and G Pasadilla (2010). The ASEAN services sector and the growth rebalancing model. *ADBI Working Paper Series No. 246*. Tokyo: ADBI. http://www.adbi.org/working-paper/2010/09/07/4059.asean.services.sector.growth.rebalancing/ [18 March 2015].

Grainger, A (2008). Customs and trade dacilitation: from concepts to implementation. *World Customs Journal*, 2(1), 17–30. http://www.worldcustomsjournal.org/media/wcj/-2008/1/customs_and_trade_facilitation_from_concepts_to_implementation.pdf [18 March 2015].

Organisation for Economic Co-operation and Development (OECD) (2014). *Southeast Asia Investment Policy Perspectives*. Paris: OECD Publishing. http://www.oecd.org/daf/inv/investment-policy/Southeast-Asia-Investment-Policy-Perspectives-2014.pdf [20 March 2015]

Tijaja, J (2013). *Standards in Global Value Chains: Rationale, Role and Implications*. Hong Kong: Fung Global Institute. http://www.fung-globalinstitute.org/en/wp-content/uploads/WP_Standards_0.pdf [20 March 2015]

Chapter 13

WATCH MANUFACTURING

Deborah Elms

13.1. BACKGROUND AND HISTORY OF THE COMPANY

The firm represented in this study was started in 1983 in Hong Kong, China. The current managing director worked for a different watch company for a decade before launching his own original equipment manufacturing (OEM) plant. In the early days, the company ordered components from global suppliers, designed products, assembled the products in Hong Kong and shipped them out to customers in the United States, Europe and elsewhere.

In 1990, the company expanded by opening a watch factory in Switzerland. A year later, the company entered discussions about taking over another Swiss factory. The merger was completed in 1992, bringing other brands into the firm and adding another factory in Switzerland.

In 1999, the company augmented its manufacturing operations by establishing a retail business in Hong Kong and in Guangdong. Its retail outlets sell not only the company's own brands, but also other Swiss watches such as Omega, Longines and Cartier. The company was then divided into three segments: manufacturing, branding and retailing.

Today, the company continues to manufacture mechanical movement watches for other brands as well as creating watch pieces with its own label. Customers in the United States, Europe, Middle East, South Africa, Asian economies or South America may require as few as 500 pieces in an order. At its peak in the mid-2000s, the Hong Kong plant turned out 80,000 pieces a month. Now, with manufacturing split increasingly with Switzerland, the

Hong Kong factory creates 50,000 pieces per month. The firm employs roughly 200 people in the factory and headquarters in Hong Kong.

The company also owns a factory that manufactures components for watches in mainland China as well as retail outlets in Shanghai. The company's component factory in the Pearl River Delta is important because prior to its acquisition 10 years ago, the manufacturing operation was struggling with delivery delays and inventory management problems. The factory produces 25,000–30,000 boxes a month. The firm keeps nearly 90 per cent of the output. The rest is delivered to a select few buyers. Its capacity is limited.

For its retail shops, the firm has five multi-brand stores in Hong Kong as well as four individual brand stores that carry lines such as Omega, Breguet and Piaget as well as their own brands. The shops in Hong Kong sell high-level multi-brand items like Cartier, Breitling, Longines, Bulgari and Frank Mueller. The firm also helps the local Hong Kong office for Cartier with retailing their products.

In China, the shops carry fashionable and mid-range models with a different price range than the Hong Kong stores. The company also sells individual brands like Gucci in three multi-brand stores.

The company's own brand is carried by 12–15 stores in total — some owned by the company and others with different owners. The company has salespeople in its branch offices in Beijing, Shanghai and Guangzhou. The Shanghai office, for example, employs two to three people to import watches from Switzerland and Hong Kong and serves as dealer for Cartier.

13.2. TRACING THE CREATION OF WATCH PRODUCTS

The company has a complex business model, as it manufactures for multiple brands and clients in different economies, as well as maintaining its own brands. A significant share of the company's business is creating products for client firms, usually by working for companies that secure brand licenses. For example, Fossil in the United States has licenses for brands like DKNY and Emporio Armani. Fossil orders watches for these licenses from many different suppliers, including the company featured in this case study. Another customer, a Japanese company, has a license for Hunting World. These client companies have no factories of their own but rely on OEM manufacturers like the Hong Kong firm in this study to make watches for different licensed brands.

13.2.1. *Sales and Design*

For OEM watches, the value chain can be considered to start from the sales relationships with clients. The sales team travels to meet specific clients in different markets to discuss emerging fashion trends since watches have gone from being time pieces to fashion accessories.[1] They also attend watch exhibitions to obtain information about trends and markets. They provide these inputs to the research and development (R&D) team who take care of engineering and design. The design team creates designs that are submitted to clients for consideration. Every year, designers create two to three new collections in line with ever changing fashion trends. Some existing collections are maintained and others are replaced, depending on sales volumes.

Because the company designs watches for its own product line as well as for other customers, a design team of 10 people works with a product development team of six persons to create new items regularly. These R&D staff members, for example, look for new materials to use like creating watch cases of stainless steel, ceramics or plastics. They search for new items in the market or new combinations. The straps, for instance, could be plastic, crocodile or stainless steel.

The team, however, does not create new watch movements because it finds the technology too complex. Quartz movements, however, could be created in-house, but mechanical movements

---

[1]Perhaps, this is why about 70–80 per cent of firm production is ladies' watches.

are not easy to develop or manufacture. In fact, only a few companies in the world can create these movements. For the Swiss factory, all movements are sourced from the same two companies in Switzerland that can create them. The two companies left to create movements are Swatch (although they largely manufacture only for their own brands) and Selecta. For the Hong Kong factory, movements may be sourced from Japan or China. Citizen and Seiko make movements, although quantities are limited and mostly used for their own brands. Increasingly, the Chinese firm of Seagao is creating movements. The quality is not Swiss, but it is improving. It used to be the case that Russian companies also made mechanical movements, but they no longer do so.

Once the client is satisfied with the design, drawings and samples, they may place an order with the company for large-scale manufacture.

13.2.2. *Materials Sourcing*

The purchasing team then goes to different suppliers to examine samples and find new sources for materials and parts, and start purchasing components. For the case study firm, the movement is generally Swiss or Japanese. Components like the case or bracelet are usually sourced from China or Japan. Other components come from China and Thailand.

13.2.2.1. *Proximity of suppliers*

The suppliers, some of which are affiliated companies, are all located near to the Hong Kong factory so that the company can communicate easily. If the company ordered cases, dials and buckles in Shanghai or elsewhere, it would be more challenging to manage. The relationship would not be worth pursuing, unless the factory had specialised skills or higher quality. As it is, the Pearl River Delta has the strengths for manufacturing, particularly for more sophisticated goods. In addition, the logistics and supply chain connections are good. The close location of parts and components suppliers allows the firm to save money on logistics and waste less time. If they encounter quality problems, the engineers from the plant can be dispatched to fix them the same day. If the suppliers were far from Hong Kong, such as in Fujian, they would have to work off plans. This is more problematic and generally means more work and difficulties for both sides.

For Chinese components, most suppliers deliver the items directly to the Hong Kong factory. These firms handle logistics and customs declarations.

It used to be the case that dials and watch boxes were ordered from Thailand. At that time, orders were given to the supplier and the company approved the samples and gave specific air shipment instructions. However, the Thai company has since relocated production to China and the packaging is now delivered directly by ground to Hong Kong.

13.2.2.2. *Synchronisation of component availability*

Inventory control of components is a challenge because the parts may not arrive at the factory at the same time. Sometimes, the case might be ready in three months, but the dials are not ready for four months. The procurement team has to make sure it synchronises the preparedness of all the components. If there is a delay in a component, say the bracelet, the watch assembly cannot be completed. The parts to be managed, overall, are relatively modest in number and include the hands (hour, minute and second hands), movements, bracelets, crowns, buckles and cases.

13.2.2.3. *Quality control of materials*

The orders are generally crafted to the specifications drawn up by the firm's client, and the fabrication is high precision. The engineering department has to follow up to make certain that all components meet the requisite criteria before proceeding to manufacture samples. Sometimes, the company sends quality control people to the suppliers to inspect on site. This is because, without a conversation with their engineers and production teams, quality issues can emerge later. If the firm waits until parts are delivered to the factory floor to undertake inspections, it is too late because it could take three months to have new parts delivered.

After the quality control process is complete, the samples can be sent to factories for approval. If everything is satisfactory, the components are delivered to the factory. The incoming quality control team examines the parts before handing them over to the production department for final assembly.

13.2.3. *Production, Packaging, Delivery*

From the time the samples are approved, it takes one to two weeks to complete an order which includes assembly and final testing. Quality control on the final product is performed in-house.

After this is finished, the watches are transferred to the shipping department for final packaging. Then, the order is sent to the client according to their shipping instructions. After that, the claims department follows up with clients for payment.

Because the factory has two facilities, clients can choose where they want watches manufactured — in Hong Kong or in Switzerland. Increasingly, clients and customers are demanding that the watches be assembled in Switzerland since these are perceived to be of higher quality. This is true even though costs of labour in Switzerland are higher.

Under Swiss law, watches must have at least 60 per cent of the value made domestically in order to be considered of Swiss origin. In practice, this means using only Swiss-made movements with either the hands or crowns also from Swiss component suppliers. The cases and bracelets can come from Hong Kong or other places, as long as the total value of imports does not exceed 40 per cent. The company's Swiss factory has less than 30 employees.

If the customer chooses to have the watch made in Hong Kong, by contrast, the complete movement does not need to come from Switzerland, and could have Japanese, Thai or Chinese parts.

13.2.4. *The Company's Own Brand Value Chain*

Besides OEM watches, the company also has its own-brand watch which is manufactured at the factory in Hong Kong. The company's research and development skills, often honed for creating OEM products, are also used in creating the company's in-house branded products. A similar value chain to that for OEM watches is involved, from material sourcing to assembly and delivery, once the in-house design is finalised. About the only thing that is different in this value chain is the lack of intensive discussions on designs with the firm's clients. Most of the components used in own-brand watches come from the Pearl River Delta in China. These parts include the case (with the glass), bracelet or straps, hands and the buckle. The movement is from Japan, and the dials (face) and watch boxes are either from China or Thailand.

Although the supplier of movements for the firm's brand is Japanese, this company has a branch office in Hong Kong. This means there is no need for the Hong Kong firm's staff to travel. The stock is kept by the Japanese company in Hong Kong and therefore, the case study firm does not import its movements directly for their own branded watch.

13.2.5. *After-sales Service*

After-sales services do not make up a big part of the revenue for the company. It does, however, maintain some stock of watch components in the Hong Kong factory in order to service and repair their high-end watches.

13.3. SERVICES IN THE VALUE CHAIN

Approximately 42 separate services enter the watch manufacture and retail value chain. These services correspond to some 80 per cent of the cost of producing the watch. Only 13 services are fully or partially outsourced. Eight of these are fully outsourced and five are partially outsourced. The services entering the value chain and information about their sourcing are listed in Appendix.

13.3.1. *In-house Services*

Most of the company's services are supplied in Hong Kong alongside the factory by around 130 of the factory's staff. An additional 70 staff can be found on the factory floor — the ones directly involved in manufacturing. Services provided in-house include sales, purchasing, design, R&D, accounting, quality control, engineering and shipping.

The reasons given for preferring where possible to supply services in-house are the level of trust, quality control and belief that such specialised services are better handled by the company. Services are outsourced where the government or exclusively mandated external entities are required to supply the service (audit, government inspections) or where networked advantages are inescapable or intrinsic to the service (insurance, IT security alarm services, financial services and certain advertising).

As for in-house supply, the company's reliance as far as possible on its own workforce means, for example, that it does its own in-house cleaning of the factory. The firm does not want to outsource this activity because external service providers may not meet certain requirements for a clean manufacturing environment. With the high value of watch components, the firm also handles security with in-house staff. Over the years, the system of reporting any theft directly to police has meant that few problems have arisen with security. Strict confidentiality agreements between the firm and their clients also preclude allowing many non-employees' access to the factory premises. Upcoming fashion development in watches under design and production must be kept secret.

Inventory and stock control are managed in-house. Parts have to be ordered in excess of the total amount anticipated for watch delivery, since some pieces will be lost or have insufficient quality. In addition, the company maintains an inventory of spare parts for customers since consumers often request new buckles, hands or crystal pieces for years or even decades after a watch has been sold.

Factory equipment and machines are also largely maintained in-house. Much is automated, such as dial assembly machines and waterproof testing machines. Most machines are not terribly sophisticated and the staff can handle routine maintenance issues.

The firm's factory in Hong Kong was the first to receive Swiss machines 8–10 years ago. These machines were necessary to reduce labour costs and keep the company going in Hong Kong.

The use of automation is necessary to stay competitive. Even in the Swiss factory, automation is important, but the kind of watches the company produces cannot be completely made by machines. Trained staff members in both locations must handle the final assembly and the placement of the movements in watch cases.

If, for some reason, the automation equipment breaks down in a way that cannot be repaired by on-site staff in Hong Kong, parts and equipment could take as long as one to two weeks to arrive from Switzerland. In the meantime, workers would take over the tasks by hand. Efficiency falls by 30 per cent, but in 8–10 years, workers have only had to step in for the equipment twice.

The equipment to make the factory a dust-free environment is also maintained in-house.

The completed watches are boxed is the client so requests. For some customers, watches are placed 10 per tray. For others, the watches are individually boxed with price tags to save the client labour costs.

For the firm's own branded models, shops typically want one to three models of roughly 30 pieces at a time. These items are delivered in plastic bags with the boxes delivered separately to the shops. The boxes are made in China and delivered to the company.

The company handles human resources in-house. It directly sources and hires regular employees like factory floor workers, drivers and so forth. For management candidates, every two to three years the company might outsource the job of finding suitable applicants.

The company has a corporate social responsibility (CSR) department that encourages the staff to engage in recreation, although the company does not have facilities on site. The firm, in general, participates in a range of CSR activities.

13.3.2. *Outsourced Services*

Legal services are not handled in-house. The marketing department works with an agency for advertising. Advertising costs for the branded product are quite high. The company has hired an entertainment personality to promote the brand.

The workshop employees are provided with uniforms. The acquisition and laundering of uniforms is outsourced.

In Hong Kong, many services are delivered through government agencies. For example, water, electricity and sewerage are all supplied by the main grid. IT and telecommunications use normal providers in Hong Kong. The government also mandates medical insurance and the company is required to pay into the pension scheme. By the nature of supply of these services, these necessarily have to be outsourced.

The company used to outsource its on-site catering service, but now it just provides an extra payment for workers to find food outside.

13.4. GOVERNMENT POLICY INTERACTION AND OTHER BUSINESS CHALLENGES

13.4.1. *Locational Advantage of Hong Kong, China*

While there used to be many watch manufacturers in Hong Kong, the case study firm is one of the few left. Some moved factories to the Pearl River Delta. In price, these factories are more competitive. However, design, quality and service are also important to customers. These elements are at present easier to maintain at a high level in Hong Kong than on the mainland. The "made in Hong Kong label" still means something to final customers.

This is true even if the products made in Hong Kong cost a bit more. The price of a watch assembled in China might be 3 USD. In Hong Kong, by contrast, the company has to pay up to USD9 for assembly. However, since the production cost of a watch is a relatively minor element of the overall price, as long as total production costs are no more than USD6 or so higher in Hong Kong than in China, the company can still make a profit.

Travel time to the supplier factories in China averages one to three hours. Given transportation issues in China, it might take the same length of time to travel between factories even if the firm relocated its Hong Kong plant to the Pearl River Delta area.

13.4.2. *Labour Issues*

The watch design workers tend to leave every two to three years for new positions elsewhere, usually into other watch or toy factories. The firm hires junior or young people out of design schools. They also check with vocational training colleges for staff. Fresh graduates work well during their brief tenure with the company.

There is pressure on labour costs as unions press for the establishment of minimum wages and standard working hours.

13.4.3. *Permits and Inspections*

The factory needed to apply for a license to operate. Most of the inspections undertaken by various government agencies are related to health and safety. For example, the labour department comes to check on working hours. Overtime permission must be approved by the government. There is no limit on the amount of overtime, but the firm needs to submit an application before it can permit workers to do overtime work. It used to be the case that many workers put in one to two hours of overtime each day, but this has become very expensive.

The fire department inspects the facilities periodically and the firm needs a separate license from the Department to operate.

13.4.4. *Incentives*

The company does not receive incentives from the Hong Kong government. But, it would support plans to bring back manufacturing to Hong Kong, particularly in high value-added activities. The company believes that manufacturing is needed to give young people a place to work. Otherwise, there are likely to be no positions available for young citizens much beyond banking and property companies. Jobs in manufacturing can be of high quality. These positions also generate home-grown skills and knowledge, as well as other jobs in the local economy and in the region.

APPENDIX

Table A.1: Services for the Manufacture and Retailing of Watches.

Stage of supply chain	Service	Corresponding CPC ver. 2 code	In-firm supply	Outsourced and to whom?	Reason(s) for house supply/ outsourcing	Remarks
Pre-Manufacture	Procurement agent for materials sourcing	85999 — Other support services n.e.c.	Yes		In-house team knows product and needs	
	Quality assurance services of materials and components	83441 — Composition and purity testing and analysis	Yes		In-house team knows product and needs	
	Freight transportation services	Division 65 — Freight transport services	Yes			
	Storage of parts and components	67290 — Other storage and warehousing services	Yes on site and in China facility			
	Conception and design of product	83920 — Design originals	Yes			
	Industrial design	83912 — Industrial design services	Yes			

	Patent acquisition	83960 — Trademarks and franchises n/a	Yes	Yes		Split responsibility depending on brand, but own brand is wholly handled in-house
Manufacture	Product administration-production management	83115 — Operations management and consulting services	Yes			
	Product administration-repair and maintenance of factory equipment	87156 — Maintenance and repair services of commercial and industrial machinery	Yes		Concern over delays and lack of outside specialised knowledge of equipment	
	Quality assurance services of manufactured product	83441 — Composition and purity testing and analysis	Yes		In-house team knows product and needs	

(Continued)

Table A.1: (*Continued*)

Stage of supply chain	Service	Corresponding CPC ver. 2 code	In-firm supply	Outsourced and to whom?	Reason(s) for house supply/outsourcing	Remarks
	Warehousing services for parts and components	67290 — Other storage and warehousing services	Yes			
	Cleaning services for factory	85330 — General cleaning services	Yes		Specialised equipment needs particular care	
	Security guards for factory and warehouses	85250 — Guard services	Yes		High value parts, components and final products need tight security	
	Government inspections on fire prevention, health hazards, labour regulations	91290 — Public administrative services related to public order and safety affairs		Yes	Government mandated	

Personnel search and referral services-recruitment of workers	85112 — Permanent placement services, other than executive search services	Yes	
Social insurance for factory workers	91320 — Administrative services related to government employee pension schemes; old-age disability or survivor's benefit schemes	Yes	
Medical services	93121 — General medical services	No	Workers receive government medical benefits
Design of packaging	83919 — Other specialty design services	Yes	

(Continued)

Table A.1: (*Continued*)

Stage of supply chain	Service	Corresponding CPC ver. 2 code	In-firm supply	Outsourced and to whom?	Reason(s) for house supply/ outsourcing	Remarks
	Packaging services	85400 — Packaging services	Yes			
Post-Manufacture	Customs-related services	85999 — Other support services n.e.c.	Yes	Yes	Depends on whether delivery to customers or own retail shops in Mainland China	
	Land transport of goods	65112 — Road transport services	Yes	Yes	Own conveyances and external suppliers	
	Repair and maintenance for fleets	87143 — Maintenance and repair services of trailers, semitrailers and other motor vehicles n.e.c.	Yes	Yes	Ad hoc in-house and regular servicing outsourced to specialists	

	Air transport services	6531 — Air transport services of freight		Yes	
	Retail services	622 — Specialised store retail trade services	Yes	Yes	Both in-house retail and other retail outlets even for own brand
	Retail administration-operation management	83115 — Operations management consulting services	Yes		
	Retail administration-operation management	83116 — Supply chain and other management consulting services	Yes		
Back Office Support	Auditing on financial accounts	Group of 822 — accounting, auditing and bookkeeping services		Yes	External auditing mandated

(Continued)

Table A.1: *(Continued)*

Stage of supply chain	Service	Corresponding CPC ver. 2 code	In-firm supply	Outsourced and to whom?	Reason(s) for house supply/ outsourcing	Remarks
	Financial services	7112 — deposit services 7113 — credit-granting services		Yes	Requires financial institutions	
	Internal auditing (including audits of financial accounts and corporate governance)	83118 — Head office services	Yes			
	Insurance services for machinery	71332 — Marine, aviation and other transport insurance services		Yes	Requires insurance providers	
	Company secretary services	83990 — All other professional, technical and business services, n.e.c.	Yes			

Corporate communications, marketing and public relations	83114 — Marketing management consulting services	Yes	Yes	Split between in-house person who works with outside advertising agency
Human resources management	83113 — Human resources management consulting services	Yes		
Human resources management-personnel search and referral services for back office staff	8511 — Personnel search and referral services	Yes		
IT and information system management	8316 — IT infrastructure and network management services		Yes	A networked service

(Continued)

Table A.1: (*Continued*)

Stage of supply chain	Service	Corresponding CPC ver. 2 code	In-firm supply	Outsourced and to whom?	Reason(s) for house supply/ outsourcing	Remarks
	Safety and security services	85230 — Security systems services		Yes	External suppliers of alarm systems	
	Safety and security services	85250 — Guard services	Yes			
	Uniform-laundry	97130 — Other textile cleaning services		Yes		
	Legal services	82120 — Legal advisory and representation services concerning other fields of law		Yes		
	Product development R&D	81129 — Research and experimental development services in other engineering and technology	Yes			

Product development R&D	81400 — Research and development originals	Yes		
Retail administration-advertising	83611 — Full service advertising		Yes	Includes hiring an entertainment personality
After-sale services-warranty and repair services	Group of 872 — Repair services of other goods	Yes		

Source: Author.

Chapter 14

MANUFACTURING OF AUTOMOTIVE COMPONENTS IN MEXICO: PERSPECTIVES FROM THREE FIRMS

Andre Wirjo, Gloria O. Pasadilla and Joel G. Bassig

14.1. INDUSTRY OVERVIEW

Mexico is now the fourth largest car exporter and occupies the eighth position in the world as a car producer (PwC Mexico, 2014a). Because car manufacturing requires efficient supply chains, the automotive parts and components industry has followed its growth, making Mexico the fifth largest producer as of 2012, behind China, Japan, USA and Germany (ProMexico, 2013). The automotive industry's contribution to manufacturing GDP of Mexico has increased from 9.4 per cent in 1989 to 19.8 per cent in 2011 and it provides employment to 13.4 per cent of total industrial labour in 2010 [Mexican Automotive Industry Association (AMIA), 2015]. Investments by automotive and auto parts manufacturers together made up 20.8 per cent of inward FDI received by Mexico in 2012 (AMIA, 2015).

Despite the fact that a car brand and model are often associated with the final assemblers, manufacturing a car actually involves many suppliers of different tiers from around the world. Figure 14.1 illustrates this complexity, showing that different parts indicated may be supplied by different suppliers. It should be noted that the list provided in the figure is non-exhaustive and there are parts

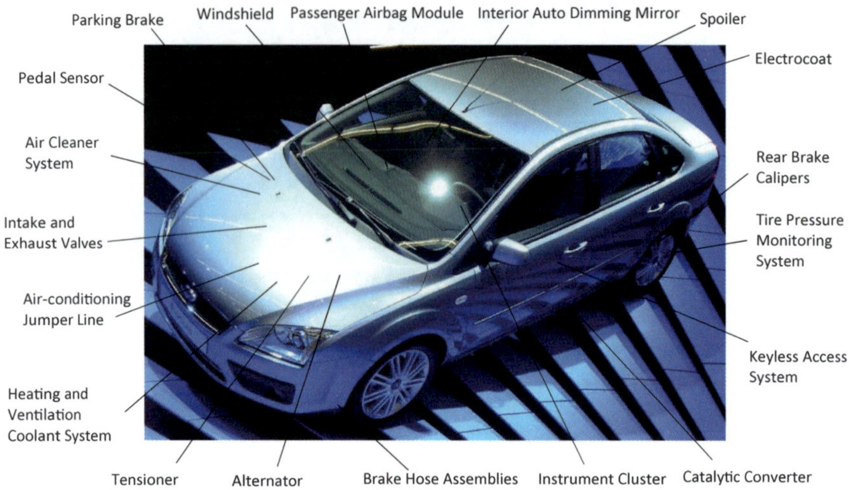

Figure 14.1: Simplified Supply Chain of a Car.

which are not identified in the figure. In turn, these suppliers have their own global value chains (GVCs) for producing and supplying parts and components to the car manufacturer. The three case study firms in this chapter are examples of tier 1 or 2 parts suppliers for the car industry.

14.2. BACKGROUND INFORMATION ON THE THREE FIRMS[1]

The three firms indicated in this report are Japanese manufacturers of various automotive parts that are based in Bajio, a region in North Central Mexico which includes parts of the states of Guanajuato, Queretaro, Aguascalientes and Jalisco. Information about these firms, henceforth referred to as Firms A, B and C, are described below.[2]

[1] All information about the firms in this case study are from their corporate websites and interviews.

[2] In this paper, Firms A, B and C are used interchangeably to refer to either the parent firm or their respective subsidiaries in Mexico.

14.2.1. *Firm A (Chassis Parts Manufacturer)*

Firm A is a designer and manufacturer of chassis parts, which are vital for vehicle safety and stability. In some instances, the firm is viewed as a tier 1 supplier because its products are directly supplied to automotive manufacturers. In other cases, it is viewed as a tier 2 supplier because the firm sells its products to tier 1 suppliers, which then incorporate them into their products before selling them to automotive manufacturers. It is headquartered in Hiroshima, Japan and has a wholly-owned subsidiary in Guanajuato.

It provides a fully integrated production starting from material molding all the way to machining process and assembly. Its involvement in chassis parts started with knuckle components but has expanded its range of chassis products to parts such as hub support, bush bracket, hanger bracket, arms and joint shaft (see Fig. 14.2). Today, the firm manufactures parts that the automotive manufacturers used to produce by themselves, a show of their confidence and trust on Firm A. Some of the firm's major customers include Mazda, Ford, Mitsubishi and Aston Martin.

14.2.2. *Firm B (Brake Hose End Fittings Manufacturer)*

Firm B is a manufacturer of brake hose end fittings and supplies them to automotive suppliers worldwide which, in turn, supplies the finished products to major vehicle manufacturers. In this regard, the firm is a tier 2 supplier of the automotive industry. The firm also manufactures stud bolts and supplies them directly to automotive manufacturers as a tier 1 supplier. It is headquartered in Tokyo, Japan and has wholly-owned subsidiaries in Thailand in addition to Guanajuato. Its Mexican subsidiary is involved in supplying to the North American, South American and European markets. The firm started with production of stud bolts, which it continues to supply directly to automotive manufacturers. However, unlike the earlier years where stud bolts made up the bulk of its sales, it now contributes only about 30 per cent of firm's total sales. Since 1964, the firm's technology has centred around

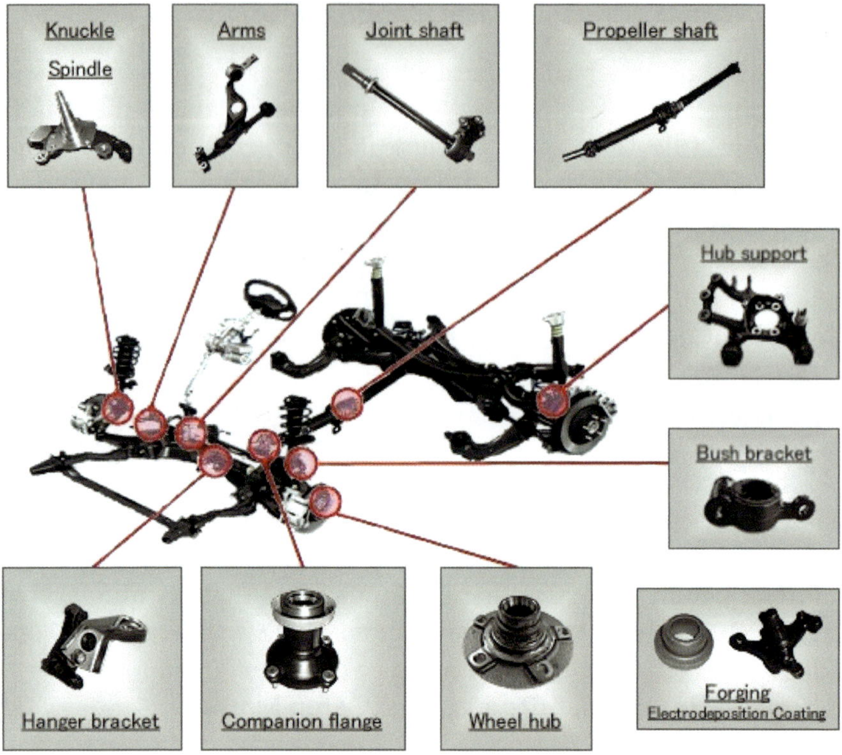

Figure 14.2: Examples of Chassis Parts Produced by Firm A.

Source: Courtesy of the firm.

cold forming process which allows most products to be manufactured in one integrated procedure. This is in contrast to conventional process where final products are usually composed of two or more parts that have to be combined together by pressing, for instance (see Fig. 14.3). In addition, this technology enables the manufacture of high-precision, high-quality near-net-shapes that were previously impossible to produce using conventional technology. The use of cold forming process, therefore, cuts the number of production steps and leads to speed enhancement as well as cost optimisation. Firm B's experience in cold forming process has allowed it to adapt the technology for the manufacture of other products such as air bag parts. The plan is to use the technology to

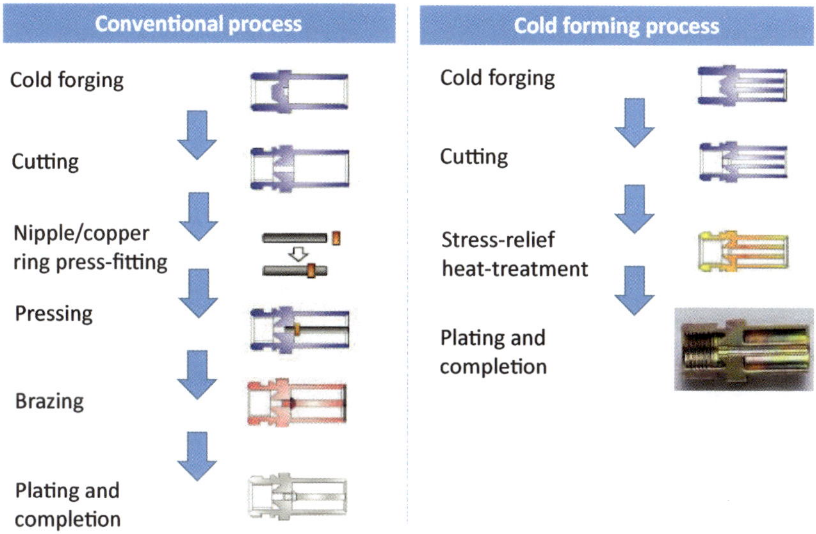

Figure 14.3: Comparison of Conventional and Cold Forming Process for the Manufacturing of Brake Hose End Fittings.

Source: Courtesy of the firm.

continue diversifying the range of products that the firm manufactures.

Specifically on brake hose end fittings, which now make up about 70 per cent of firm's sales, the firm produces many different versions of it, among them female type, male type, banjo type, block type and variant type (see Fig. 14.4). After they are assembled to both sides of a rubber hose, the hose would be installed on the brake caliper. These brake calipers are subsequently installed on all the wheels and play an important role in braking.

14.2.3. *Firm C (Antenna Manufacturer)*

Firm C is a specialised supplier of automotive antenna which serves almost all major vehicle manufacturers and automotive suppliers worldwide either as a tier 1 or tier 2 supplier. It is headquartered in Japan and has wholly-owned subsidiaries in nine economies around the world focusing on functions such as research and

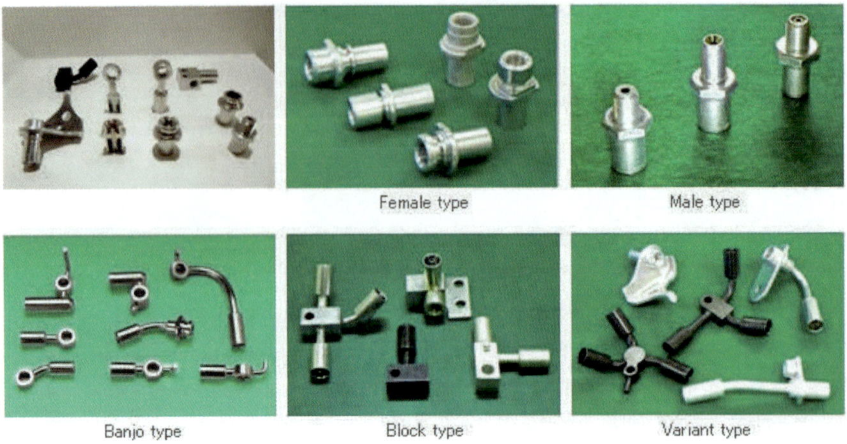

Figure 14.4: Different Brake Hose End Fittings Manufactured by Firm B.
Source: Courtesy of the firm.

development (R&D), sales and manufacturing. Its Mexican subsidiary is located in Queretaro and was established 27 years ago in 1988. As an indication of its global footprint, more than two-thirds of Firm C's consolidated net sales in 2014 were made outside Japan and covered Asia, Europe, North America and South America.

The firm has been producing antennas since its founding and has expanded the type of antennas that it manufactures over the years. Among the different types of antenna that the firm currently manufactures and supplies are fin-type/low-profile antenna (LPA), rod-type antenna, screen-type antenna, film-type antenna, in-dash–type antenna, in-spoiler–type antenna and other integrated/hidden-type antenna.

These antennas serve different functions and can be mounted in different parts of a car depending on its type. The fin-type antennas (see Fig. 14.5), for instance, supports all kinds of radio application such as analogue amplitude modulation (AM) and frequency modulation (FM), digital audio broadcasting (DAB) as well as satellite broadcasting. They can also be used in telematics application such as global positioning system (GPS). On the other hand, the on-dash–type antennas can be used for dedicated short-range

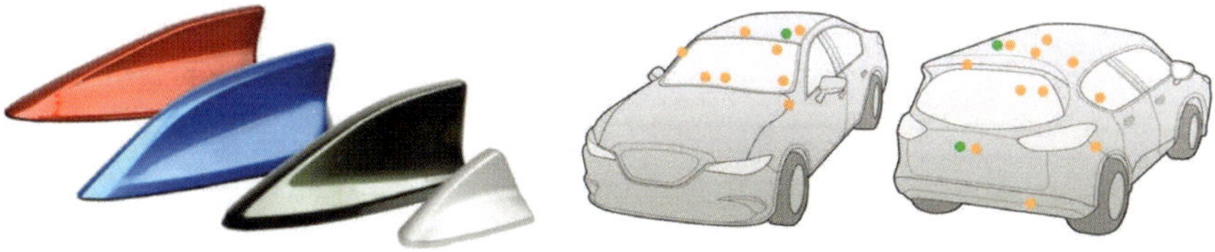

Figure 14.5: Fin-type Antennas Produced by Firm C and their Mounting Positions.

Note: Possible mounting positions of fin-type antennas are shown by green dots, while that for other type of antennas are shown by yellow dots.
Source: Courtesy of the firm.

communication (DSRC) and support technologies such as electronic toll collection (ETC). The firm indicated that it continues to develop new antennas in response to diversifying needs. Indeed, its engineers work closely with engineers of automotive companies to develop antennas that suit their needs.

In addition to antennas, Firm C also supplies products and peripheral technologies associated with automotive antennas such as cable routing in cars, coaxial cables, copper clad aluminium (CCA) cables, shielded twisted pair (STP) cables, connectors assembly, amplifiers and noise filters.

14.3. DESCRIPTION OF THE VALUE CHAIN

This case study will focus on the three firms' Mexican subsidiaries, which were established in different years and for various reasons. The subsidiaries of Firms A and B are relatively younger, having been established in 2013 and 2012, respectively. In contrast, Firm C's subsidiary was set up in 1988.

All the subsidiaries were established with the objective of accessing markets in North America and South America, the immediate one being Mexico itself. Indeed, many of their client firms are located in the same Bajio region where they are based. However, there exist other more idiosyncratic reasons for the establishment of these subsidiaries. Firm A, for example, wanted to leverage on the low production cost provided by Mexico while Firm B shared that its Mexican subsidiary was also established with the purpose of experimenting with new production methods which are relatively more challenging to apply in its Japan HQ and Thai subsidiary due to the presence of well-established procedures that have stood the test of time.

The product of focus and dimension of the value chain covered by this study differ between firms and are summarised in Table 14.1 below. This is to be expected considering that these firms, as indicated in the earlier section, produce very different parts and components despite all of them being in the automotive industry. Furthermore, in the interest of accuracy and capturing more recent and relevant information, while the dimension of value chain for Firms A and B begins with their subsidiary's establishment stage and ends with the provision of after-sales parts/services to the client firms that for Firm C begins with pre-manufacturing stage.

Figure 14.6 gives a general representation of some of the important steps involved at each phase of the value chain. It should be noted, however, that each firm may not necessarily undertake all the steps listed.

14.3.1. *Establishment*

Firms A and B utilised several services activities during the establishment stage of their Mexican subsidiary including company registration, business consultancy and legal services. Firm A fully outsourced these activities while Firm B either fully or partially outsourced them to third-party providers for various reasons such as efficiency (outsourced firms are more familiar with local rules and regulations); or economies of scale (e.g., banking and financial services); or legally mandated (e.g., licensing from the government).

The establishment phase of Firm C is not covered in this report as its subsidiary was set up 27 years ago and what the firm experienced at that time would perhaps no longer reflect the current situation faced by newly established firms or subsidiaries.

Table 14.1: Product of Focus and Dimension of Value Chain by Firm.

		Dimension of value chain covered				
	Product of focus	Establishment	Pre-manufacturing	Manufacturing	Post-manufacturing	After-sales
Firm A	Part of suspension systems of car (i.e., knuckle, hub support and bush bracket)	✓	✓	✓	✓	✓
Firm B	Brake hose end fittings	✓	✓	✓	✓	✓
Firm C	Fin-type/low-profile antenna (LPA)		✓	✓	✓	✓

Source: Authors.

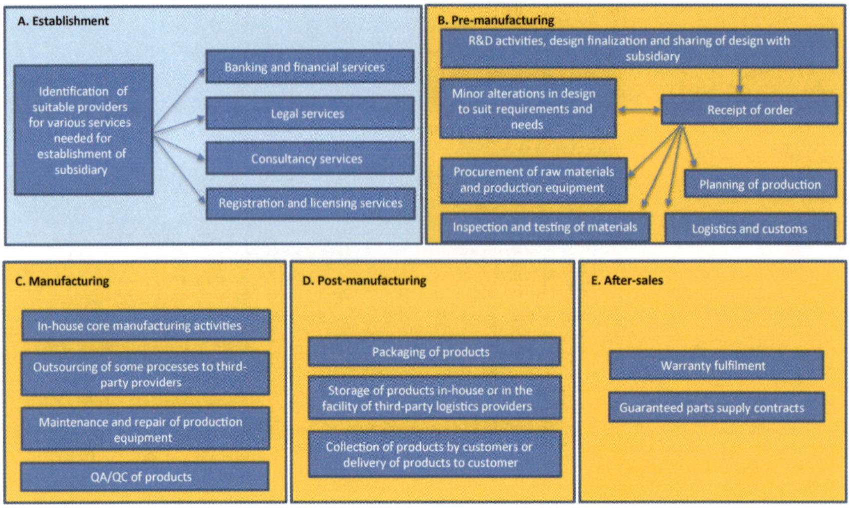

Figure 14.6: Dimension of the Value Chain Covered by the Case Study.

Note: Establishment stage is only applicable to Firms A and B.

Source: Authors.

14.3.2. *Pre-manufacturing*: *Design and Materials Sourcing*

14.3.2.1. *Firm A* (*chassis parts manufacturer*)

Firm A's pre-manufacturing stage begins when the lead automotive manufacturer invites the participation of the firm's engineers in the design of a new car model approximately 3 years before its launch into the market. Firm A indicated that out of the 10-man design team led by the car manufacturer, three people would usually come from them. These engineers are based at the premise of the automotive manufacturer and would provide their expertise on chassis parts and components that the firm will eventually produce and supply to the manufacturer. As long as the design has not been finalised, Firm A's engineers would have a free hand on the design and specifications of parts such as hub support and knuckle. However, once the design has been finalised, it would become the property of the lead automotive manufacturer and only it has the rights to modify the design. As a supplier, firm has to abide by the agreed design and specifications.

All finalised designs are shared by the automotive manufacturer through its supplier portal. Regular access to the portal is critical for all suppliers including Firm A because it is only from the portal that the firm would be able to foresee the weekly demand for its parts and components and hence respond accordingly. Sometimes, when demand is extraordinarily high, Firm A's Mexican subsidiary seeks the support of its manufacturing facilities in Japan to meet the demand of its client firms.

Changes to the designs, if any, could also be seen via the portal. Minor changes are usually handled by Firm A's subsidiary but for cases beyond the capability of the local team (e.g., design change that necessitates an installation of new production line) it has to request for assistance from its Japan headquarters which, besides approving the production, may also send the appropriate number of staffs who can temporarily help out.

Raw materials are sourced from either Japan or Korea and they come in processed form of iron and steel. Firm A disclosed that it would prefer to source raw materials locally but, so far, has not been able to identify local suppliers who can produce the same high quality materials. Upon arrival, its in-house staffs test the raw materials but, at times, their suppliers send their own staffs to test the materials on-site. For materials imported from Japan, its headquarters may also be involved in the testing activities.

14.3.2.2. *Firm B (brake hose end fittings manufacturer)*

In contrast to Firm A whose involvement starts from the pre-launch of the new car model, the pre-manufacturing stage of Firm B's Mexican subsidiary usually begins when its customers place orders for brake hose end fittings with the firm. Due to the firm's reputation and long-term relationships with existing manufacturers, orders are placed directly with them without need to participate in competitive biddings. Unlike Firm A where its engineers have a role in product designs, Firm B's customers usually specify the product designs.

Upon receipt of the customer's order, staff at its Mexican subsidiary shares the product designs and specifications with firm's R&D department in Japan. There are currently no such departments in its two subsidiaries in Thailand and Mexico although Firm B has plans to establish an R&D department in Mexico in the future. The role of the R&D department is to design moulds which are then used in the production stage. For products that are entirely new, the R&D department is also involved in the design of production machinery as well as production line. Even though the product design is provided by the customer, Firm B's R&D department can make design suggestions if they affect the manufacturing process. Communications between customer and R&D department are mediated by Firm B's Mexican subsidiary. However, customer at times also liaises directly with firm's R&D department in Japan, for instance when the customer also has its design team in Japan. Once the mould design has been finalised and the moulds produced, they are sent to firm's Mexican subsidiary for installation in the production machinery. Besides installation, Firm B's engineers at its Mexican subsidiary are also involved in process improvements.

Depending on the type of material inputs, sourcing and procurement are handled by Firm B's in-house staffs, its HQ or third-party trading companies. The steel used for production is currently imported from Japan. Firm B has tried to source steel locally in Mexico but found that its quality does not meet the firm's requirements. The logistics of getting material inputs into the facility is fully outsourced to third-party providers and include services such as customs clearance, loading and transport. Composition and purity testing of these inputs are performed in-house but may also be outsourced to its affiliates or third-party providers if the firm does not have in-house equipment.

In preparation for the manufacturing stage, Firm B trains many of its local production staff such as machine operators in its facility in Japan for 6 months. In addition, to ensure that processes in the new facility start off correctly, Firm B now has five Japanese staff on site to provide expertise and training on activities such as plant management, cold forming process technology, plating technology and QA/QC procedures. Specifically for training pertaining to external certification, external trainers come to provide training to firm's staffs.

Customers require that firm's manufacturing facility has the necessary certifications to produce parts and components. To this end, Firm B has completed CQI-11 Plating System Assessment program, which was developed by the Automotive Industry Action Group (AIAG) to ensure adherence to plating procedures and systems.[3] This has then led to the firm passing the Nissan Alliance Supplier Evaluation Standard (ASES), which is designed by Nissan to evaluate if a firm qualifies to become a suitable supplier.[4] At the moment, the firm is in the process of obtaining ISO/TS16949 certification which defines the quality management system requirements for the design and development, production and, when relevant, installation and service of automotive-related products.[5]

14.3.2.3. *Firm C* (*antenna manufacturer*)

The pre-manufacturing stage of Firm C's Mexican subsidiary begins when its customers place direct orders for antennas with the firm. In some cases, it could also involve the customers requesting for quotations from the firm or submissions of bid in response to call by customers. However, prior to describing this part of the value chain in greater detail, it is worth taking a few steps back to understand the processes that happen before this stage.

Firm C's actual involvement in the manufacturing process of an automotive antenna begins much earlier through R&D activities

---

[3]More details about CQI-11 Plating System Assessment can be obtained at: http://cqi-support.de/en/cqi-standards/cqi-11.

[4]More details about Nissan ASES can be obtained at: http://www.nissan-global.com/EN/QUALITY/PRODUCTS/GLOBAL_SP/GUARANTEE/ (Nissan Motor Corporation, 2015).

[5]More details about ISO/TS16949 can be obtained at: http://www.iso.org/iso/catalogue_detail?csnumber=52844.

carried out by its engineers based in the United States, United Kingdom and Japan. The activities carried out by these engineers could generally be divided into two parts: (1) internal research and (2) working with their counterparts in the client firms. As antenna experts, they constantly carry out internal research and collaboration to improve existing antenna technology.

More significantly, Firm C's engineers also work with their counterparts in the automotive companies to develop antennas that meet their requirements. In fact, depending on the travel distance between the firm and their customers' premises, Firm C's engineers, like Firm A, could be based in their customers' facility as guest/resident engineers. The fin-type antenna is a good example of an antenna that was conceived through the close collaboration between firm's engineers and that of its customers. Considering that Firm C's engineers work with different customers, there is a need to maintain secrecy and ensure that proprietary information of one customer is not shared unintentionally with another customer. Firm C signs non-disclosure agreement (NDA) with each customer to protect such confidential information.

Once the antenna design has been finalised, its drawings and schematics would be passed to firm's manufacturing facility where mass production engineers prepare for large-scale production of the antenna. Although not formally part of the design team, it is possible for mass production engineers to provide feedback to firm's HQ and obtain its approval to manufacture a specific type of antenna that is adapted to local context after conducting field test and obtaining the nod of its customers. Due to the present infrastructure in Mexico, this revised design would still be able to perform its function well but at the same time, could be produced at a lower cost than the original version. Usually, its customers appreciate the revised, lower cost design which are adapted to the local context.

Firm C's Mexican subsidiary has its own procurement department. This department generally handles the sourcing and procurement of material inputs, which can come from either small or large firms, domestic firms or multinationals. However, Firm C also has affiliates in other economies and if the suppliers for certain inputs happen to be based in the same economy where its affiliates are (e.g., Japan), then it may ask its affiliate to procure on behalf of the Mexican subsidiary. Firm's supplies come from different parts of the world including Japan, Malaysia, the United States and Europe. The logistics of getting material inputs into the facility is fully outsourced to third-party providers, which provides bundled services which include customs clearance, loading and transport. Composition and purity testing of these inputs are performed in-house although they may be outsourced to its affiliates or third-party providers if the firm does not have the requisite in-house equipment.

14.3.3. *Manufacturing Process*

14.3.3.1. *Firm A (chassis parts manufacturer)*

Except for basic machining and fabrication activities which are usually bundled with the steel materials as well as painting work which it outsources to a local firm, all manufacturing activities including forging and thermal treatment are performed in-house. Similar to sourcing of raw materials, firm indicated that it would have outsourced certain activities to local providers but at the moment, it has not been able to identify qualified providers. Maintenance and repair services of machineries are carried out by Firm A's own staffs but are also outsourced to third-parties such as equipment providers if they could not be maintained and repaired internally.

Once manufacturing has been completed, products are tested to ensure that they are in good, working conditions. While these are usually conducted in-house, some customers also prefer to test the products themselves.

14.3.3.2. *Firm B (brake hose end fittings manufacturer)*

The production process of brake hose end fittings by Firm B involves three main procedures; namely, cold forming process, machining and plating. Cold forming process is the firm's mainstay technology and therefore fully conducted in-house. The firm also carries out machining and plating in-house but in certain cases, may outsource the activities. For instance, they are sometimes carried out by the customers if they order non-finished products from the firm. In other words, firm's role in the manufacturing process is only at the cold forming stage. Heat treatment, if needed, is outsourced to its partners because of lack of in-house equipment.

Firm B initially planned to outsource plating activities in Mexico. However, it found local plating activities to be limited, expensive and do not meet quality requirements. Therefore, it decided to undertake in-house plating activities for its own products as well as offer the services to other customers.

Firm B's philosophy that quality must be 100 per cent perfect and the fact that it manufactures brake hose end fittings, which play an important role in ensuring the safety of automotives and their users, means that it is necessary for the firm to develop rigorous QA/QC procedures. Firm wants that all products (i.e., 100 per cent) are automatically inspected by machines followed by manual inspection. The largest share of its direct labour is, in fact, involved in the manual inspection. Furthermore, randomly sampled products are inspected using higher-end equipment.

14.3.3.3. *Firm C (antenna manufacturer)*

As with Firms A and B, Firm C outsources certain non-proprietary steps such as fabrication and basic machining, usually to the suppliers of its material inputs. Thus, many of the firm's material inputs are already in processed form when they enter its facility. Otherwise, all final products are manufactured in-house. Maintenance and repair services of factory equipment are carried out by firm's in-house staff or outsourced to third-parties such as equipment providers if they could not be performed internally.

Products are tested and certified upon completion of manufacturing. These are usually conducted in-house although some customers may require them to be done by its certified third-party testers.

14.3.4. *Post-manufacturing*

Post-manufacturing, in this case study, refers to packaging, handling and transport of products to customers. Firm A does the packaging services in-house. Third-party logistics providers are then engaged to transport the products to the customers, which are mostly based domestically.

Firms B and C, likewise, undertake the packaging services in-house but are also sometimes provided by third-party logistics providers or done by the customers themselves. If done by third-party logistics providers, packaging is bundled with other services like customs clearance, loading and transport. Products are then stored either in the firm's facility or leased warehouse within the industrial park before being transported to its customers.

For Firm C products sold domestically, they are usually sold ex-works and transport services are provided by its customers via milk-run system, where customers come to collect the products themselves from Firm C's facilities.

14.3.5. *After-sales Services*

The exact provision of after-sales services by each firm varies but, in general, it consists of the following: (1) guaranteeing that replacement parts will be available for a specified number of years (10–20 years) after mass production has been completed (it should be noted that besides automotive manufacturers, replacement parts may also be sought by car dealers and repair shop), (2) replacing parts during the warranty period via the automotive manufacturers in case of any defects in its products and (3) standing ready to send its engineers to the automotive manufacturers' facilities to resolve any technical issues.[6]

[6] The last one applies only to Firm A.

14.4. SERVICES ALONG THE VALUE CHAIN

14.4.1. *Services Identification and Value Contribution*

The total number of services identified in each firm's value chain differs and varies between 49 and 61 services (see Fig. 14.7). These

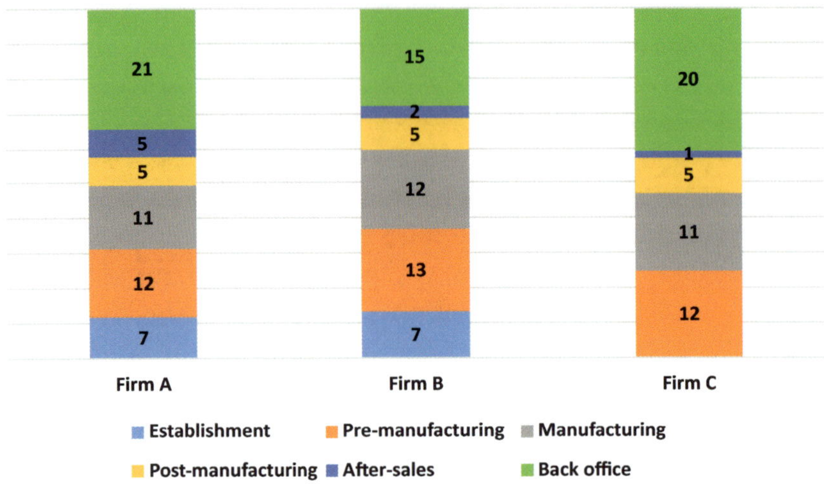

Stage	Examples of key services
Establishment	• Company registration and licensing services • Business consultant services
Pre-manufacturing	• Engineering services • Design services • Procurement services
Manufacturing	• Production management services • QA/QC services
Post-manufacturing	• Packaging services • Transport/logistics services
After-sales	• Inventory management services
Back office	• Headquarter services • Financial services • Legal services

Figure 14.7: Breakdown of Number of Services by Stage and Examples of Key Services.

Source: Authors.

services are categorised according to the various stages within the chain: (1) services during establishment, (2) pre-manufacturing services, (3) services during manufacturing, (4) post-manufacturing services, (5) after-sales services and (6) back office services. It is worth noting in the comparison across the three firms though that the exact dimension of the value chain differs, that is, Firm C has no establishment phase while the other two have.

Although information on which are the high value-added services inputs is not available, it could be surmised that engineering services are among the critical ones across all three firms. The role of engineering services in the firms' core activities such as those pertaining to R&D (both development and utilisation) and quality assurance and control is invaluable. They work closely with their counterparts from R&D departments (in the parent firm) and/or client firms to design components that would become an integral part of the final car model. Thereafter, they need to adapt proprietary technology to produce these components at the right specifications and quality. Often, these may include designing new machinery and re-configuring production lines in addition to good production planning. Engineering services would also be needed to establish the proper QA/QC procedures to ensure that supplied components are of the highest standards.

The importance of services to the case study firms can also be inferred from the share of employees devoted to service activities. Firm A shared that approximately a third of its on-site employees are involved in the provision of services such as plant management, procurement services, production planning, etc. The remaining two-thirds are operators who are directly involved in the manufacturing process on the production floor. In the case of Firms B and C, the share of services employees is approximately 42 and 20 per cent, respectively. Given that some of the staffs categorised as manufacturing employees could also be involved in activities such as testing and QA/QC, the share of employees involved in services provision in these firms could well be higher than this estimate.

Where it is possible to obtain information on costs, they provide another perspective through which the relevance of services to firm is clearly exhibited. Firm B, for instance, noted that direct manufacturing cost which includes cost of raw materials, machines and direct labour made up only about half of firm's total operational cost. This indicates that the cost of everything else (back office, logistics, etc.) which are mostly services in nature made up the other half.

In contrast, services made up only approximately 15 per cent of Firm C's total cost, which is relatively low compared with other case study firms which disclosed estimates of its cost structure. Several reasons could explain this observation. First, Firm C has a large direct labour cost. With 80 per cent of its employees directly involved in manufacturing on the production floor, the share of services significantly shrink. Second, Firm C has minimal expenditure on services generally because (1) the cost of logistics are usually covered by its suppliers in the case of material inputs and by its customers in the case of finished products, (2) the firm does not need to send its engineers to provide after-sales services and (3) it does not have to undertake much advertising and marketing because of its business-to-business (B2B) business model. Third, the share of services does not include the value-added that are embedded in material inputs purchased by Firm C which comes in processed form and therefore already contains a significant amount of services value added which are not considered in this study.

Looking ahead, the share of services in each firm's value chain is expected to remain significant because of various reasons. First, one of the primary reasons for the establishment of these subsidiaries is to allow the firms (and ultimately its client automotive firms) to access the NAFTA market at preferential rates. To achieve this, there must be sufficient transformation of imported inputs and/or regional value content[7] and value addition through services is one option to do so. Second, while significant share of their raw

[7]More information on NAFTA Rules of Origin can be obtained at: http://tcc.export.gov/Trade_Agreements/All_Trade_Agreements/NAFTA_Annex_401_5.asp#P39_3158. Rules pertaining to firm's products can be found in Section XVII, Chapter 87.

materials are currently imported, firms generally have plans to source more from local suppliers. The savings from sourcing materials locally could reduce the cost of raw materials on one end and, concurrently, increase the share of services on the other end. In addition, some of the raw materials obtained by firms are already in processed form. As such, sourcing them locally means that services associated with the processing activities will also be captured locally. Last but not least, Firms B and C have plans to undertake R&D activities in the near future. In addition, Firm B is currently expanding its facility and when construction is ready, the expanded facility is expected to host a section that will focus on new processes where engineering services are likely to be intensely used.

14.4.2. *Outsourcing, Bundling and other Aspects of Services Supply*

Further disaggregating the services identified in these value chains and classifying them into whether they are provided in-house or outsourced showed that each firm has different mix of in-house and outsourced services (see Table 14.2 and Appendix).[8]

Services provided entirely in-house are generally those that are considered firms' core activities such as R&D, production management, head office and QA/QC services.[9] All firms indicated that these services are conducted in-house because they involve propri-

---

[8]Some of the reasons for firms to outsource services include (1) government services, such as visa and immigration services as well as company registration and licensing services; (2) required by laws and regulations, such as external auditing services by third-party providers; (3) lack of expertise or specialization in-house to provide certain services, such as legal services and medical services; (4) need access to the best services, such as maintenance and repair services for some equipment; (5) lack of feasibility to supply service in-house, such as banking services and utilities services; (6) economies of scale, such as transport/logistics services; (7) need for strong relationships with government agencies, such as customs clearance services; and (8) network economies, such as security services.

[9]The only reason where QA/QC is outsourced is the lack of expertise/equipment and this may be the case if the equipment is rarely used and capital expenditure on it is deemed unnecessary by firm.

Table 14.2: Number of In-house and Outsourced Services by Firm.

	Firm A	Firm B	Firm C
In-house	26	19	25
Partially outsourced	7	32	19
Fully outsourced	72	40	44

Note: Services are defined as follows: (1) in-house if it is provided by the case study firm itself or its affiliates, (2) fully outsourced if it is provided fully by third-party providers and (3) partially outsourced if it is provided partially by the firm its affiliates and partially by third-party providers.
Source: Authors.

etary technology and/or critical to ensure that the products meet the highest standards of quality required by the client firms.

Firm B further added that it initially tried to outsource plating activities. However, it found that the cost was very high and the quality of plating was not up to the standards required by the firm. Firm decided to undertake its own plating activities and, at the same time, provides plating services to other manufacturers.

Other reasons for keeping certain services in-house, according to Firm C, include the use of common IT portal and sales reporting template across subsidiaries.

On outsourced services, Firm A is unique (when compared to Firms B and C) in that it fully outsources majority of its non-core activities to third-party providers which include numerous business processes such as accounting and human resources. While it is not surprising to observe the outsourcing of these activities, it is not often that one finds firms that fully instead of partially outsource them (see Box 1 for some of the motivations behind these decisions). Firm A disclosed that two staff members from its third-party provider of human resources management are based on-site as though it were its own HR department. Confidentiality agreements are signed between the firm and its third-party providers to prevent leakage of corporate information (e.g., salaries it offers) considering that these providers also service other firms.

Box 14.1. Third-party Providers of Non-core Activities

Operating in a foreign economy is a challenge for many firms because it requires extensive knowledge of that economy's regulations in different areas such as company registration, customs and labour. Often, the amount of resources allocated to understanding them are so massive that the firm decides either not to set up in the economy or if it has set up, to revise its timeline from establishment to operation. Realizing these opportunities, many third-party providers are now offering services whereby they will handle the non-core but necessary functions of the firm while leaving the firm to focus on core activities such as production and quality control. Among the possible benefits of this arrangement is increased efficiency and productivity of foreign firms' operations.

One provider of these services in Mexico is the North American Production Sharing, Inc. (NAPS). It has serviced more than 100 firms including one case study firm in this report since its establishment in 1991. As can be seen in Table 14.3 below, the range of services provided by NAPS is very broad and can be tailored to suit the needs of individual client firm. At the same time, these services allow its client firms to focus on improving efficiency in its production and operations.

Table 14.3: Sample Functions Provided by NAPS.

Area	Activities
Pre-establishment support	• Site selection • Organizational structure
Human resources	• Personnel recruitment • Supervision
Accounting	• Tax compliance • Payroll processing
Customs	• Regulatory compliance • Shipping
Environmental	• Certifications • Audits

Due to the nature of outsourced services in general and the extent through which they are outsourced, many of these services are bundled. For instance, one of the third-party providers actually handles activities ranging from company registration and obtaining the necessary government licensing to recruitment and legal matters. Another example where bundling is observed is in the supply of raw materials to the firm. When they are delivered to the firm, these raw materials are already in processed form in that the suppliers have performed basic machining and fabrication works on them. In addition, transport/logistics provider rarely provides only point-to-point transportation services but also customs clearance, freight insurance and even loading and unloading of materials inputs.

14.5. POLICIES AFFECTING THE VALUE CHAIN

A critical component of this study is the analysis of how policies influence the value chain and identify areas of improvements in the process. The firms identified a number of policy issues in different areas including human capital and labour mobility.

14.5.1. *Frequent Revision of Tax Laws and Regulations*

In an effort to minimise tax evasion and at the same time boost tax revenues, the Mexican government has been revising its tax laws and regulations, with the last reform being signed by the President in December 2013 and changes generally effective 1 January 2014. Firm B has been trying to comply with these revisions but indicated that the constant revisions have made it challenging for its staff to understand and follow because time and effort are required to comply with the laws and regulations. There were instances when Firm B thought that it has complied with everything only to find that the laws and regulations have been revised yet again.

An example of a recently instituted regulation is the one pertaining to the requirement to maintain electronic accounting records and submission of general ledger to the tax authorities on a monthly basis. In addition, digital invoices of transactions must be issued by all taxpayers including the firm (Deloitte, 2015; Ernst & Young, 2014). Although it has good intent, the government appeared to have been unprepared for the wealth of information that they received. Overwhelmed with data, the government is trying to change the regulations again.

14.5.1.1. *Recent changes to IMMEX*

The increased level of competition at the global markets motivated the Mexican government to publish the Decree for the Promotion of the Manufacturing, Maquila and Export Services Industry (IMMEX Decree) in November 2006. One of the main objectives of the Decree is to increase the competitiveness of its export sector by simplifying the compliance regime and reducing firms' operational costs. Among the defining characteristics of the IMMEX Program is one that allows maquiladoras to temporarily import material inputs, machinery and equipment without paying the general import tax, value added tax (VAT) and in some cases, countervailing duties, as long as these inputs are intended for use in the production, transformation and repair of goods for exports.[10] Firms only need to pay tax if the finished products are sold domestically in Mexico.

The 2013/2014 tax reform, however, has included a revision of the IMMEX Program. Specifically, the VAT and excise tax exemptions on the temporary import of goods have been abolished. This means that VAT rate of 16 per cent have to be paid upfront at point of entry. The maquiladora then needs to obtain certification from the Mexican tax authorities. If it is certified, the VAT that is technically imposed on the imported goods will be eliminated by a full tax credit (Deloitte, 2014).

[10] This is applicable on the condition that imported goods are used to make the finished products and exported out within 18 months from the time of entry of inputs (PwC Mexico, 2014a).

The revision essentially changes government's fundamental take on the activities of maquiladoras by assuming that finished products are sold domestically and it is firms' responsibility to prove otherwise. As such, the process of getting reimbursement becomes very bureaucratic and firms have to produce many documents to get its reimbursement request processed. The process is so tedious that Firm B has to hire a full-time additional headcount just to handle reimbursement matters. Firm further added that although it was stated on paper that firms with certification need to wait for a certain number of days to get reimbursed, firm has waited beyond that to obtain its reimbursement.

Interestingly, the revision of the IMMEX Program also has unintended consequences on entities such as Firm A. The fact that majority of Firm A's clients are based domestically means that it has to eventually pay tax for the finished products that it supplies to these clients and hence does not benefit from the program. However, the way the IMMEX Program was originally implemented essentially means that the firm could reduce its cash outflows for taxation purposes during importation stage, hence allowing it to explore opportunities such as purchasing more productive resources including raw materials. The recent revision, unfortunately, means that it can no longer do so.

14.5.1.2. *Reimbursement pertaining to construction*

On a related note of bureaucratic reimbursement process, Firm B also indicated that the construction of its facility was taxed, and although it can be waived, many documents are needed to obtain the waiver of construction tax.

14.5.2. *Minimal and Conflicting Incentives for Investment Promotion*

Many governments have rolled out the red carpet to attract foreign investments into their economies. The Mexican government is not an exception. The IMMEX Program is one such program and in addition to the benefits mentioned in the previous section, there are also income tax benefits including an additional tax deduction which is equal to 47 per cent of specific benefits given to employee including contributions to pension and retirement funds, overtime payments, Christmas bonuses, food coupons and those pertaining to vacation (Deloitte, 2014).

It also introduced Sectoral Promotion Programs (PROSEC) in order to promote the manufacture of products in certain industry such as electrical and electronics, chemicals as well as automotive.[11] Under PROSEC, eligible firms can import goods such as material inputs and capital assets at preferential general import duty rates if they would be used in the manufacturing process. Unlike IMMEX where the finished products must be exported, PROSEC is indifferent as to whether the firm exports or not.

The government has also set up the Integral Foreign Trade Information System (SIICEX) website to facilitate access to information such as laws and regulations and most importantly, trade treaties and agreements that have been concluded by Mexico. It should be acknowledged that the market potential made possible by North American Free Trade Agreement (NAFTA) has served as a strong incentive for firms to base their facilities in Mexico.

Despite these incentives, Firm B noted that it has received minimal support from the Mexican government to set up its manufacturing facility. So far, it has only received travel grants for the firm's Mexican engineers to attend training in Japan.

14.5.3. *Availability of Qualified Suppliers and Human Capital*

The manufacturing of automotive components usually requires several processes which are shown in Table 14.4. Depending on various reasons such as economies of scale and availability of equipment, firms tend to outsource some of these processes to external providers. However, Firm A shared that so far, it only

---

[11] For a complete listing of industry covered by PROSEC, please refer to: http://www.economia.gob.mx/industry/foreign-trade-instruments/prosec.

Table 14.4: Possible Processes Involved in the Manufacturing of Different Automotive Parts.

Suspension system	Body	Braking systems
• Machining	• Die-cutting	• Machining
• Forging	• Sheet metal	• Die-cutting
• Welding	• Electrostatic paint	• Forging
• Pressure injection	• Surface treatment	• Welding
• Thermal treatment	• Plastics	• Sintering
• Surface treatment	• Welding	• Thermal treatment
		• Surface treatment
Transmission system	Traction system	Cooling system
• Machining	• Machining	• Machining
• Forging	• Forging	• Die-cutting
• Die-cutting	• Die-cutting	• Sheet metal
• Pressure injection	• Welding	• Welding
• Sintering	• Pressure injection	• Mechanic assembly
• Mechanic assembly	• Mechanic assembly	• Thermal treatment
• Thermal treatment	• Thermal treatment	• Surface treatment
	• Surface treatment	• Plastics
Steering system	Safety system	Electric and electronic components
• Pressure injection	• Machining	• Machining
• Mechanic assembly	• Die-cutting	• Die-cutting
• Surface treatment	• Mechanic assembly	• Welding
	• Thermal treatment	• Mechanic assembly
	• Surface treatment	• Plastics
	• Plastics	
Exhaust system	Rims and tires	Fuel consumption
• Machining	• Machining	• Machining
• Die-cutting	• Die-cutting	• Extrusion
• Welding	• Sheet metal	• Sheet metal
• Pressure injection	• Thermal treatment	• Welding
• Mechanic assembly	• Surface treatment	• Surface treatment
• Thermal treatment		• Die-cutting
• Surface treatment		• Stamping
		• Plastics

Source: Promexico.

outsourced painting activities to a local firm in Mexico. The other processes are either performed in-house or were carried out by suppliers of its raw materials in Japan and Korea. The same can be said for Firm B which outsources minimal processes locally. One of the primary reasons for these decisions is the absence of qualified providers for such processes in Mexico. All three firms also noted that it is a challenge to obtain good quality inputs locally despite continuous attempts to do so.

Pertaining to the hiring of staff, Firm A noted that while it faced no significant problem in hiring, it still needs to send its employees for additional training to bring them to the level required by the firm. Indeed, since establishment, firm has organised several training courses for supervisors and operators in Japan. The areas of training for supervisors are very broad and range from human resources and labour relations to quality control and plant management. In addition, firm also provides on-the-job training for its operators through the use of skill maps, a chart-like tool whereby the level of technical skills that each staff has attained is displayed to assist in the training process.

On the other hand, Firm B mentioned that it encounters challenges in recruiting managers with the right qualifications in Mexico. The fact that many firms are setting up in the area compounds the difficulty from competition in recruiting talent. While the main reason for firm to have five Japanese staff in its facility is to ensure that everything starts off properly, it is also partially due to firm's inability to find competent managers locally. Eventually, Firm B hopes to reduce the number of foreign staff considering that it is also not easy to persuade them to relocate to Mexico even with strong incentives.

Firm C, it appears, has similar personnel difficulties faced by both Firms A and B, especially because it is seeking approval from its headquarters to set up an R&D department in Mexico which can adjust its products to the local conditions. As a supplier to automotive firms, Firm C believes that having an R&D facility in Mexico would go a long way in facilitating collaborations in product development with automotive firms. But in its headquarters reckoning, the relatively limited availability of skilled labour may put Mexico at a disadvantage as a regional R&D centre relative to other neighbouring economies like Brazil and Chile.

The Mexican government recognises the presence of education-industry gap and in an effort to minimise it, has created Colegio Nacional de Educacion Profesional Tecnica (CONALEP)/National College of Technical Education whose objective is to be a centre for training, evaluation and certification of skills required by the industry.[12] Since its inception in 1978, CONALEP has been initiating reforms to ensure that its educational model remains consistent and relevant to industry's needs. Today, CONALEP is a federalised institution comprising of 30 state colleges located in major cities and industrial areas in Mexico. Despite the progress, there have been feedbacks that CONALEP's curriculum can be further improved in order to produce technicians for tier 1 and 2 automotive suppliers.

14.5.4. *International Labour Mobility*

Ley Federal del Trabajo/The Federal Labour Law is the primary law governing employment in Mexico since 1970s.[13] One of the Articles of the Law that usually pose a challenge to most firms is Article 7, which requires that at least 90 per cent of a firm's employees must be Mexican nationals except for directors, administrators and other manager-level employees.[14] Case study firms indicated that they generally do not face any issues in adhering to the Article

[12] More information about CONALEP can be obtained at: http://www.conalep.edu.mx/qspropuesta/Paginas/default.aspx (CONALEP, 2015).

[13] More information about labour and employment laws in Mexico can be obtained at: http://www.naalc.org/migrant/english/pdf/mgmexfwg_en.pdf (Commission for Labor Cooperation, 2015) and http://www.acc.com/accdocket/online exclusives/foreign-nationals-mexico.cfm?makepdf=1 (Schiaffino, 2015).

[14] The Federal Labour Law was last revised substantially in 2012 to promote business competitiveness (https://www.dlapiper.com/en/singapore/insights/publications/2012/12/mexico-enacts-amendments-to-federal-labor-law/). However, the 90 per cent rule appears unchanged.

and the Law as a whole because they bring in far less foreign workers than they are allowed.

However, there are situations when adherence may be a challenge. One such instance is when Firm A needed to install a new or overhaul its production line. Depending on the complexity of the tasks, firm may need to host a significant number of Japanese engineers for an extended period which creates the possibility that the share of Mexican employees could temporarily fall below the 90 per cent threshold level. To overcome this problem, Firm A only applies working visa for several of its foreign engineers and asks the rest of the team to enter Mexico on Visitor Permit.[15] Yet, this solution is not foolproof and comes with its own set of restrictions and challenges.

First, holders of Visitor Permit are technically not allowed to undertake paid activities in Mexico and this means that engineers who come in on Visitor Permit are not eligible for all the benefits that come along with having a working visa including being paid for the services they render. Second, the Visitor Permit is only valid for up to 180 days and because it cannot be renewed, holders will need to leave Mexico upon its expiry. If the tasks take more than 6 months to complete, engineers with Visitor Permit have to exit Mexico and possibly return to Japan before re-entering Mexico again. This raises the cost of the firm and adds additional burdens to the engineers considering the travel distance between Mexico and Japan.

An alternative solution would be to change the status of the engineers from being holders of Visitor Permit to that of Visa de Residente Temporal/Temporary Resident Visa but this cannot be done domestically in Mexico.[16] As a solution, Firm A sends its engineers to go to the Mexican Consulates in the nearest economy, the United States, to apply for the change in status.

---

[15]More information about Mexico's immigration policies can be obtained at: http://www.mexperience.com/living/immigration-mexico.php (Mexican Visas and Immigration, 2015).

[16]There are two exceptions where a foreign national can exchange his or her Visitor Permit to Resident Visa in Mexico: (1) if the foreign national has a close family in Mexico; and (2) if the foreign national applies for residency on humanitarian grounds.

Although none of the firms in this case study has encountered this before, other sources mentioned that it may no longer be possible to do so in the Mexican Consulates located in economies neighbouring Mexico. Applicants have to return to their home economy and collect them at the Mexican Consulates there. To add to the regulatory burden, while issuance of Temporary Resident Visa is done by Mexican Consulates, applicants still need to register at the local immigration office within 30 days upon arrival in Mexico so as to acquire the visa card.

14.5.5. *Logistics, Infrastructure and Security Concerns*

Logistics plays an important role in the smooth functioning of the GVCs. It ensures that material inputs reach the firms' facility in time for manufacturing activities to be carried out and thereafter, allows products to be sent to customers. Realizing the importance of transportation and logistics for Mexico, the Transport and Communications Investment Program 2013–2018 was launched by the President in 2013 and total investment was estimated at USD 100 billion (PwC Mexico, 2014b). As an indication that the investment has improved the economic infrastructure, Firm B noted that ports in Mexico are not as saturated as those in the United States such as Los Angeles and that, in fact, during the recent labour strike in Los Angeles ports, the firm exported products from Japan to the United States via Mexico. This triangulation proved to be cheaper and more efficient compared with airfreighting the products directly to the United States.

Despite the improvements over the years, however, Mexico ranks at the bottom three among the top 10 car-producing economies in the world in terms of overall LPI score as well as selected logistics index components such as infrastructure, international shipments and logistics quality and competence (Table 14.5). Trucking products from Mexico to the Mexico–United States border remains expensive, perhaps partly due to protected competition in freight transport.[17]

---

[17] Domestic transportation of freight (with the exception of messenger or package delivery services) is reserved for Mexican nationals only.

Table 14.5. Logistics Performance Index and Selected Components in the Top 10 Car-producing Economies in the World (2013).

No.	Economy	Overall LPI	Infrastructure	International shipments	Logistics quality and competence
1	China	3.53	3.67	3.50	3.46
2	United States	3.92	4.18	3.45	3.97
3	Japan	3.91	4.16	3.52	3.93
4	Germany	4.12	4.32	3.74	4.12
5	Korea	3.67	3.79	3.44	3.66
6	India	3.08	2.88	3.20	3.03
7	Brazil	2.94	2.93	2.80	3.05
8	Mexico	3.13	3.04	3.19	3.12
9	Thailand	3.43	3.40	3.30	3.29
10	Canada	3.86	4.05	3.46	3.94

Source: Ranking of top 10 car-producing economies in the world in 2013 is obtained from PwC Mexico report entitled "Doing Business in Mexico — Automotive Industry". The scores for overall LPI, infrastructure, international shipments and logistics quality and competence are obtained from World Bank (2014, http://d21a6b425f3bbaf58824-9ec594b5f9dc5376fe36450505ae1164.r12.cf2.rackcdn.com/LPI_Report_2014.pdf). The overall LPI score is the weighted average of the scores of an economy on six key dimensions: (1) efficiency of the clearance process by border control agencies including customs, (2) quality of trade- and transport-related infrastructure, (3) ease of arranging competitively priced shipments, (4) competence and quality of logistics services including transport operator and customs brokers, (5) ability to track and trace consignments and (6) timeliness of shipments in reaching destination within scheduled or expected delivery time.

Besides infrastructure and logistics limitations, security concerns in some of the ocean-bordering states such as Nayarit, Jalisco, Colima, Michoacán and Guerrero can also affect the firms' activities.[18] An example is the inability of Firm C to gain access to its inputs at the port due to closure of customs in Jalisco a few months ago as a result of attack by gangsters. The tightening of security has also slowed down transportation of goods since there are more inspections by the police and military on the highways. It does not

[18] More details on the state of security in these states can be obtained at: http://travel.state.gov/content/passports/english/alertswarnings/mexico-travel-warning.html (Bureau of Consular Affairs, 2015).

help that the number of highways linking the ports to Firm C's facility are limited.

14.5.6. *Moving Forward*

This case study has attempted to give the perspectives of firms manufacturing different automotive components on the importance of services to its operations in Mexico. It has also identified the policies that are impacting its operations and hence, its overall competitiveness. Without doubt, policies and, consequently, policymakers have a significant role in supporting businesses, specifically in ensuring that their own economies have the right environment in helping businesses to grow.

The inadequate availability of qualified providers of manufacturing-related processes limits firms' outsourcing abilities. Likewise, the relatively inadequate supply of skilled labour could upset possible plans of expansion such as building R&D departments. While the firms provide training to its employees and on-the-job training is the norm in most firms, minimizing the gap between education and industry would still help the firms by reducing the resources that are currently allocated to staff training and use the savings to explore options to improve its operations.

Most importantly, this specific case study has shown that government regulations and procedures can have large implications on the firms. A case in point is the IMMEX program whose revision of tax collection period essentially meant bigger cash outflow for the firm upfront when raw materials arrive at the Mexican port compared to before the revision. Other cases include the fact that Visitor Permit could not be renewed domestically in Mexico and the change of status from being holders of Visitor Permit to that of Temporary Resident Visa has to be done in Mexican Consulates located in the economy of origin. Firms would benefit if policies take into considerations the policies' impacts on affected firms.

APPENDIX

Summary of Services Utilised and Provided by All Three Firms

Table A.1: Establishment Phase.

Service		Corresponding CPC ver. 2 code	Firm A	Firm B	Firm C
1	Business liaison services (e.g., services from trade development centres or SME centres)	9113 — Public administrative services related to the more efficient operation of business	Fully outsourced, efficiency; economies of scale	Fully outsourced, lack of expertise	
2	Company registration and licensing services	91138 — Public administrative services related to general economic, commercial and labour affairs	Fully outsourced, efficiency; economies of scale; government services; strong relationship with government agencies	Partially outsourced, government services; strong relationship with government agencies	
3	Information and statistical services	9113 — Public administrative services related to the more efficient operation of business	Fully outsourced, efficiency; economies of scale	Partially outsourced, government services	
4	Visa and immigration services for foreign investors/ employees (if any)	91290 — Public administrative services related to other public order and safety affairs	Fully outsourced, efficiency; economies of scale; government services; strong relationship with government agencies	Fully outsourced, strong relationship with government agencies; government services	

(Continued)

Table A.1: (*Continued*)

Service		Corresponding CPC ver. 2 code	Firm A	Firm B	Firm C
5	Business consultant services	8311 — Management consulting and management services	Fully outsourced, efficiency; economies of scale; lack of expertise		
		83129 — Other business consulting services	Fully outsourced, efficiency; economies of scale; lack of expertise	Fully outsourced, lack of expertise	
6	Banking and finance services	71121 — Deposit services to corporate and institutional depositors	Fully outsourced, efficiency; economies of scale; not possible to supply in-house	Fully outsourced, not possible to supply in-house	
		71135 — Non-mortgage loan services for business purposes	Fully outsourced, efficiency; economies of scale; not possible to supply in-house	Fully outsourced, not possible to supply in-house	
7	Legal services	82130 — Legal documentation and certification services	Fully outsourced, efficiency; economies of scale; lack of expertise	Fully outsourced, lack of expertise	

Source: Authors.

Table A.2: Pre-manufacturing Stage Including Sourcing and Importation of Raw Materials.

Service	Corresponding CPC Ver. 2 Code	Firm A	Firm B	Firm C
8 Product research and development	81129 — Research and experimental development services in other engineering and technology	In-house	Partially outsourced, lack of expertise	In-house
	81400 — Research and development originals	In-house	In-house	In-house
	83700 — Market research and public opinion polling services	In-house		In-house
9 Product design services	83920 — Design originals	In-house	In-house	In-house
	83912 — Industrial design services	In-house	In-house	In-house
10 Engineering services	83310 — Engineering advisory services	In-house	In-house	In-house
	8332 — Engineering services for specific projects	In-house	In-house	In-house

(*Continued*)

Table A.2: *(Continued)*

Service	Corresponding CPC ver. 2 code	Firm A	Firm B	Firm C
11 Sales and marketing services	83700 — Market research and public opinion polling services		In-house	In-house
	8311 — Management consulting and management services (may include customer relationship management)		In-house	In-house
12 Procurement services	83116 — Supply chain and other management consulting services	In-house	Partially outsourced, efficiency; lack of expertise	In-house
	85999 — Other support services n.e.c.			In-house
13 Customs clearance services and logistics of raw materials	67110 — Container handling services	Fully outsourced, efficiency; economies of scale; strong relationship with government agencies; government services	Fully outsourced, efficiency; strong relationship with government agencies; government services	Fully outsourced, efficiency; strong relationship with government agencies; government services

		85999 — Other support services n.e.c.	Fully outsourced, efficiency; economies of scale; strong relationship with government agencies; government services	Fully outsourced, efficiency; strong relationship with government agencies; government services	Fully outsourced, efficiency; strong relationship with government agencies; government services
14	Technical testing of raw materials	83441 — Composition and purity testing and analysis services	Partially outsourced, raw materials providers	Partially outsourced, lack of expertise/ equipment; not possible to supply in-house	Partially outsourced, not possible to supply in-house due to lack of equipment
15	Transport services of raw materials	651 — Land transport services of freight	Fully outsourced, efficiency; economies of scale	Fully outsourced, efficiency; economies of scale	Fully outsourced, efficiency; economies of scale
		652 — Water transport services of freight	Fully outsourced, efficiency; economies of scale	Fully outsourced, efficiency; economies of scale	Fully outsourced, efficiency; economies of scale
		6531 — Air transport services of freight	Fully outsourced, efficiency; economies of scale	Fully outsourced, efficiency; economies of scale	Fully outsourced, efficiency; economies of scale
		67910 — Freight transport agency services and other freight transport services	Fully outsourced, efficiency; economies of scale	Fully outsourced, efficiency; economies of scale	Fully outsourced, efficiency; economies of scale

(*Continued*)

Table A.2: (*Continued*)

Service		Corresponding CPC ver. 2 code	Firm A	Firm B	Firm C
16	Freight insurance of raw materials	71333 — Freight insurance services	Fully outsourced, efficiency; economies of scale; not possible to supply in-house	Fully outsourced, not possible to supply in house	Fully outsourced, not possible to supply in house
17	Storage and warehousing services of raw materials	67290 — Other storage and warehousing services	Partially outsourced, efficiency; economies of scale	Partially outsourced, leased warehouse for overcapacity	In-house
18	Training services for staffs	92919 — Other education and training services, n.e.c.	In-house	Partially outsourced, lack of expertise	Partially outsourced, not possible to supply in-house; need access to best services
19	Construction services	54 — Construction services	Fully outsourced, lack of expertise; need access to best services	Fully outsourced, lack of expertise; need access to best services	
20	Installation services of equipment	873 — Installation services (other than construction)	In-house	Partially outsourced, lack of expertise; by equipment suppliers	Partially outsourced, lack of expertise; by equipment suppliers

Source: Authors.

Table A.3: Manufacturing Stage.

Service	Corresponding CPC ver. 2 code	Firm A	Firm B	Firm C
21 Production administration — production management	83115 — Operations management consulting services	In-house	In-house	In-house
22 Maintenance and repair of factory equipment	87156 — Maintenance and repair services of commercial and industrial machinery	Partially outsourced, lack of expertise; by equipment suppliers	Partially outsourced, lack of expertise; by equipment suppliers	Partially outsourced, to equipment suppliers
23 Utilities (electricity, gas and water supply)	691 — Electricity and gas distribution (on own account)	Fully outsourced, not possible to supply in-house	Fully outsourced, not possible to supply in-house	Fully outsourced, not possible to supply in-house
	692 — Water distribution (on own account)	Fully outsourced, not possible to supply in-house	Fully outsourced, not possible to supply in-house	Fully outsourced, not possible to supply in-house
24 Water purification services	8344 — Technical testing and analysis services		In-house	
25 Manufacturing services provided in-house and by suppliers of activities such as basic machining	886 — Basic metal manufacturing services	Fully outsourced, efficiency; economies of scale; lack of in-house equipment; services provided as part of raw materials	Usually in-house but may be outsourced for specific cases (e.g., if customers order non-finished products)	Partially outsourced, comes together with raw materials that are in processed form

(Continued)

Table A.3: (*Continued*)

Service	Corresponding CPC ver. 2 code	Firm A	Firm B	Firm C
	887 — Fabricated metal product, machinery and equipment manufacturing services	Fully outsourced, efficiency; economies of scale; lack of in-house equipment; services provided as part of raw materials	Usually in-house but may be outsourced for specific cases (e.g., if customers order non-finished products)	Partially outsourced, comes together with raw materials that are in processed form
	885 — Rubber, plastic and other non-metallic mineral product manufacturing services			Partially outsourced, comes together with raw materials that are in processed form
26 Engineering services during manufacturing	83310 — Engineering advisory services	In-house	In-house	In-house
	8332 — Engineering services for specific projects	In-house	In-house	In-house
27 Warehousing services for intermediate goods	67290 — Other storage and warehousing services	Partially outsourced, efficiency; economies of scale	Partially outsourced, leased warehouse for overcapacity	In-house

28	Quality control and assurance as well as compliance with ISO	8344 — Technical testing and analysis services	In-house	Partially outsourced, lack of expertise/equipment; not possible to supply in-house	In-house
29	Product testing to obtain certification at export market	8344 — Technical testing and analysis services	Partially outsourced, customer requirement; performed by customers	Partially outsourced, lack of expertise/equipment; not possible to supply in-house	Partially outsourced, required by some customers
30	Sewage water treatment services	94110 — Sewerage and sewage treatment services	Fully outsourced, lack of expertise; performed by industrial park operator	In-house	Fully outsourced, lack of expertise; performed by estate manager
		943 — Waste treatment and disposal services	Fully outsourced, lack of expertise; performed by industrial park operator	Partially outsourced, for hazardous waste (i.e., those containing oil); required by laws and regulations	Fully outsourced, lack of expertise; performed by estate manager
31	Specialised cleaning services for machines and equipment	85340 — Specialised cleaning services	In-house	Partially outsourced, lack of expertise; by equipment suppliers	Partially outsourced, to equipment suppliers

(Continued)

Table A.3: *(Continued)*

Service	Corresponding CPC ver. 2 code	Firm A	Firm B	Firm C
32 Waste collection and recycling services	894 — Materials recovery (recycling) services, on a fee or contract basis	Fully outsourced, lack of expertise	Fully outsourced, lack of expertise	
	942 — Waste collection services	Fully outsourced, lack of expertise	Fully outsourced, lack of expertise	Fully outsourced, lack of expertise; performed by estate manager

Source: Authors.

Table A.4: Post-manufacturing Stage.

Service	Corresponding CPC ver. 2 code	Firm A	Firm B	Firm C
33 Packaging services	83919 — Other specialty design services	In-house	Partially outsourced, may be provided by logistics firms or firm's customers	Partially outsourced, required by some customers to use certain logistics company which also provides packaging services
	85400 — Packaging services	In-house	In-house	Partially outsourced, required by some customers to use certain logistics company which also provides packaging services
34 Warehousing services for products	67290 — Other storage and warehousing services	Partially outsourced, efficiency; economies of scale	Partially outsourced, leased warehouse for overcapacity	In-house
35 Customs clearance services and logistics of products	67110 — Container handling services	Fully outsourced, efficiency; economies of scale; strong relationship with government agencies; government services	Partially outsourced, efficiency; strong relationship with government agencies; government services	Fully outsourced, efficiency; strong relationship with government agencies; government services

(Continued)

Table A.4: (*Continued*)

Service	Corresponding CPC ver. 2 code	Firm A	Firm B	Firm C
	85999 — Other support services n.e.c.	Fully outsourced, efficiency; economies of scale; strong relationship with government agencies; government services	Partially outsourced, efficiency; strong relationship with government agencies; government services	Fully outsourced, efficiency; strong relationship with government agencies; government services
36 Transport services of products	651 — Land transport services of freight	Fully outsourced, efficiency; economies of scale	Fully outsourced, efficiency; economies of scale	Fully outsourced, efficiency; economies of scale
	652 — Water transport services of freight	Fully outsourced, efficiency; economies of scale	Fully outsourced, efficiency; economies of scale	Fully outsourced, efficiency; economies of scale
	6531 — Air transport services of freight	Fully outsourced, efficiency; economies of scale	Fully outsourced, efficiency; economies of scale	Fully outsourced, efficiency; economies of scale
	67910 — Freight transport agency services and other freight transport services	Fully outsourced, efficiency; economies of scale	Fully outsourced, efficiency; economies of scale	Fully outsourced, efficiency; economies of scale
37 Freight insurance of products	71333 — Freight insurance services	Fully outsourced, efficiency; economies of scale; not possible to supply in-house	Fully outsourced, not possible to supply in house	Fully outsourced, not possible to supply in house

Source: Authors.

Table A.5: After-sales Services.

Service	Corresponding CPC ver. 2 code	Firm A	Firm B	Firm C
38 Travel services for engineers and other staffs pertaining to after-sales services	8551 — Reservation services for trans portation	Fully outsourced, efficiency; economies of scale		
	85521 — Reservation services for accommodation	Fully outsourced, efficiency; economies of scale		
39 Accommodation services	6311 — Room or unit accommodation services for visitors	Fully outsourced, efficiency; economies of scale		
	63220 — Room or unit accommodation services for workers in workers hostels or camps	Fully outsourced, efficiency; economies of scale		
40 Visa and immigration services for staffs	91290 — Public administrative services related to other public order and safety affairs	Fully outsourced, efficiency; economies of scale; government services; strong relationship with government agencies		
41 Warehousing services for after-sales inventories	67290 — Other storage and warehousing services	In-house	In-house	
42 Inventory management services	83116 — Supply chain and other management consulting services	In-house	In-house	In-house

Source: Authors.

Table A.6: Business Processes (Back Office Support).

Service	Corresponding CPC ver. 2 code	Firm A	Firm B	Firm C
43 Company registration and licensing services (obtaining permit to operate)	91138 — Public administrative services related to general economic, commercial and labour affairs			Fully outsourced, government services
44 Government licensing and inspections on fire prevention, health hazards, environmental protection and other aspects	91133 — Public administrative services related to mining and mineral resources, manufacturing and construction	Fully outsourced, efficiency; economies of scale; government services; lack of expertise	Partially outsourced, lack of expertise	Fully outsourced, government services
	91290 — Public administrative services related to other public order and safety affairs	Fully outsourced, efficiency; economies of scale; government services; lack of expertise	Partially outsourced, lack of expertise	Fully outsourced, government services
45 Headquarter services	83118 — Head office services	In-house	In-house	In-house
46 Management services	83111 — Strategic management consulting services	In-house		In-house

47	Accounting, auditing and bookkeeping services	82210 — Financial auditing services	Fully outsourced, efficiency; economies of scale; required by laws and regulations	Fully outsourced, required by laws and regulations	Fully outsourced, required by laws and regulations
		8222 — Accounting and bookkeeping services	Fully outsourced, efficiency; economies of scale	In-house	In-house
48	Financial services	71121 — Deposit services to corporate and institutional depositors	Fully outsourced, efficiency; economies of scale; not possible to supply in-house	Fully outsourced, not possible to supply in-house	Fully outsourced, not possible to supply in-house
		71140 — Financial leasing services	Fully outsourced, efficiency; economies of scale; not possible to supply in-house	Fully outsourced, not possible to supply in-house	Fully outsourced, not possible to supply in-house
		71313 — Group pension services	Fully outsourced, efficiency; economies of scale; not possible to supply in-house	Fully outsourced, not possible to supply in house	Fully outsourced, not possible to supply in-house

(*Continued*)

Table A.6: (*Continued*)

Service	Corresponding CPC ver. 2 code	Firm A	Firm B	Firm C
	83112 — Financial management consulting services	Fully outsourced, efficiency; economies of scale; not possible to supply in-house		Fully outsourced, not possible to supply in-house
49 Legal services	82120 — Legal advisory and representation services concerning other fields of law	Fully outsourced, efficiency; economies of scale; lack of expertise	Fully outsourced, lack of expertise	Fully outsourced, lack of expertise
	82130 — Legal documentation and certification services	Fully outsourced, efficiency; economies of scale; lack of expertise	Fully outsourced, lack of expertise	Fully outsourced, lack of expertise
50 Insurance services (commercial life and accident/health insurance, property insurance for the factory compound, product quality insurance, management liability insurance)	7131 — Life insurance and pension services	Fully outsourced, efficiency; economies of scale; not possible to supply in-house	Fully outsourced, better portfolio; not possible to supply in-house	Fully outsourced, not possible to supply in-house

	7132 — Accident and health insurance services	Fully outsourced, efficiency; economies of scale; not possible to supply in-house	Fully outsourced, better portfolio; not possible to supply in-house	Fully outsourced, not possible to supply in-house
	71334 — Other property insurance services	Fully outsourced, efficiency; economies of scale; not possible to supply in-house	Fully outsourced, better portfolio; not possible to supply in-house	Fully outsourced, not possible to supply in-house
	71335 — General liability insurance services	Fully outsourced, efficiency; economies of scale; not possible to supply in-house	Fully outsourced, better portfolio; not possible to supply in-house	Fully outsourced, not possible to supply in-house
	714 — Reinsurance services	Fully outsourced, efficiency; economies of scale; not possible to supply in-house		
51 Human resources services	91320 — Administrative services related to government employee pension schemes; old-age disability or survivors' benefit schemes, other than for government employees	Fully outsourced, efficiency; economies of scale; lack of expertise; government services	Partially outsourced, lack of expertise; government services	Partially outsourced, government services

(Continued)

Table A.6: *(Continued)*

Service	Corresponding CPC ver. 2 code	Firm A	Firm B	Firm C
	91330 — Administrative services related to unemployment compensation benefit schemes	Fully outsourced, efficiency; economies of scale; lack of expertise; government services		Partially outsourced, government services
	8511 — Personnel search and referral services	Fully outsourced, efficiency; economies of scale	Partially outsourced, efficiency	Partially outsourced, efficiency; cost consideration
	8512 — Labour supply services	Fully outsourced, efficiency; economies of scale; network economies	Partially outsourced, efficiency; network economies	Partially outsourced, network economies
	83113 — Human resources management consulting services	Fully outsourced, efficiency; economies of scale	In-house	Partially outsourced, lack of expertise
52 Business and management consultancy services	8311 — Management consulting and management services	In-house		Fully outsourced, lack of expertise

53	Corporate communications and public relationship	83114 — Marketing management consulting services	In-house		In-house
		83121 — Public relations services	In-house	In-house	
54	Courier, postal and local delivery services	681 — Postal and courier services	Fully outsourced, not possible to supply in-house	Fully outsourced, not possible to supply in-house	Fully outsourced, not possible to supply in-house
55	Information technology (IT) services	8313 —IT consulting and support services	Fully outsourced, lack of expertise	Partially outsourced, lack of expertise	Partially outsourced, lack of expertise
		8314 —IT design and development services	Fully outsourced, lack of expertise	Partially outsourced, lack of expertise	Partially outsourced, lack of expertise
		83151 — Website hosting services	Fully outsourced, lack of expertise	Partially outsourced, lack of expertise	In-house
		8316 — IT infrastructure and network management services	Fully outsourced, lack of expertise	Partially outsourced, lack of expertise	Fully outsourced, lack of expertise

(*Continued*)

Table A.6: *(Continued)*

Service	Corresponding CPC ver. 2 code	Firm A	Firm B	Firm C
56 Telecommunication services	841 — Telephony and other telecommunications services	Fully outsourced, not possible to supply in-house	Fully outsourced, not possible to supply in-house	Fully outsourced, not possible to supply in-house
	84210 — Internet backbone services	Fully outsourced, not possible to supply in-house	Fully outsourced, not possible to supply in-house	Fully outsourced, not possible to supply in-house
	8422 — Internet access services	Fully outsourced, not possible to supply in-house	Fully outsourced, not possible to supply in-house	Fully outsourced, not possible to supply in-house
57 Uniform design, alteration and laundry	83919 — Other specialty design services	In-house	Partially outsourced, cost consideration	In-house
	97130 — Other textile cleaning services	Fully outsourced, efficiency; economies of scale	Fully outsourced, for specific clothings; cost consideration	
58 Transport services for staffs	641 — Local transport and sightseeing transportation services of passengers	Fully outsourced, efficiency; economies of scale	Fully outsourced, efficiency; cost consideration	

59	Property management services	72212 — Non-residential property management services on a fee or contract basis	Partially outsourced, efficiency; lack of expertise	Fully outsourced, efficiency cost consideration	Fully outsourced, efficiency; lack of expertise
60	Real estate services	72112 — Rental or leasing services involving own or leased non-residential property	In-house		
61	Medical services	93121 — General medical services	Fully outsourced, lack of expertise		Partially outsourced, lack of expertise
62	Catering services	63393 — Other contract food services	Fully outsourced, lack of expertise		Fully outsourced, lack of expertise
63	Security services	85230 — Security systems services	Fully outsourced, lack of expertise	Partially outsourced, lack of expertise	Fully outsourced, lack of expertise
		85250 — Guard services	Fully outsourced, lack of expertise; network economies	Partially outsourced, lack of expertise; network economies	Fully outsourced, lack of expertise; network economies
64	Cleaning services	853 — Cleaning services	Fully outsourced, lack of expertise		Fully outsourced, lack of expertise

Source: Authors.

REFERENCES

Alvarez, L, T Escribano and KT Hardy (2012). Mexico enacts amendments to Federal Labor Law. DLA Piper, 3 December 2012. https://www.dlapiper.com/en/singapore/insights/publications/2012/12/mexico-enacts-amendments-to-federal-labor-law/ [8 July 2015].

Bureau of Consular Affairs (2015). Mexico Travel Warning. U.S. Department of State. http://travel.state.gov/content/passports/english/alerts warnings/mexico-travel-warning.html [8 July 2015].

Commission for Labor Cooperation (2015). Foreign Worker's Guide to Labor and Employment Laws in Mexico. http://www.naalc.org/migrant/english/pdf/mgmexfwg_en.pdf [8 July 2015].

CONALEP (2015). What is the CONALEP — National College of Technical Professional Education. http://www.conalep.edu.mx/English/Paginas/default.aspx [8 July 2015].

CQI-11 Plating System Assessment (2015). TopQM-Systems. http://cqi-support.de/en/cqi-standards/cqi-11 [14 July 2015].

Deloitte (2015). Mexico Highlights 2015. Mexico City: Deloitte, 2015. http://www2.deloitte.com/content/dam/Deloitte/global/Documents/Tax/dttl-tax-mexicohighlights-2015.pdf [8 July 2015].

Deloitte (2014). Taxation and Investment in Mexico 2014. Mexico City: Deloitte, 2014. http://www2.deloitte.com/content/dam/Deloitte/global/Documents/Tax/dttl-tax-mexicoguide-2014.pdf [8 July 2015].

Ernst & Young (2014). Mexican tax authorities issue rules for electronically filing accounting records. 11 July 2014. http://www.ey.com/GL/en/Services/Tax/International-Tax/Alert--Mexican-tax-authorities-issue-rules-for-electronically-filing-accounting-records [14 July 2015].

ISO/TS 16949:2009. ISO. http://www.iso.org/iso/catalogue_detail?csnumber=52844 [14 July 2015].

Mexican Automotive Industry Association (2015). Mexican Automotive Industry — PowerPoint Presentation for Mexico Automotive Summit on 24 September 2015. http://mexicoautomotivesummit.com/highlights.html [8 July 2015].

Mexican Visas and Immigration (2015). Mexperience. http://www.mexperience.com/living/immigration-mexico.php [8 July 2015].

Nissan Motor Corporation (2015). Quality initiatives. http://www.nissan-global.com/EN/QUALITY/PRODUCTS/GLOBAL_SP/GUARANTEE/ [14 July 2015].

North American Production Sharing, Inc. http://www.napsintl.com/ [1 September 2015].

ProMexico (2013). The Auto Parts Industry. Mexico City: ProMexico, May 2013. http://www.oesa.org/Doc-Vault/Knowledge-Center/Intl-Markets-and-Trade-Content/ProMexico-Auto-Parts.pdf [8 July 2015].

PwC Mexico (2014a). Doing Business in Mexico — Automotive Industry. Mexico City: PwC, September 2014. http://www.pwc.de/de/internationale-maerkte/assets/doing-business-mexico-automotive.pdf [8 July 2015].

PwC Mexico (2014b). Insights of Transportation & Logistics Sector in Mexico. Mexico City: PwC, September 2014. http://www.pwc.com/es_MX/mx/knowledge-center/archivo/2014-09-transportation-and-logistics.pdf [8 July 2015].

Schiaffino, M (2015). Important Considerations When Employing Foreign Nationals in Mexico. Association of Corporate Counsel. http://www.acc.com/accdocket/onlineexclusives/foreign-nationals-mexico.cfm?makepdf=1 [9 July 2015].

World Bank (2014). Connecting to Compete — Trade Logistics in the Global Economy. Washington. World Bank. http://d21a6b425f3b-baf58824-9ec594b5f9dc5376fe36450505ae1164.r12.cf2.rackcdn.com/LPI_Report_2014.pdf [8 July 2015].

Chapter 15

MANUFACTURING OF TELECOMMUNICATIONS EQUIPMENT

Huani Zhu and Gloria O. Pasadilla

15.1. INDUSTRY OVERVIEW

Telecommunications equipment manufacturers refer to firms that manufacture hardware and devices related to telecommunications such as modems, circuit-switches systems, routers and base transceiver stations. In the early days, telecoms equipment mainly referred to those equipment used in a telephone network. Nowadays, the term includes a much broader range of equipment with more sophisticated devices and more embedded functions as technology evolves.

As shown in Fig. 15.1, main participants in the telecommunications industry include component suppliers, telecoms equipment makers, telecommunications network service providers, customers and regulators. Telecoms equipment manufacturer's source raw materials from component suppliers, conduct their own research and development (R&D), create and design their devices and establish their distribution channels. This case study focuses on a Chinese telecoms equipment manufacturer based in Shenzhen, China, which has a growing international business.

Telecoms equipment manufacturing industry is dynamic — it has experienced robust growth in the past decade driven by factors such as economic growth, technological innovation, increased

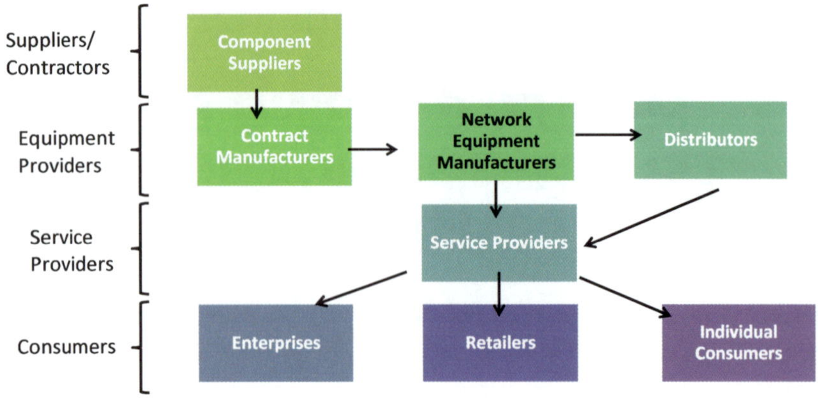

Figure 15.1: Telecommunications Industry Supply Chain.
Source: Adapted from Reyes *et al.* (2002).

demand for communication services, and growth of internet and wireless communication. At the same time, transformation has been taking place towards greater usage of wireless network, especially on mobile handsets. Telecoms equipment makers also evolved from simply manufacturing switches to developing their own-developed equipment and integrating their service offerings with their products. The industry is relatively concentrated with only a few established players (see Fig. 15.2) primarily because of high capital investment, in particular in R&D which is critical for anticipating technological changes and maintaining its market position.

The business model of telecoms equipment vendor is highly technology driven. Firms compete through their technology innovation and capacity for sustained deployment of new products and services. Furthermore, the shortened life span of telecoms products brings additional pressure on equipment manufacturers to coordinate research, design and production in order to ensure a sustainable product portfolio. The industry has also evolved towards greater emphasis on services especially because of the software-based functions of hardware telecommunication products. In addition, new product creation also gives rise to new opportunities to develop new services in various contexts, such as integration services and management of big data.

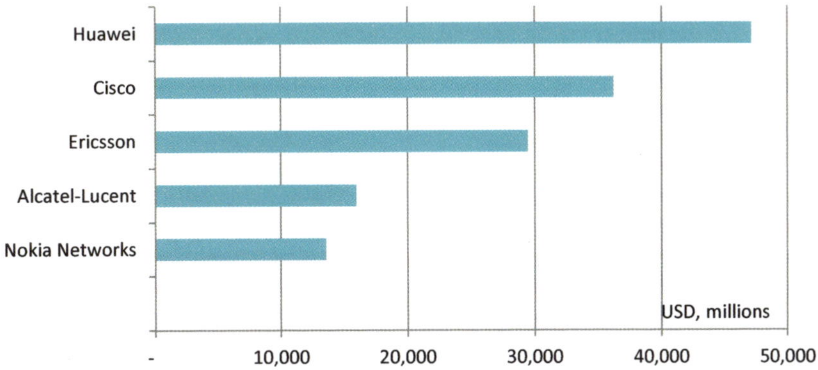

Figure 15.2: Top Telecoms Equipment Manufacturers Based on FY 2014 Revenue.
Source: Company annual reports; Infonetics Research; IMF IFC database, APEC Policy Support Unit estimates.

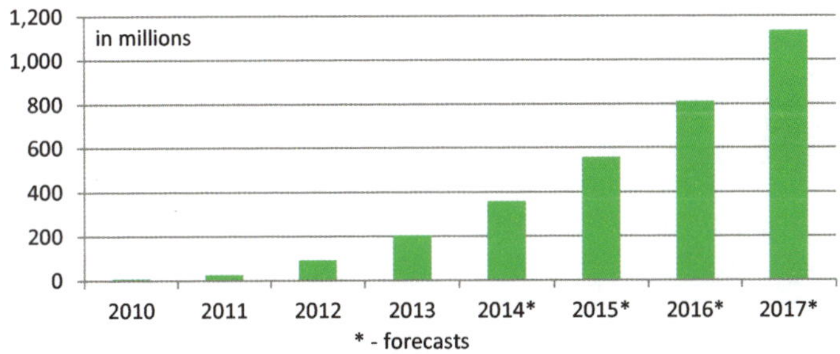

Figure 15.3. Number of 4G Long Term Evolution (LTE) Connections Worldwide.
Source: The Mobile Economy 2014, GSMA.

Global telecommunications market is growing fastest in Asia Pacific[1] and other emerging markets. Moreover, the focus of investments has started to shift towards construction and optimization of mobile broadband as the deployment of 3G and 4G networks is expected to reach greater depths[2] (see Fig. 15.3).

[1]In 2009, Asia Pacific accounted for the world's largest share of telecommunications investments, surpassing North America and Western Europe.

[2]Cisco Visual Networking Index: "Global Mobile Data Traffic Forecast Update, 2014–2019".

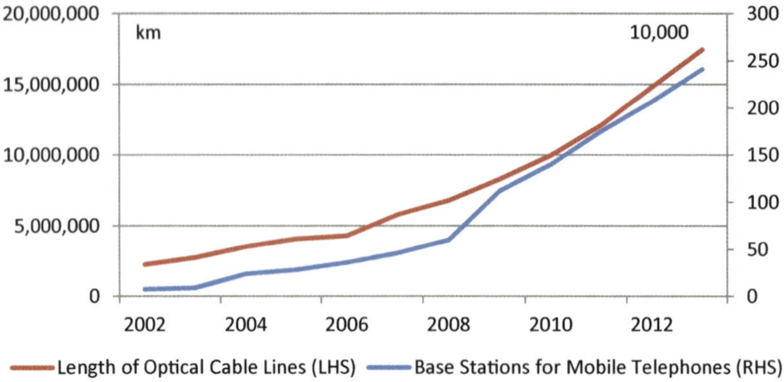

Figure 15.4: Indicators of Telecommunication Capacity of China.

Source: China Statistical Yearbooks.

For the case study firm that is discussed subsequently, growth of the Chinese telecommunications market is important since it obtains roughly half of its revenue from the domestic market. Its telecoms equipment sales growth benefited from investments made by telecoms service network providers. These investments are evident in improved telecommunications infrastructure and capacity as shown in Fig. 15.4.

15.2. BACKGROUND INFORMATION ON THE FIRM[3]

The case study firm is an integrated telecoms equipment manufacturer with a significant presence in both the Chinese domestic market as well as the international market. The predecessor of the firm was established in 1985 based on companies under the former Ministry of Aerospace Industry. The firm started as an original equipment manufacturer by taking in orders to manufacture small electronic goods. In the late 1990s, the case study firm was restructured as a publicly listed enterprise and decided to specialise in telecoms equipment industry. As of 2014, the same state-owned

[3]Information on the firm has been sourced from its corporate website and annual reports.

corporation remained the largest shareholder of the case study firm with around 31 per cent of the shareholdings.

Headquartered in Shenzhen, the firm has more than 100 subsidiaries globally across 6 continents. Majority of the subsidiaries are sales offices of its telecommunications products, distribution points of contact and engineering services stations. The firm employs roughly 65,000 staff globally, with around 60,000 based in China. Its global production sites include China, India, Pakistan, Venezuela, Indonesia and the United States. Locating factories in developing economies with relatively cheap labour give the firm competitive advantage. Locating close to the local market likewise helps it minimise logistic expenses.

The firm leveraged its growth on a strong emphasis on R&D. The firm has consistently invested around 10 per cent of its revenue in R&D, transforming strategic research direction into product development ideas. It established 18 R&D centres in China, France, the United States and Sweden with over 30,000 research professionals employed — representing more than one-third of total employees. It also actively collaborates with Chinese research institutes and universities and participates in some of technology research projects funded by the government. According to patent filling statistics from the World Intellectual Property Organization (WIPO), the firm has always been one of the top three enterprises that filed the most patents since 2010.

With products and services being used in more than 160 economies, the firm's principal activities include designing, developing, producing, distributing, installing and maintaining a wide range of telecoms equipment and systems. As shown in Table 15.1, its products fall into three broad categories: (i) carriers' network; (ii) handset terminals and (iii) telecoms software systems, services and other products. Products under carriers' networks, with the highest operating profit margin, have consistently been contributing around half of the total revenue.

The firm has sustained a strong foothold in the Chinese telecoms equipment market for several reasons. First, it has long established relationships with leading telecoms providers in China.

Table 15.1: Selected Products and Services Offered by the Firm.

Handset	Terminal	Smart Phone; Pad; Data Card; Convergence Terminal
Carriers' Networks	Wireless	GSM systems; CDMA systems; 4G LTE; Base Stations
	Fixed Access	Integrated Service Access Platform; DSL Systems
	Core Network	Voice Communication; Packet Core; Convergence User Data
	Bearer	Optical Transmission; Data Communication; Routers; Ethernet Switch
Telecoms Software Systems, Services and Others	Cloud Computing and IT	Cloud Computing Infrastructure; Operating Support Systems; Server; Storage
	Services	Maintenance Support Services; IT Integration Services
	Energy and Infrastructure	Telecoms Power System; Telecoms Tower; UPS (Inverter Power Supply)

Source: Company website and annual reports.

Second, it has developed a broad range of products based on common technology platforms with integrated end-to-end solutions, leveraging on its technical know-how and R&D. Third, it has maintained a low-cost structure evidenced by its stable operating profit margins in major segments over the years.

For the purpose of identifying a specific value chain, this study focuses on a set of products under the core network group that provide voice communication services, integrated core network gateway and centralised management of systems. The firm enjoys great success in their LTE network products by signing more than 140 LTE/EPC contracts around the world, breaking industry performance records and winning award for its product at the LTE

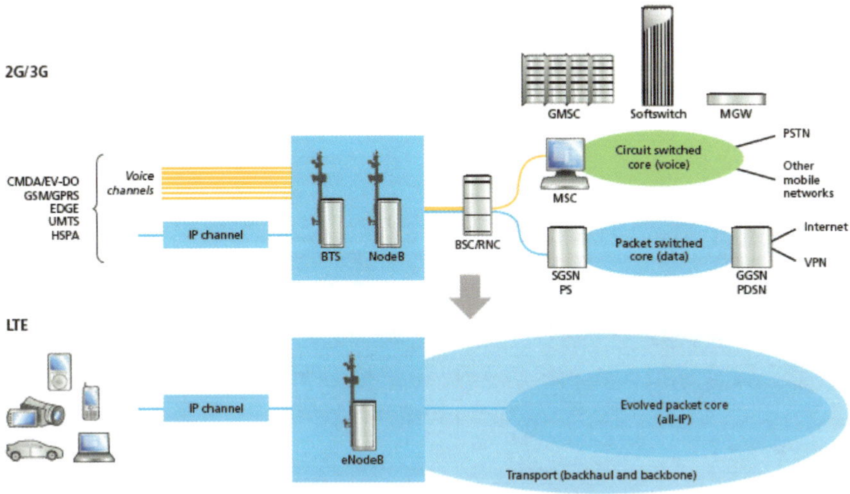

Figure 15.5: Evolution from Separate Circuit Switched (CS) and Packet Switched (PS) Core Sub Domains to One Common IP Core.

Source: Courtesy of Alcatel Lucent.

World Summit in 2014.[4] Figure 15.5 demonstrates the architecture of evolved packet core (EPC), a new all-IP mobile core network for the LTE that unifies two distinct mobile core sub-domains into a single IP domain.

15.3. DESCRIPTION OF THE VALUE CHAIN

Telecoms equipment providers normally sell their products by grouping them for an end-to-end solution to enable them to gain extra revenues from services, installation and deployment. In general, the case study firm has separate production lines for its products under carriers' networks and handset terminals as the products from the former category rely more on R&D and bear a greater proportion of core technologies. This study focuses on a group of wireless communications products under the carriers' networks category. These products have consistently generated

[4] Based on the firm's press releases; 26th June 2014.

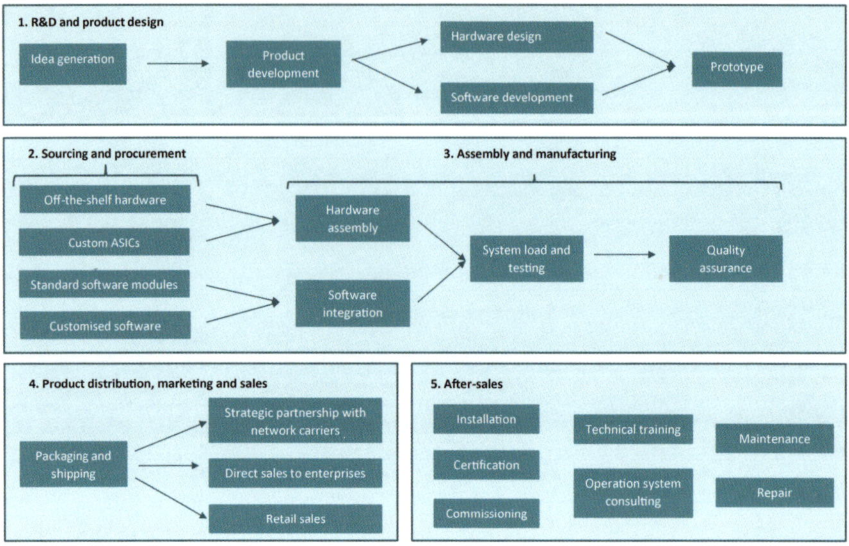

Figure 15.6: The Case Study Firm's Value Chain.

Source: Authors.

more than half of the firm's revenues. As the rapid deployment of 4G networks and the firm's active involvement in pre-5G and 5G development, these wireless products also provide enormous potential for the firm going forward.

Figure 15.6 presents the supply chain of the firm's wireless products. It involves five main stages, viz. — (i) R&D and product design; (ii) sourcing and procurement; (iii) assembly and manufacturing; (iv) product distribution, marketing and sales; and (v) after-sales services.

15.3.1. *R&D and Product Design*

The firm's supply chain starts from R&D. Products are conceptualised mainly from two sources: international standards (global system for mobile communications (GSM), core-division multiple access (CDMA) systems; 4G LTE) and interaction with clients, including telecom carriers around the globe. The firm noted that there are increasing numbers of idea generation coming from introduction of international standards. However, far from a passive recipient of international standards, the firm has become actively involved in the establishment of industry standards. Such engagement ultimately helps the firm to develop products that meet industry standards in a timely and cost-effective manner.

The firm's 18 R&D centres around the globe are highly specialised in a way that each of them concentrates on a specific product family. These R&D centres collaborate via technology sharing and transfer. In general, those R&D centres based in developed economies focus more on developing technological innovation since these economies are at the technological frontier. This knowledge is then transferred back to the R&D centres based in China, which focus more on product development. The case study firm always retains its own core technologies protected by patents in order to maintain and enhance its competitive strengths.

The case study firm starts from modifying the existing products as technology development in telecoms equipment industry continuously evolves. After an idea is generated, the firm will develop product architecture, design software and hardware, create prototype and conduct testing. When a concept of a new product is developed, the firm gathers a team from different divisions to evaluate the feasibility of the product before formally launching a project. The feasibility study includes input from technical assessment, financial analysis and market research. Decisions for the project to proceed are made by high-level management in every stage during the product development process.

The firm has an in-house team taking care of patent application and acquisition since the firm needs to retain its core competency by maintaining confidentiality. The same team is also in charge of licensing services for the right to use trademarks, franchises and other intellectual property products from other companies. For instance, when the firm's core network products use software developed by another company or to use a particular telecoms technology, it needs to obtain licenses.

15.3.2. *Sourcing and Procurement*

Despite having separate production lines for carriers' networks products and handset terminals, the firm has a centralised procurement policy for both units in order to gain economies of scale. Required inputs for manufacturing carriers' networks product typically include: off shelf hardware, standard software modules, custom software and custom application-specific integrated circuits (ASICs). Electrical and electronic components like circuit boards and ASICs are the main raw materials the firm uses to manufacture their products.

The firm has developed and formulated a set of internal qualification and bidding procedures to determine a potential supplier's product quality and price before appointing it as an approved supplier. After forecasting the required amount of raw materials and components, the firm invites qualified suppliers to submit their bids. Successful bidders will then enter into a Purchase Framework Agreement which specifies product types, quantities and prices, quality specifications, delivery schedules, locations and modes of delivery and other contract details. When a negotiation with a supplier takes place, the component prices are normally determined on an arm's length basis and on normal commercial terms.

The firm obtains around 20 per cent of its purchases from its five largest suppliers as of 2014. In order to prevent over-reliance on a particular supplier, the firm tries to have at least two suppliers for each type of raw material, with some exception.

Transportation of raw materials is mostly outsourced to third-party logistics providers. Except when the material is supplied in-house, short-distance transportation is supplied either in-house or from affiliated companies.

15.3.3. *Assembly and Manufacturing*

Products are mainly manufactured in-house with production organized based on product lines. While the firm used to manufacture, assemble and test their products primarily in China, it now has an expanding list of overseas manufacturing centres. Besides being nearer to the end market, considering rising Chinese wages, another advantage of establishing overseas production base is cheap labour costs.

For production base located in China, the firm covers most of the activities in-house, from managing warehouse, manufacturing and product testing to supporting activities such as security, maintenance and specialized cleaning of manufacturing equipment. One exception is the mandated tests on products required by the authorities such as US Federal Communications Commission, Certification and Accreditation Administration of China or independent safety science company like Underwriters Laboratories for the UK market. Such tests normally concerns security issues or electromagnetic compatibility or radiation. The firm normally outsources this to a third party. For overseas manufacturing centres, supporting services are largely outsourced to third parties.

Once the manufacturing process is completed, the responsibility of ensuring product quality normally rests on the quality control department based in the Shenzhen headquarters.

15.3.4. *Product Distribution, Marketing and Sales*

The firm designs and packs most of its products in-house while outsources transportation services to its affiliated companies or third-party logistics providers. The firm has an associate[5] that specialises in end-to-end supply chain services including transporting goods and handling customs-related services.

The firm has separate sales channels for its telecoms equipment products and handset terminals since these two product categories are very different in terms of complexity, core technologies involved and target customers. Telecoms equipment is distributed mainly through business-to-business (B2B) channel. For instance, the firm sells its products through telecoms systems contracts. Under such sales arrangement, products from several segments

---

[5] Associate refers to a company for which the case study firm only possess a minority stake in the ownership of the company.

can be sold at the same time and it gives the firm opportunities to provide additional services such as customization, installation and integration.

For core network products, the firm manages sales in-house given its well-established network of sales offices. Such arrangement allows the firm to maintain close relationship with major customers, gain knowledge about customers' demand and subsequently anticipate their needs. Subsidiaries with sales function in various regions participate in the tendering process organized by telecoms services providers on the firm's behalf. These subsidiaries are responsible for negotiations and contracting if their bidding was successful. Similarly, for sales to international markets, the firm's overseas sales offices liaise with local telecoms services providers to analyse their requirements before designing and proposing customized solutions. Ideally the firm prefers to establish direct relationship with the network service providers. In certain jurisdictions where there is restriction on foreign companies' participation in local telecommunications sector, the firm enters into bidding by setting up cooperative arrangements with local partners.

15.3.5. *After-sales Services*

Due to the sophisticated nature of telecoms equipment, after-sales services play an important role for the case study firm. Customer support services that the firm provides include installation, certification and commissioning services, maintenance and repair services, technical training and operation system consulting services. Most of the firm's after-sales services are provided in-house, except in cases when the required service falls outside of the firm's core operation. For example, when the firm needs to build a base station (as part of its product delivery) in a developing economy, it typically outsources the construction to a third-party construction company. For repair services, the decision to supply in-house or outsource depends on the complexity of the problem. For overseas clients, locally based specialists take charge of standard maintenance and simple repair services. For more sophisticated breakdowns, the faulty component is normally shipped back to China while the firm sends the spare parts to the customer. In order to ensure timely repair services, the firm usually keeps spare parts in the warehouse alongside other inventories.

15.4. SERVICES ALONG THE VALUE CHAIN

A total of 72 services have been identified in the case study firm's value chain, which can be divided into the following stages: (i) R&D and product design; (ii) sourcing and procurement; (iii) assembly and manufacturing; (iv) product distribution, marketing and sales; (v) after-sales and (vi) business processes (Fig. 15.7).

It is not difficult to infer that the most important service inputs are the ones in the earliest and the latest stages, viz. (i) R&D and product design; and (v) after-sales. Services that support the R&D and product development activities are vital because telecoms equipment vendors build their core competence and develop their competitive advantage based on technological innovation. After-sales

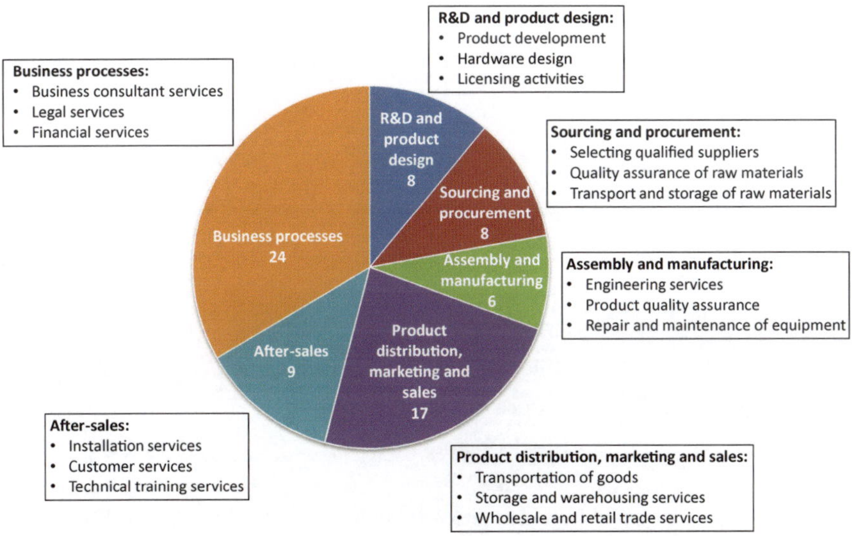

Figure 15.7: Breakdown of Services by Stage.

Source: Authors.

services provide more value-added opportunities for the firm beyond selling the equipment. On the other hand, that the manufacturing stage and business processes consist of the most number of services imply that services are indeed integrated pervasively into its manufacturing value chain.

The importance of services can also be demonstrated from the case study firm's employee headcount. While employees devoted to manufacturing activities constitute around one fifth of total headcount, service-oriented segments such as customer service, marketing and sales make up another 18 per cent and 17 per cent of the workforce, respectively. The largest proportion of employees, around 36 per cent, works on R&D-related activities, which is again heavily reliant on services. In total, service-related employees account for roughly 70 per cent of the firm's total workforce.

The 72 services identified in the value chain can be further disaggregated into 126 detailed services. Out of the 126 services, our analysis estimates that 46 services are supplied in-house, 53 are partially outsourced and 27 are fully outsourced as shown in the Table 15.2.

There are several reasons why the firm choose to outsource its services. In general they can be grouped into the following

Table 15.2: Number of Services by Source.

	Sources of Services	Number of Services (Percentage of Total)
Fully in-house	In-house	35 (28%)
	In-house and affiliated companies	8 (6%)
	Affiliated companies	3 (2%)
Partially outsourced	In-house and third parties	31 (25%)
	Affiliated companies and third parties	11 (9%)
	In-house and affiliated companies and third parties	11 (9%)
Fully outsourced	Third parties	27 (21%)

Source: Authors.

categories: (i) the services lie outside the firm's core operations, such as advertising and utilities supply; (ii) as the firm continuous to expand internationally, it outsources some of the services for its overseas operation, such as storage of products and installation of equipment; (iii) some of the government mandated testing was outsourced; (iv) the firm lacks expertise in providing specialised professional services such as market research; (v) to gain economies of scale in area such as transportation.

Besides its core services in areas such as R&D, product design, patent acquisition, quality assurance, maintenance and repair services, the firm also supplies a considerable amount of supporting services in-house such as packaging, retail trade and human resources. This is partly due to economies of scale given the firm's sizeable presence in its headquarter, which comprises more than 85 per cent of the workforce. However, this may change going forward as Chinese wages continue to rise and the firm face greater pressure to remain competitive.

15.5. POLICIES AFFECTING THE VALUE CHAIN

The following section identifies policies that affect the firm's value chain. While telecoms equipment industry is known for its stringent standards and registration regulation, that is, procedures to follow and requirements to be fulfilled by equipment manufacturers seeking to register their telecoms equipment for sale and use, the firm nevertheless does not experience any challenges in registering or obtaining certifications of their products. Majority of the policy challenges identified by the firm are linked to its international expansion.

15.5.1. *Cyber-security Concerns*

Rapid development of technology and the increasing complexity of the supply chain increased security vulnerabilities given the fact that technology is designed, developed and deployed in and from various geographic locations.

The case study firm identified cyber security as an issue of great concern to them. The firm's effort to expand sales in developed markets such as North America, Europe and Australia has faced setbacks after reports identified foreign, especially Chinese communications equipment, as posing a perceived potential threat to their cyber security, especially if the network is used for carrying government communications. Hence, under a security exemption, either the firm's telecom equipment are prohibited from government procurement in these economies or banned from bidding for government-funded broadband network.

As a result, the case study firm faces considerable obstacle in expanding into certain overseas markets. Although it can sell its handset terminals abroad, its more profitable system products for core network are not welcomed.

15.5.2. *Anti-dumping*

The firm faced anti-dumping investigation from some regulators about alleged unfair competition through its heavily subsidized activities, driving big price wedge between Chinese telecoms equipment producers and their local counterparts. The case was subsequently settled but it dented the firm's confidence in a possible expansion project and launching of more products and services, as well as technology investment in the region.

15.5.3. *Local Content Requirements*

The firm indicated that some emerging economies have protectionist policies in place to force localisation of production. In one ASEAN economy, the regulations require importers of 4G LTE smartphones to produce, starting in 2017, 30 per cent "local content" in the territory. Such regulation implies that, for the case study firm to comply with the regulations, its subsidiary in the economy will have to do the assembly of various components and parts that the company will export to its own subsidiary. Since assembly cost is lower in China, such arrangement would result in higher production costs.

Another pending local content requirement being contemplated in this economy would apply not only to hardware products but also to software and R&D and design development. If this policy is approved, the case study firm is likely to face problems related to the difficulties finding enough qualified researchers locally.

15.5.4. *Intellectual Property*

Safeguarding intellectual property is vital for telecoms equipment manufacturers since this gives them a competitive advantage. Intellectual property disputes between telecoms equipment makers are common because they would use each other's licensed technology in their own products occasionally. In cases when they adopt other's patents, the case study firm pays or establishes agreements with the patent holder. Companies might accuse others of demanding excessive prices for their patents or over-asserting the essentialness of their patents which affect any potential patent infringement decision by the courts.[6]

The case study firm has an in-house legal team dedicated to intellectual property management. Normally the firm establishes licensing agreements with major players in the industry for the rights to use each other's patents in their products. A fair price is agreed upon most of the time since the firm indicated that their royalty payment cost is manageable.

15.5.5. *Labour Mobility*

Due to the complexity of telecoms equipment, engineering resources often need to travel to the customer site to resolve technical issues

[6] The firm identifies Europe as the main market where it faces thorny intellectual property issues. A court case last year led to an injunction restricting the case study firm from selling equipment that could infringe on the disputed patent. Other lawsuits are afoot involving rival firms in 4H and camera technology patents. Continuous legal battles increase the firm's legal costs, affect the firm's corporate image and put the firm in an unfair competition if the anti-competitive antitrust complaint is found to be credible.

as part of its after-sales services. Because of visa issues, the firm cannot make use of their vast pool of experienced technical specialists in China. If engineers from its Chinese headquarters are required, the visa requirements and the length of time waiting for their issuance affect the timeliness of their repair service.

Visa applications also affect the firm's ability to participate in overseas tendering process. For instance, when their potential customer from overseas calls for a tender, bidders need to be present on site. When a project is sizeable, a team with personnel from various divisions is normally required to be present to assist the proposal submission and to demonstrate what the firm is capable of doing. Again, the visa restrictions for Chinese nationals and the uncertainty of obtaining the required visa on time, make it difficult for the case study firm to deploy the necessary number of persons to the tendering, thus hampering their ability to exhibit the firm's full capabilities.

15.6. CONCLUSION

This case study demonstrates the importance of embedded services in the manufacturing of telecoms equipment. Services evidently play a vital role due to the significance of R&D and innovation in the telecoms equipment manufacturing along with the development of more value-added services especially in the after-sales market.

On the policy front, some of the issues in telecommunications are thorny in nature due to the complex interplay of politics and economics. Bilateral, regional or multilateral agreements to open each other's market in telecom and cooperate in R&D and technical standard-setting will foster healthy competition and encourage knowledge transfer.

The restriction on movement of specialised personnel due to visa issues impedes the firm's ability to provide the best after-sales services in a timely manner. This might negatively affect the firm's customer satisfaction level and hinder future business opportunities. From the end user's point of view, such restriction on labour mobility does not best represent their economic interests either because the delay in repair services might be costly to their operation.

Similarly for the protectionist policies such as local content requirement, the final consumers are worse off since those policies push up the production costs which are ultimately passed on to end consumers. Therefore, it is in the economy's own interest to formulate policies that balance between supporting local manufacturing industry and consumers' welfare.

APPENDIX

Services Along the Value Chain

Table A.1: R&D and Product Design.

Service	Corresponding CPC ver. 2 code	Supplied in-house	Outsourced to affiliated companies and reasons	Outsourced to third-party suppliers and reasons	Bundled
1 Product development/ R&D	81129 — Research and experimental development services in other engineering and technology	Yes	Yes (Set up R&D centres in developed economies to take advantage of the frontier technology)	No	
	81400 — Research and development originals	Yes	No	No	
	91114 — Government services to research and development	Yes	No	No	

2	Market research	83700 — Market research and public opinion polling services	Yes	No	Yes (for handset terminals only; reason: efficiency and lack of expertise)
3	Conception and design of product	83920 — Design originals	Yes	Yes	Yes (for handset terminals only; reason: efficiency)
		83939 — Other scientific and technical consulting services n.e.c.	No	No	Yes (only once; reason: lack of expertise)
4	Hardware design	83912 — Industrial design services	Yes	No	No
		83919 — Other specialty design services	Yes	No	No
5	Software development	8314 — IT design and development services	No	No	Yes (reason: efficiency and cost consideration)
6	Patent acquisition	83960 — Trademarks and franchises	Yes	No	No

(Continued)

Table A.1: (*Continued*)

	Service	Corresponding CPC ver. 2 code	Supplied in-house	Outsourced to affiliated companies and reasons	Outsourced to third-party suppliers and reasons	Bundled
7	Licensing services	73330 — Licensing services for the right to use R&D products	Yes	No	No	
		73340 — Licensing services for the right to use trademarks and franchises	Yes	No	No	
		73390 — Licensing services for the right to use other intellectual property products	Yes	No	No	
8	Prototype testing	83443 — Testing and analysis services of integrated mechanical and electrical systems	No	No	Yes (for government mandated testing; reason: efficiency)	

Source: Authors.

Table A.2: Sourcing and Procurement.

Service		Corresponding CPC ver. 2 code	Supplied in-house	Outsourced to affiliated companies and reasons	Outsourced to third-party suppliers and reasons	Bundled
9	Assessing and selecting qualified suppliers (organising tendering, verifying supplier qualification, negotiation)	85999 — Other support services n.e.c.	Yes	No	No	
10	Customer-related services for raw materials imports	67110 — Container handling services	No	No	Yes	
		85999 — Other support services n.e.c.	Yes	Yes	No	

(*Continued*)

Table A.2: (*Continued*)

Service		Corresponding CPC ver. 2 code	Supplied in-house	Outsourced to affiliated companies and reasons	Outsourced to third-party suppliers and reasons	Bundled
11	Raw material management	83190 — Other management services, except construction project management services	Yes	No	No	
12	Quality assurance services of raw materials	83441 — Composition and purity testing and analysis services	Yes	No	No	

13	Transport services of raw materials	6511 — Road transport services of freight	No	No	Yes (reason: economies of scale)
		6512 — Railway transport services of freight	No	No	Yes (reason: economies of scale)
		6521 — Costal and transoceanic water transport services of freight	No	No	Yes (reason: economies of scale)
		6531 — Air transport services of freight	No	No	Yes (reason: economies of scale)

(*Continued*)

Table A.2: (*Continued*)

Service		Corresponding CPC ver. 2 code	Supplied in-house	Outsourced to affiliated companies and reasons	Outsourced to third-party suppliers and reasons	Bundled
14	Repair and maintenance for fleets (if self-owned)	87143 — Maintenance and repair services of trailers, semitrailers and other motor vehicles n.e.c.	Yes	Yes (for short distance transportation only)	No	
15	Freight insurance of raw materials	71333 — Freight insurance services	No	No	Yes (reason: lack of expertise)	
16	Storage of raw materials	67290 — Other storage and warehousing services	Yes	Yes (for overseas operation; reason: cost consideration)	No	

Source: Authors.

Table A.3: Assembly and Manufacturing.

Service		Corresponding CPC ver. 2 code	Supplied In-House	Outsourced to affiliated companies and reasons	Outsourced to third-party suppliers and reasons	Bundled
17	Product management	83115 — Operations management consulting services	Yes	No	No	
		83117 — Business process management	Yes	No	No	
18	Product quality assurance and compliance with standards	83441 — Composition and purity testing and analysis services	Yes	No	Yes (for government mandated testing; reason: efficiency)	
		83443 — Testing and analysis services of integrated mechanical and electrical systems	Yes	No	Yes (for government mandated testing; reason: efficiency)	

(*Continued*)

Table A.3: (*Continued*)

Service		Corresponding CPC ver. 2 code	Supplied In-House	Outsourced to affiliated companies and reasons	Outsourced to third-party suppliers and reasons	Bundled
19	Warehousing services for intermediate goods	67290 — Other storage and warehousing services	Yes	No	Yes (for overseas operation only; reason: efficiency	
20	Installation of production equipment	87320 — Installation services of industrial, manufacturing and service industry machinery and equipment	No	No	Yes (reason: lack of expertise)	
		87360 — Installation services of electrical machinery and apparatus n.e.c.	No	No	Yes (reason: lack of expertise)	

21	Engineering services	83310 — Engineering advisory services	Yes	Yes	Yes
		83325 — Engineering services for telecommunications and broadcasting projects	Yes	Yes	Yes
22	Repair and maintenance services of machines and equipment	87156 — Maintenance and repair services of commercial and industrial machinery	Yes	No	Yes (reason: efficiency)
		87130 — Maintenance and repair services of computers and peripheral equipment	Yes	No	Yes (reason: efficiency)
		87152 — Maintenance and repair services of electrical machinery and apparatus n.e.c.	Yes	No	Yes (reason: efficiency)

Source: Authors.

Table A.4: Product Distribution, Marketing and Sales.

Service		Corresponding CPC ver. 2 code	Supplied in-house	Outsourced to affiliated companies and reasons	Outsourced to third-party suppliers and reasons	Bundled
23	Design of Packages	83919 — Other specialty design services	Yes	No	No	
24	Packaging Services	85400 — Packaging services	Yes	No	No	
25	Cargo handling services for delivery to wholesaler/retailer	67110 — Container handling services	No	Yes	Yes	
26	Customs-related services of products exports	85999 — Other support services n.e.c.	No	Yes	Yes	

27	Transport of goods	6511 — Road transport services of freight	No	Yes	Yes
		6512 — Railway transport services of freight	No	No	Yes
		6521 — Costal and transoceanic water transport services of freight	No	Yes	Yes
		6531 — Air transport services of freight	No	No	Yes
28	Freight insurance	71333 — Freight insurance services	No	No	Yes
29	Storage and warehousing services for finished goods	67290 — Other storage and warehousing services	Yes	No	Yes (for overseas operation only)

(Continued)

Table A.4: (*Continued*)

	Service	Corresponding CPC ver. 2 code	Supplied in-house	Outsourced to affiliated companies and reasons	Outsourced to third-party suppliers and reasons	Bundled
30	Wholesale trade services	6128 — Wholesale trade services on a fee or contract basis, of machinery, equipment and supplies	No	No	Yes (form strategic partnership with network carriers; reason: economics of scale)	
31	Retail trade services: by store	6218 — Non-specialized store retail trade services, of machinery, equipment and supplies	Yes	No	Yes	
		6228 — Specialized store retail trade services, of machinery, equipment and supplies	Yes	No	Yes	

32	Retail trade services: by internet or mail-order	6238 — Mail order or Internet retail trade services, of machinery, equipment and supplies	Yes	Yes	Yes
33	Retail trade services: by other non-store retail trade services	6248 — Other non-store retail trade services, of machinery, equipment and supplies	Yes	No	No
34	Retail services on a fee or contract bases	6258 — Retail trade services on a fee or contract basis, of machinery, equipment and supplies	Yes	No	No

(Continued)

Table A.4: (*Continued*)

Service		Corresponding CPC ver. 2 code	Supplied in-house	Outsourced to affiliated companies and reasons	Outsourced to third-party suppliers and reasons	Bundled
35	Retail operation management	83115 — Operations management consulting services	Yes	No	No	
		83116 — Supply chain and other management consulting services	Yes	No	No	
36	Retail site development for new shops	83911 — Interior design services	Yes	Yes	No	
		Group of 546 — Installation services	Yes	Yes	No	
		Group of 547 — Building completion and finishing services	Yes	Yes	No	
37	Security services (cash delivery)	85240 — Armoured car services	Yes	No	No	
		85250 — Guard services	Yes	No	No	

38	Leasing/rental services	Group of 731 — Leasing or rental services concerning machinery and equipment without operator	No	Yes	Yes
		Group of 732 — Leasing or rental services concerning other goods	No	Yes	Yes
		71140 — Financial leasing services	No	Yes	Yes
39	Retail Advertising	83611 — Full service advertising	No	No	Yes (reason: lack of expertise)
		83620 — Purchase or sale of advertising space or time, on commission	No	No	Yes (reason: lack of expertise)
		83612 — Direct marketing and direct mail services	No	No	Yes (reason: lack of expertise)

Source: Authors.

Table A.5: After-sales.

Service		Corresponding CPC ver. 2 code	Supplied in-house	Outsourced to affiliated companies and reasons	Outsourced to third-party suppliers and reasons	Bundled
40	Installation services of equipment	87340 — Installation services of radio, television and communication equipment and apparatus	Yes	Yes (deploy local technical specialist; reason: efficiency)	Yes (for overseas operation)	
41	Certification and commissioning services of equipment	83443 — Technical testing and analysis services of integrated mechanical and electrical systems	Yes	No	No	
42	Customer services: general and technical support	85931 — Telephone call centre services	Yes	No	No	

43	Customer services: loyalty program	95999 — Other services provided by membership organizations n.e.c.	Yes	No	No
44	Maintenance and repair services of equipment	87153 — Maintenance and repair services of telecommunication equipment and apparatus	Yes	Yes (for overseas operation; reason: efficiency)	No
45	Inventory management of spare parts for replacement	67290 — Other storage and warehousing services	Yes	Yes	Yes
46	Auxiliary services to ensure successful delivery of products	532 — Civil engineering works	Yes	Yes	Yes (for overseas operation; reason: lack of expertise)
		85999 — Other support services n.e.c.	Yes	Yes	Yes

(Continued)

Table A.5: (*Continued*)

Service		Corresponding CPC ver. 2 code	Supplied in-house	Outsourced to affiliated companies and reasons	Outsourced to third-party suppliers and reasons	Bundled
47	Technical training	92919 — Other education and training services, n.e.c.	Yes	No	No	
48	Operating system consulting	8316 — IT infrastructure and network management services	Yes	No	Yes (when the project is highly complexed or the firm is short-staffed)	

Source: Authors.

Table A.6: Business Processes (Support Services).

Service		Corresponding CPC ver. 2 code	Supplied in-house	Outsourced to affiliated companies and reasons	Outsourced to third-party suppliers and reasons	Bundled
49	Company registration, licensing and business liaison services	91133 — Public administrative services related to the more efficient operation of business	Yes	No	No	
50	Government inspections on fire prevention, health hazards, environmental protection and other aspects	91133 — Public administrative services related to mining and mineral resources, manufacturing and construction	Yes	No	No	
		91137 — Public administrative services related to multipurpose development projects	Yes	No	No	

(*Continued*)

Table A.6: (*Continued*)

	Service	Corresponding CPC ver. 2 code	Supplied in-house	Outsourced to affiliated companies and reasons	Outsourced to third-party suppliers and reasons	Bundled
51	Visa and immigration services for employees	91290 — Public administrative services related to other public order and safety affairs	Yes	No	No	
52	Business consultant services	8311 — Management consulting and management services	Yes	Yes	Yes (reason: lack of expertise)	
		83129 — Other business consulting services	Yes	No	Yes (reason: lack of expertise)	
		83118 — Head office services	Yes	No	No	
53	Banking and finance services	71121 — Deposit services to corporate and institutional depositors	Yes	Yes	No	
		71135 — Non-mortgage loan services for business purposes	Yes	Yes	No	

54	Legal services	82130 — Legal documentation and certification services	Yes	No	Yes (reason: lack of expertise)
		82120 — Legal advisory and representation services concerning other fields of law	Yes	No	No
55	Auditing on financial accounts	8221 — Financial auditing services	Yes	Yes	Yes
		8222 — Accounting and bookkeeping services	Yes	Yes	Yes
56	Corporate finance services	71511 — Mergers and acquisition services	Yes	No	Yes
		71521 — Corporate finance and venture capital services	Yes	No	Yes

(Continued)

Table A.6: *(Continued)*

Service		Corresponding CPC ver. 2 code	Supplied in-house	Outsourced to affiliated companies and reasons	Outsourced to third-party suppliers and reasons	Bundled
57	Insurance and pension services	71313 — Group pension services	Yes	Yes	Yes	
		71321 — Accidental death and dismemberment insurance services	Yes	Yes	Yes	
		71322 — Health insurance services	Yes	Yes	Yes	
		71701 — Services of holding equity of subsidiary companies	Yes	Yes	Yes	
58	Insurance services for machinery	71332 — Marine, aviation and other transport insurance services	Yes	Yes	Yes	

59	Corporate communications, marketing and public relationship	83114 — Marketing management consulting servicesf831	Yes	No	No
		83121 — Public relations services	Yes	No	No
60	Courier, postal and local delivery services	6811 — Postal services	No	Yes	Yes
		6812 — Courier services	No	Yes	Yes
		6813 — Local delivery services	No	Yes	Yes
61	Estate management	72112 — Rental or leasing services involving own or leased non-residential property	Yes	No	No

(Continued)

Table A.6: (*Continued*)

Service		Corresponding CPC ver. 2 code	Supplied in-house	Outsourced to affiliated companies and reasons	Outsourced to third-party suppliers and reasons	Bundled
62	Human resources management; Personnel search and referral services for back office staff	83113 — Human resources management consulting services	Yes	No	No	
		8511 — Personnel search and referral services	Yes	No	No	
		8512 — Labour supply services	Yes	No	Yes	
63	IT and information system management, consulting and support, with webpage development	83132 — Information technology (IT) support services	Yes	No	Yes	
		8315 — Hosting and information technology (IT) infrastructure provisioning services	Yes	No	Yes	
		8316 — IT infrastructure and network management services	Yes	No	Yes	

64	Safety and security services	85230 — Security systems services	Yes	No	No
		85250 — Guard services	Yes	No	Yes (for overseas operation only)
65	Telecommunication services	Group of 841 — Telephony and other telecommunications services	Yes	No	Yes
		Group of 842 — Internet telecommunication services	Yes	No	Yes
66	Installation services for equipment and related wiring	873 — Installation services (other than construction)	Yes	No	Yes

(*Continued*)

Table A.6: (*Continued*)

Service		Corresponding CPC ver. 2 code	Supplied in-house	Outsourced to affiliated companies and reasons	Outsourced to third-party suppliers and reasons	Bundled
67	Utilities supply: electricity, gas and water	8631 — Support services to electricity transmission and distribution	Yes	No	Yes	
		86320 — Gas distribution services through mains (on a fee or contract basis)	No	No	Yes	
		86330 — Water distribution services through mains (on a fee or contract basis)	No	No	Yes	
		86340 — Distribution services of steam, hot water and air-conditioning supply through mains (on a fee or contract basis)	No	No	Yes	

68	Worker-related services: catering , dormitory, medical services, retail services, transportation services	63393 — Other contract food services	Yes	No	Yes
		63220 — Room or unit accommodation services for workers in workers hostels or camps	Yes	No	No
		93121 — General medical services	No	No	Yes
		6212 — Non-specialized store retail trade services, of food, beverages and tobacco	No	No	Yes
		96520 — Sports and recreational sports facility operation services	Yes	No	Yes
		64114 — Local special-purpose scheduled road transport services of passengers	No	No	Yes

(Continued)

Table A.6: (*Continued*)

Service	Corresponding CPC ver. 2 code	Supplied in-house	Outsourced to affiliated companies and reasons	Outsourced to third-party suppliers and reasons	Bundled
69 Social insurance for factory workers	91320 — Administrative services related to government employee pension schemes; old-age disability or survivors' benefit schemes, other than for government employees	Yes	No	Yes	
	91330 — Administrative services related to unemployment compensation benefit schemes	Yes	No	Yes	

70	Gardening of the premises; general and specialized cleaning services for machines and equipment	85970 — Landscape care and maintenance services	No	No	Yes (reason: economies of scale)
		85330 — General cleaning services	No	No	Yes (reason: economies of scale)
		85340 — Specialized cleaning services	Yes	No	Yes
71	Sewage water treatment services	94110 — Sewerage and sewage treatment services	No	No	Yes (reason: economies of scale)
72	Waste collection and recycling services	9421 — Collection services of hazardous waste	Yes	No	Yes
		94239 — General waste collection services, other	Yes	No	Yes
		94229 — Collection services of non-hazardous recyclable materials, other	Yes	No	Yes

Source: Authors.

REFERENCES

Alcatel Lucent (2009). Introduction to evolved packet core. http://www3.alcatel-lucent.com/wps/DocumentStreamerServlet?LMSG_CABINET=Docs_and_Resource_Ctr&LMSG_CONTENT_FILE=White_Papers/Intro_EPC_wp_0309.pdf [June 2015].

Cisco (2015). Cisco visual networking index: global mobile data traffic forecast update, 2014–2019. http://www.cisco.com/c/en/us/solutions/collateral/service-provider/visual-networking-index-vni/white_paper_c11-520862.html [June 2015].

Deals Street Asia (2015). Indonesia smartphones requires 30% local content starting 2017. http://www.dealstreetasia.com/stories/indonesia-smartphones-requires-30-local-content-starting-2017-8925/ [July 2015].

Icegate Solutions Inc. (2007). Telecommunications equipment value chain study. http://profnik.moodlehub.com/pluginfile.php/24570/mod_resource/content/1/Background_Documents/0107861e.pdf [May 2015].

Infonetics Research Scorecard (2013). Telecommunications equipment vendor leadership scorecard. https://www.infonetics.com/download.asp?id=37 [June 2015].

Reuters (2013). U.S. law to restrict government purchases of Chinese IT equipment. http://www.reuters.com/article/2013/03/28/us-usa-cybersecurity-espionage-idUSBRE92Q18O20130328 [July 2015].

Reuters (2015). Indonesia plays hardball with smartphone manufacturers. http://www.reuters.com/article/2015/04/02/us-indonesia-mobile-phone-manufacturers-idUSKBN0MT2GF20150402 [July 2015].

Reyes, P, M Raisinghani and M Singh (2002). Global supply chain management in the telecommunication industry: the role of information technology in integration of supply chain entities. *Journal of Global Information Technology Management*. http://business.baylor.edu/Pedro_Reyes/publications/Global%20supply%20chain%20management%20in%20the%20telecommunications%20industry.pdf [June 2015].

The Independent (2012). China telecoms giant could be cyber-security risk to Britain. http://www.independent.co.uk/news/uk/politics/china-telecoms-giant-could-be-cybersecurity-risk-to-britain-8420432.html [July 2015].

Wall Street Journal (2014). EU, China complete deal on telecoms equipment. http://www.wsj.com/articles/eu-china-reach-deal-on-telecoms-equipment-1413797081 [July 2015].

Wikinvest, Networking & Communication Equipment. http://www.wikinvest.com/industry/Networking_%26_Communication_Equipment [June 2015].

Chapter 16

MANUFACTURING OF PRINTED CIRCUIT BOARDS IN CANADA

Ben Shepherd

16.1. INDUSTRY OVERVIEW

Printed circuit boards (PCBs) are used in a wide variety of electrical equipment and electronics goods, including consumer electronics such as computers. PCBs provide mechanical support for electronics components, and connect them using conductive tracks and other features. The essence of producing a PCB is to take a conductive material, typically copper (see also the case of aluminium for quantum computing applications in the following paragraphs), and laminate it onto a non-conductive substrate. PCB features are then etched onto the copper using chemicals. Figure 16.1 shows a typical small PCB.

World exports of PCBs amounted to $39.5bn in 2014, with Asia as the main centre of export production, in particular China (Fig. 16.2). Importing is more geographically dispersed, with Europe and North America figuring prominently, but again Asia remains dominant (Fig. 16.3). Extensive two-way trade in similar products therefore characterises the market.

PCBs are electronics components that serve as inputs into a large number of value chains, both local and global in scope. PCB manufacturers occupy a mid-range position in terms of typical electronics value chains. They are downstream from designers, who create the patterns of circuitry that manufacturers etch onto the PCBs. But they are upstream from final assemblers, who bring together PCBs with electronics components such as integrated circuits, resistors and capacitors to create finished goods.

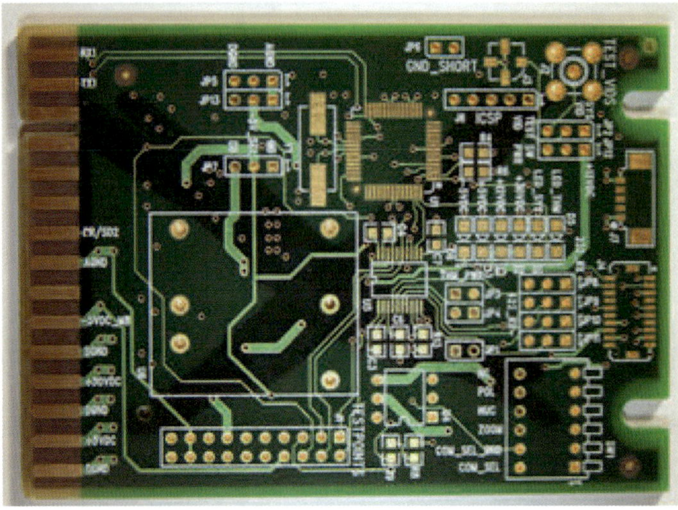

Figure 16.1: A Typical Small PCB.

Source: Case study firm.

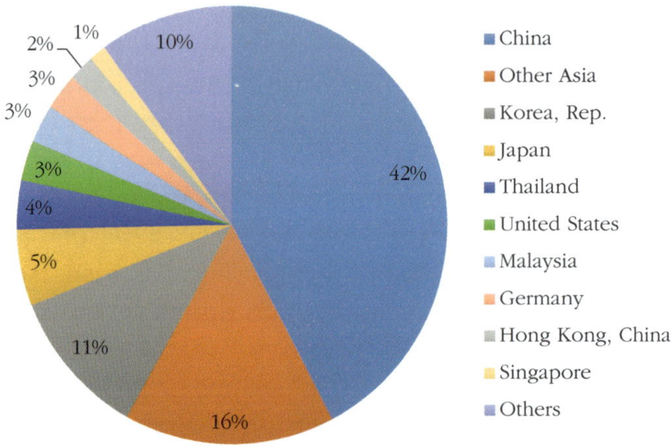

Figure 16.2: Market Shares of the Top 10 PCB Exporters, 2014, Per cent.

Source: UN Comtrade via World Integrated Trade Solution (WITS); author's calculations.

PCBs are widely used products, but although demand is increasing overall (Fig. 16.4),[1] production has steadily been shifting from developed to developing economies, particularly in China. In

[1] The 2014 figure is likely understated due to reporting lags in WITS.

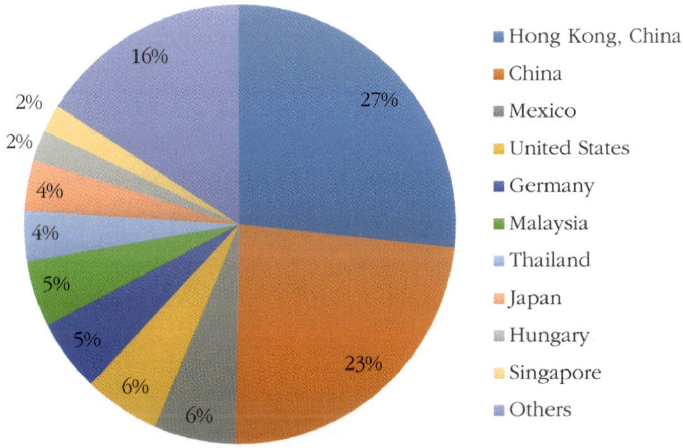

Figure 16.3: Market Shares of the Top 10 PCB Importers, 2014, Per cent.

Source: UN Comtrade via WITS; author's calculations.

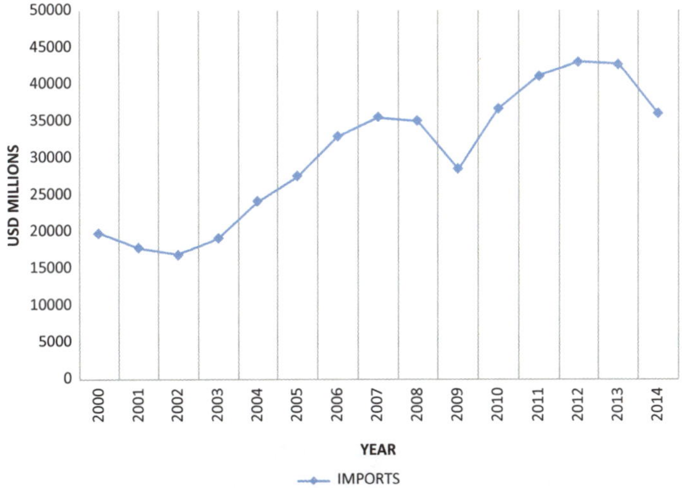

Figure 16.4: Total World Imports of PCBs, 2000–2014.

Source: UN Comtrade via WITS; author's calculations.

North America, for example, 257 out of a total of 567 PCB manufacturers have closed over the last 10 years. The industry is widely regarded as being in the sunset stage in most segments, so production costs and lead times are important determinants of commercial success. There is also a perception in some quarters that more stringent environmental regulations[2] in developed economies are in part responsible for the migration of the industry toward locations where environmental rules are not enforced with the same rigor.

16.2. BACKGROUND INFORMATION ON THE FIRM

Against this background, the case study firm, a Canadian company, has been doing business since 1985 in Richmond, British Columbia. Having started in supporting the graphic arts industry, it branched into PCB production during its first two years of operations. It now works out of a 7000 sq.ft. facility near Vancouver International Airport. All production equipment is owned by the firm as a historical legacy, not leased.

The firm is a small business, currently employing 10 staff. Market conditions are difficult, as noted previously: over the last five years, sales have decreased at a rate of about 6 per cent. However, the company's website is very effective in generating sales: orders increased by 57 per cent year on year through this channel. The case study firm specialises in three lines of business. The first is the production runs of proven PCB designs. The second is the production of new designs, typically for equipment, based on information supplied by the customer. The third is the production of aluminium-based PCBs for cutting edge quantum computing applications. In the first two business lines, the case study firm's short lead times — three and five days, respectively — help it compete with overseas producers, which need to factor in much longer transport times in addition to production times.

The case study firm's production runs for these two business lines are typically small. For larger orders, it uses offshore

[2]PCB manufacturing uses strong chemicals for the etching process. Those chemicals have to be handled and disposed of in particular ways. Also, there is residue (waste) produced from etching. In this regard, stringent environmental regulations can become a factor for relocation decisions of PCB manufacturing firms.

suppliers. Typically, the case study firm's customers are technology companies and start-ups. The case study firm accompanies them from the early stages of development, up to the point where they reach a level of sophistication and scale where they can deal with offshore producers directly.

For the production of standard circuit boards and new designs, the case study firm typically deals with Canadian firms. Most business comes through established relationships that have been in place for some time, or through the company's website, which allows potential customers to upload a design and receive an estimate. A common example of a device in which the case study firm's PCBs are used for testing equipment is the oil sector, which is an important industry for Canada. In many cases, these probes are destroyed during use, so there is consistent demand for production.

The most innovative part of the case study firm's business is its production of aluminium-based PCBs for quantum computing applications in commercial businesses. Whereas traditional computers function on the basis of bits — which can either take the value of one or zero — quantum computers use "qubits", which can be one, zero or any quantum superposition of those two states. In theory, quantum computing offers the potential for significant gains in computing power and speed over traditional computers, although the technology is still in the development stage. Major breakthroughs are likely in the coming years. The case study firm appears to have been the first company globally to manufacture aluminium PCBs for use in quantum computing applications. Aluminium PCBs have important properties relative to the traditional copper alternative, and they are exploited in applications such as quantum computing. For example, aluminium PCBs can be used in low-temperature superconductive environments, as well as in applications where the radiation absorption properties of aluminium are important.

The case study firm's involvement in quantum computing through the production of aluminium PCBs is more international in focus than its core businesses, even though the market is very much a niche — it is partly for that reason that larger companies have not entered it. The case study firm deals with world-renowned commercial and research organisations. The value chain for production of quantum computer applications is more internationalised than, for example, the energy value chain referred to earlier. The case study firm deals with international clients and ships its products overseas: it is an SME that is actively involved in exporting in this part of its business. To the best of the company's knowledge, it is currently the only provider of aluminium circuit boards worldwide.

At various points in its history, the case study firm has received assistance from the Business Development Bank of Canada (BDC) enabling the current team to acquire the business three years ago.

16.3. DESCRIPTION OF THE VALUE CHAIN

Although the three main business lines at the case study firm are distinct in respect of some important characteristics, they also have important similarities that make it possible to analyse their operation from a value chain perspective in a unified way. That is the approach taken in this section — to discuss the case study firm's operations in a general way, entering into specifics in relation to the role of services in the value chain, and in particular the decision to source services in-house or through a third-party provider.

As a general proposition, services are important intermediate inputs at a variety of points in the value chain. Overall, the company estimates that services account for around 14 per cent of total costs. Some services are provided in-house, while third parties supply others. As a small business, the case study firm constantly faces the trade-off of the cost involved in using outside suppliers versus the cost and difficulty of acquiring the necessary knowledge and practice in house.

The following sections describe the case study firm's production process, from customer conception through to delivery and after sales follow up, and also discusses the role of general support services and back office functions. Full details of the firm's use of services by Central Product Classification (CPC) code are provided in the tables at the end of the case study. In total, the firm uses 40 different types of services according to that classification at all points during its production process. Seventeen of those services are supplied in-house, 13 are supplied by unrelated third parties, and 10 are supplied through a combination of those two means. As these numbers indicate, services are obviously important to the firm's operations, and its ability to link with suppliers and customers. Services are, therefore, an integral part of the value chain in this case, even though there is relatively heavy reliance on in-house supply due to the company's small size.

16.3.1. *Pre-manufacturing: Design, Materials Sourcing and Logistics*

In all cases, it is the customer who undertakes the design of the PCB layout. The customer submits the design to the firm, which provides an estimate for manufacture. If the estimate is accepted, there is a purchase order for a given number of units, and an order certification.The firm only provides input on the design if there is a blatant error, or if there is some technical problem with the design that makes it impossible to manufacture.

The only exception to this approach, and it is a partial one, is for the aluminium-based PCBs. They were developed in-house, using the case study firm's expertise, including conducting experiments and tests. So in this one particular case, in-house research and development (R&D) services were important. Although the technology is finding cutting edge applications, the case study firm has not been able to protect it by patents because NASA scientists originally developed the methods in the 1960s, and thus are in the public domain.

Materials are sourced from a variety of locations. One important input is copper, which comes mostly from the United States (80 per cent) and Canada (20 per cent). Aluminium is sourced from Europe. Chemicals used in the photolithography process are sourced from Asia.

Services are important at this stage in the value chain. For instance, a customs clearance agent is used to facilitate the flow of raw materials from overseas, with advantages in terms of compliance and risk management. Other services are provided in-house, such as quality assurance based on ISO standards, and storage and inventory management. The decision to store goods in-house is based largely on the company's scale of operations: there is a clear advantage to keeping operations under one roof, and not outsourcing this particular function. It is also part of the company's strategy of ensuring fast order turnaround — a matter of just a few days in most cases.

16.3.2. *Manufacturing Process*

Production takes place in-house, at least until the customer reaches a level of scale and sophistication were outsourcing needs to be contemplated. In that case, offshore facilities in Chinese Taipei are used: the design, which has already been produced in small numbers by the case study firm and has proven itself, is sent to the offshore partner, the article is produced at scale, and it is sent back to the customer. In the production process, the case study firm brings together raw materials and the customer's design to make the final product, typically first as a prototype, and then in whatever number the customer orders.

Scale plays a key role in the provision of services during the manufacturing phase of the value chain. The production process needs to be controlled so that uncertainties as to quality and execution can be minimised. It is, therefore, important to have access to expertise, and the key determinant of the case study firm's sourcing decision is whether or not the relevant knowledge exists in-house. If it does, the scale of the business means that there is a clear preference for leveraging internal capabilities. For instance, the case study firm has a chemist to assist with the production process itself, and this person is a member of staff, so the provision of production management services and engineering advisory services is in-house. Similarly, all repair and maintenance are done in-house, as is product testing and analysis.

Respecting environmental regulations in relation to chemicals used in the production of PCBs, as well as waste products, is a key issue for the case study firm. The company uses a combination of in-house expertise and third-party provision to ensure that waste materials are sent to the appropriate locations for disposal or processing. Some of the waste is used by a third party to reclaim copper, which is then of commercial value. Wastewater is evaluated as to composition by a third-party laboratory, to ensure compliance with domestic laws and regulations.

In terms of administrative services, they are mostly provided in-house. The main exception to that approach is for payroll: a third-party provider takes care of calculating wages, benefits, and tax and social security withholding amounts.

16.3.3. *Post-manufacturing*

This phase of the chain refers to activities such as delivery and sales. As in other stages, the case study firm's small size means that it relies on in-house expertise whenever possible, but uses third-party providers whenever it is cost effective and efficient to do so. This part of the process involves a number of services elements, which are discussed separately.

Packaging of PCBs takes place in house, and shipping documentation is also prepared within the firm. After that point, however, third-party service providers intervene. Given the short lead times that are an important part of its competitive advantage, the case study firm uses courier services to ship goods to the customer. A third party also supplies customs clearance services.

16.3.4. *Post-sales Services*

The case study firm's PCB is intermediate goods for the production of other electrical goods and electronics products. Once received by the customer, they are combined with other inputs to make a more advanced product, or even a finished consumer good. Some of the case study firm's products find industrial applications, as already mentioned, for example in the energy industry in Canada.

Of course, post-manufacturing support is important for the case study firm's customers. The case study firm retains an image of each and every PCB that is produced. If a difficulty arises, the customer sends an image of the area of the PCB with a problem. The firm examines the issue, and takes responsibility if necessary. Alternatively, the company may repair or replace the units in question. All of these activities are undertaken in-house.

16.3.5. *Back Office, Utilities and General Services*

In addition to production-specific processes, the case study firm also requires services inputs to keep the business operating properly, and to ensure compliance with relevant reporting requirements and financial obligations. Again, it uses a mix of in-house provision wherever possible, and third-party provision when the cost and efficiency benefits are significant.

An example of the ways in which these two strategies work together is in relation to accounting. Day to day accounting work and monthly summaries are produced in-house. However, year-end work — including tax compliance — is outsourced to a third-party provider. Similarly, a mix of in-house and third-party provision is used for insurance: medical insurance for employees is organised in-house, and a broker is retained for general insurance purposes to ensure that optimal insurance values and premiums are obtained. Insurance is waived on incoming shipments and instead the most reliable couriers are used — a factor that reduces shipment costs by as much as 30 per cent.

It was noted earlier that BDC played a role — including the provision of financing — in enabling the current team to acquire the case study firm. In addition, it also provides business-consulting services. BDC goes over the company's financials and provides feedback, which is then incorporated into operations as appropriate by management.

Marketing is important for the case study firm, given the competitive nature of its industry. As previously noted, its website is an important channel for generating business, including in the cutting edge area of aluminium PCBs. Development and maintenance of the site are outsourced to a Canadian firm. More generally, IT services are both provided in house — the case study firm employees developed the company's network, for example — and outsourced, such as website development and e-mail.

Even though the case study firm is a small business, R&D plays an important role. As previously noted, it appears to be the first manufacturer globally of aluminium PCBs used in quantum computing applications, for example. It has engaged with third-party service providers as part of its R&D efforts, including the University of British Columbia, and the National Research Council of Canada. The Canadian government directly supports R&D activities through cash payments, but the case study firm's customers have also been active supporters in this area.

16.4. LOOKING FORWARD: CHALLENGES AND OPPORTUNITIES FOR THE VALUE CHAIN

As noted at the outset, the PCB industry in North America has been shrinking in terms of the number of operators, in the face of increasingly strong competition from overseas, particularly Asia. The case study firm's sales have been declining, although should quantum computing take off as a technology, the company is well positioned to take advantage of new market opportunities, as it has already established links with key organisations in the sector.

The case study firm deals with international suppliers and customers in all aspects of its business, and in some cases offshores production to Chinese Taipei once a certain volume is reached. However, its value chain is perhaps more Canada-focused than others considered in this project. Nonetheless, the company very much fits the standard view of a value chain participant, namely a specialist in performing a particular task that is combined with others in complex ways to produce a final product. It is important to highlight that the case study firm deals with global companies, including Canada-based companies that are foreign invested, as well as multinationals like UPS and FedEx for courier and delivery services. The world market for PCBs is growing, and although there is a structural shift in production to low cost economies in Asia — perhaps also due to more general movement of the electronics cluster to that part of the world — there is still room for niche operators like the case study firm to enjoy considerable success.

In terms of the challenges faced by the business, a key one relates to environmental compliance. This area is believed to be — along with cost differentials — a major advantage for some Asian producers, where regulations are not as stringent as they are in Canada. As a small business, the case study firm needs to invest considerable resources in staying abreast of constantly changing environmental regulations. It uses chemicals that have environmental impacts and need to be disposed of appropriately. Environment Canada issues rules for individual chemicals, so companies like the case study firm constantly need to go back through their production processes to see if they use each one, and then comply with the relevant regulation. The case study firm is involved in an industry task force that is working to streamline the process as it applies to PCB manufacture. Limiting compliance costs is a key concern going forward.

Another issue the case study firm faces is in dealing with large, lead companies. Given their scale, approaches can sometimes appear bureaucratic to a much smaller and more nimble operator. Although the case study firm deals directly with some international clients, its most obvious path to internationalisation as a small business is through developing linkages with lead firms that export. To do that, it will need to address the issue of how best to interface with them, given differences in business practices and outlooks. Related to this issue is the development of new sectors of activity, like quantum computing. The technology offers real potential, and the case study firm has developed a competitive advantage. However, if it is to move to scale, it will be important to develop strong links with key operators, including service companies that are potentially important players in this emerging value chain.

APPENDIX

Table A.1: Establishment Phase.

Stage/category	Service	CPC ver. 2 code	Please mark (X) if the service is used in your supply/value chain	Supplied in-house	Supplied by another company in the group	Outsourced to third parties
1. Government services (licensing etc.)	Business liaison services (e.g., services from trade development centres or SME centres)	Class of 9113 — Public administrative services related to the more efficient operation of business	X			X
	Company registration and licensing services	91138 — Public administrative services related to general economic, commercial and labour affairs	X	X		
	Information and statistical services	Class of 9113 — Public administrative services related to the more efficient operation of business	X	X		

(Continued)

Table A.1: *(Continued)*

Stage/category	Service	CPC ver. 2 code	Please mark (X) if the service is used in your supply/value chain	Supplied in-house	Supplied by another company in the group	Outsourced to third parties
2. Other services (professional etc.)	Banking and finance services	71121 — Deposit services to corporate and institutional depositors 71135 — Non-mortgage loan services for business purposes	X			X
	Legal services	82130 — Legal documentation and certification services	X	X		X

Source: Author.

Table A.2: Raw Materials/Inputs Pre-production Phase.

Stage/category	Service	CPC ver. 2 code	Please mark (X) if the service is used in your supply/value chain	Supplied in-house	Supplied by another company in the group	Outsourced to third parties
1. Procurement of raw materials	Procurement agent for raw material sourcing	85999 — Other support services n.e.c.	X	X		
	Customs-related services for raw materials imported	85999 — Other support services n.e.c.	X			X
	Quality assurance services (of raw materials)	83441 — Composition and purity testing and analysis services	X	X		
2. Logistics	Freight transportation services (of raw materials) by road, rail, sea or air	Division: 65 — Freight transport services	X			X
3. Storage	Storage of raw materials — general storage	67290 — Other storage and warehousing services	X	X		
4. Product design	Conception and design of product	83920 — Design originals	X	X		
	Industrial design	83912 — Industrial design services	X	X		

Source: Author.

Table A.3: Manufacturing Phase.

Stage/category	Service	CPCver.2 code	Please mark (x) if the service is used in your supply/value chain	Supplied in-house	Supplied by another company in the group	Outsourced to third parties
1. Production administration	Production administration — Production management	83115 — Operations management consulting services	X	X		
	Production administration — Repair and maintenance of factory equipment	87156 — Maintenance and repair services of commercial and industrial machinery	X	X		
	Production administration — Quality assurance and compliance with ISO	83441 — Composition and purity testing and analysis services	X	X		
2. Services supporting factory daily operations	Cleaning services of factory	85330 — General cleaning services	X	X		
	Engineering Services	83310 — Engineering advisory services Class: 8332 — Engineering services for specific projects	X	X		X

Security guards for factory and warehouses	85250 — Guard services	X			X
Sewage water treatment services	94110 — Sewerage and sewage treatment services	X	X		
Repair and maintenance services of machines and equipment	87156 — Maintenance and repair services of commercial and industrial machinery	X	X		
Waste collection and recycling services	Class: 9421 — Collection services of hazardous waste 94229 — Collection services of non-hazardous recyclable materials, other 94239 — General waste collection services, other	X	X		X

(*Continued*)

Table A.3: (*Continued*)

Stage/category	Service	CPC ver. 2 code	Please mark (X) if the service is used in your supply/value chain	Supplied in-house	Supplied by another company in the group	Outsourced to third parties
3. Services from government regulation requirements	Government inspections on fire prevention, health hazards, environmental protection and other aspects	91133 — Public administrative services related to mining and mineral resources, manufacturing and construction 91290 — Public administrative services related to other public order and safety affairs	X	X		
4. Worker-related services	Personnel search and referral services — Recruitment of factory workers	85112 — Permanent placement services, other than executive search services	X	X		

	Social insurance for factory workers	91320 — Administrative services related to government employee pension schemes; old-age disability or survivors' benefit schemes, other than for government employees 91330 — Administrative services related to unemployment compensation benefit schemes	X	X		X

Source: Author.

Table A.4: Delivery and Sales/Post-manufacturing Phase.

Stage/category	Service	CPC ver. 2 code	Please mark (X) if the service is used in your supply/value chain	Supplied in-house	Supplied by another company in the group	Outsourced to third parties
1. Packaging and labelling	Packaging services	85400 — Packaging services	X	X		
2. Delivery to wholesaler/retailer	Customs-related services	85999 — Other support services n.e.c.	X			X
3. Sales services	Retail trade services — By Internet or mail-order	Group of 623 — Mail order or Internet retail trade services	X	X		

Source: Author.

Table A.5: Back Office, Utilities and General Services.

Stage/category	Service	CPC ver. 2 code	Please mark (X) if the service is used in your supply/value chain	Supplied in-house	Supplied by another company in the group	Outsourced to third parties
1. Finance	Auditing on financial accounts	Group of 822 — Accounting, auditing and bookkeeping services	X	X		X
	Financial services	71121 — Deposit services to corporate and institutional depositors 71313 — Group pension services 71701 — Services of holding equity of subsidiary companies	X			X
	Internal auditing (including audits of financial accounts and corporate governance)	Group of 822 — Accounting, auditing and bookkeeping services 83118 — Head office services	X	X		X

(*Continued*)

Table A.5: (*Continued*)

Stage/category	Service	CPC ver. 2 code	Please mark (X) if the service is used in your supply/value chain	Supplied in-house	Supplied by another company in the group	Outsourced to third parties
	Insurance services for machinery	71332 — Marine, aviation and other transport insurance services	X	X		X
2. General management	Business and management consultancy services	Class of 8311 — Management consulting and management services	X			X
	Company secretary services	83990 — All other professional, technical and business services, n.e.c.		X		
	Corporate communications, marketing and public relationship	83114 — Marketing management consulting services 83121 — Public relations services	X			X
	Courier, postal and local delivery services	Group of 681 — Postal and courier services	X			X

Estate management	72112 — Rental or leasing services involving own or leased non-residential property	X			X
Human resources management	83113 — Human resources management consulting services	X	X		
Human resources management — Personnel search and referral services for back office staff	Class of 8511 — Personnel search and referral services	X	X		
I.T. and information system management, consulting and support, with webpage development	Class of 8313 — Information technology (IT) consulting and support services Class of 8314 — design and development services Class of 8316 — IT infrastructure and network management services	X	X		X

(*Continued*)

Table A.5: (*Continued*)

Stage/category	Service	CPC ver. 2 code	Please mark (X) if the service is used in your supply/value chain	Supplied in-house	Supplied by another company in the group	Outsourced to third parties
	Safety and security services	85230 — Security systems services 85250 — Guard services	X			X
	Telecommunication services	Group: 841 — Telephony and other telecommunications services 84210 — Internet backbone services 84221 — Narrowband Internet access services 84222 — Broadband Internet access services	X			X
	Uniform — Laundry	97130 — Other textile cleaning services	X			X

3. Legal	Legal services	82120 — Legal advisory and representation services concerning other fields of law 82130 — Legal documentation and certification services	X		X
4. R&D	Product development/R&D	81129 — Research and experimental development services in other engineering and technology 81400 — R&D originals	X	X	X
5. Utilities	Electricity supply	Class of 8631 — Support services to electricity transmission and distribution	X		X
	Gas supply	86320 — Gas distribution services through mains (on a fee or contract basis)	X		X
	Water supply	86330 — Water distribution services through mains (on a fee or contract basis)	X		X

Source: Author.

Table A.6: Post-sales Services.

Stage/category	Service	CPC ver. 2 code	Please mark (X) if the service is used in your supply/value chain	Supplied in-house	Supplied by another company in the group	Outsourced to third parties
1. Sales/ After-sales Customer services	Customer services — Complains and compliments handling	85931 — Telephone call centre services	X	X		
	Customer services — Technical support	85931 — Telephone call centre services	X	X		
	Customer services — Warranty and repair services	Group of 872 — Repair services of other goods	X	X		

Source: Author.

Chapter 17

WINE INDUSTRY IN CHILE

Karina Fernandez-Stark and Penny Bamber

17.1. INDUSTRY OVERVIEW

Over the past 20 years, the wine industry has undergone significant changes that have fundamentally altered its competitive landscape. The 'democratisation' of wine associated with the rise of growing demand from a broader consumer-base, combined with the rise of new producers and a concentration of sales through supermarkets has shifted the focus from a producer-driven industry focused on differentiated supply to one led by demand. While natural endowments such as climate and *terroir*[1] remain the most important factors for success, the ability to innovate in production, consistently achieve economies of scale and align research and development strategies with marketing goals have become essential drivers of competitiveness (Cusmano *et al.*, 2010).

During this time, international trade of wine has grown in terms of total volumes and value and diversified into a number of new geographic markets. Total exports grew threefold from US$10.2 billion in 1995 to US$35 billion in 2013 (UN Comtrade, 2015). The geographic composition of trade has also changed — both in terms of demand and supply. While 'Old World' suppliers in Europe, including France, Italy and Spain, previously dominated the industry with over 90 per cent market share, today several 'New World' producers, including Australia, Chile, New Zealand,

[1]The combination of conditions of the soil, climate and topography.

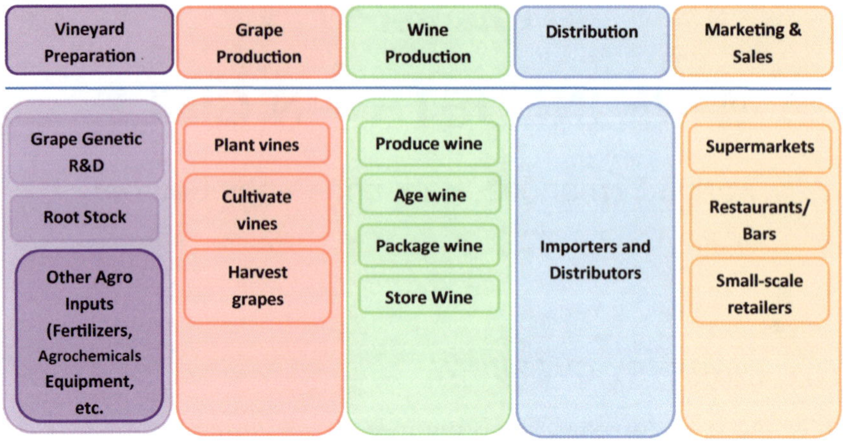

Figure 17.1: Simplified Wine Value Chain.

Source: Authors.

South Africa and the United States have emerged as important exporters. Once driven by European demand, the industry is now characterised by strong markets in the Asia-Pacific region, particularly in China and Korea (UN Comtrade, 2015) (Fig. 17.1).

The evolution of Chile's wine sector, which has been largely export-oriented since the 1990s,[2] is illustrative of these changing dynamics in the global industry. Since 1995, the economy's wine exports have grown by a factor of 10, from US$182 million to US$2 billion in 2013, simultaneously tripling its market share to reach 6 per cent. Although primarily known for its 'value for money' model, the per unit value has increased slowly but steadily over this time, particularly in the bottled wine segment (UN Comtrade, 2015). With over 117 thousand hectares under production and almost 200 exporting wineries (Farinelli, 2012), the industry expects to increase its bottled wine sales 25 per cent in value by 2020 (Todovinos.cl, 2014). This growth has been driven by several factors including technological and organisational changes and product standardisation in the industry, focusing on

[2]The share of Chilean wines sold abroad increased from 7 per cent in 1989 to 63 per cent in 2002 (Visser, 2004).

more easily identifiable varieties and innovating in marketing (Cusmano *et al.*, 2010).

17.2. BACKGROUND INFORMATION ON THE FIRM[3]

The firm represented in this case is a traditional, family-owned Chilean wine company with over 60 years of experience in the sector. The firm exports primarily bottled and boxed wine and a very small quantity wine in bulk that is bottled or packaged in the destination market. The firm exports a range of wines in the varietal, premium, super premium and ultra premium market segments and is primarily focused on the small group of grape varieties that Chile is well known for — Cabernet Sauvignon, Carmenere, Merlot, Chardonnay and Sauvignon Blanc. Its wines have been recognised for their high quality with awards from international competitions and, today, the winery is trying to position itself in the high-end market and will soon launch an icon wine.

With operations in both Chile and Argentina, the winery has its own grape production and winemaking operations. The vineyard has several hundred hectares under vines and the grapes produced in these in-house operations are typically used for the firm's finer wines, in which the vineyard requires greater control over quality. For its lower priced wines or varieties, the vineyard tends to source from grape outgrowers.

The firm's operations over the past two decades have been transformed by two key principles — modernisation and sustainability. New technologies were incorporated into the cellar to optimise quality control and production efficiencies; the firm obtained a range of certifications required for key global markets, including ISO 9000, 14000, Hazardous Analysis and Critical Control Points and British Retail Consortium certifications, while also adopting a code of conduct for labour operations; at the same

[3]All information about the firm in this case study is from its corporate website as well as from the interview.

time, a range of environmental measures such as water recycling and use of lighter bottles were adopted to help reduce the firm's carbon footprint. These efforts have supported strong growth. Between 2003 and 2013, sales more than tripled by volume and value, and the firm expanded its geographic reach into a number of new markets, particularly in Asia, including China, Japan and Singapore.

17.3. DESCRIPTION OF THE VALUE CHAIN

This case study will focus on the firm's Chilean operations, which accounts for over 90 per cent of the firm's exports. For the purposes of this case study, five main stages of the wine global value chain are discussed: vineyard preparation, and grape production (pre-production activities); wine making (production activities); and distribution, marketing, and sales (Fig. 17.2).

17.3.1. *Pre-production: Vineyard Preparation and Grape Production (Viticulture)*

17.3.1.1. *Vineyard preparation*

During this stage of the chain, the variety to be planted is carefully selected based on the *terroir*. This is the natural environment

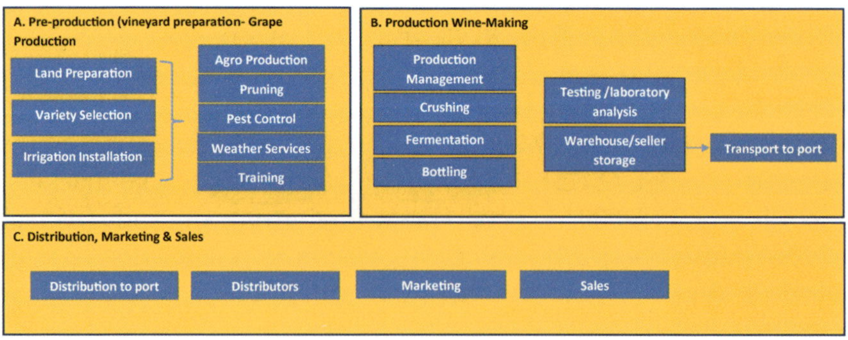

Figure 17.2: Dimension of the Value Chain Covered by Case Study.
Source: Authors.

comprised of the soil, climate and topography. In this stage, many tests are performed analysing the most suitable varieties for wine production. In addition, a number of inputs are purchased to establish the vineyard including poles, wires, and rootstock amongst others. This requires significant investment, with no returns for approximately 3 years until the vines become productive.

17.3.1.2. *Grape production*

In this stage of the chain, the grapes are cultivated and harvested. Some of the key activities are canopy management, pruning, irrigation control, spraying and harvesting. Over the past few years, there has been an increase in the mechanisation of many of these activities. This stage is also characterised by strict quality control in order to obtain the best grape possible to prepare the desired wine. The grape production of this winery follows two models: in the first model, the wine maker owns the grape production, while, in the second scheme, the wine maker sources grapes or grape juice from independent outgrowers. However, in the outgrower system, the wine maker is also involved in the cultivation process to control the grape quality.

17.3.2. *Production*: *Wine-making* (*Viniculture*)

The wine production stage is comprised of a number of processes. This process starts with the crushing of the grapes, followed by fermentation in which the temperature is controlled and all chemical reactions are completed. After that, the grapes are pressed to extract the juice, which is then blended to create the desired wine type. Finally the wine is bottled or stored before bottling. During this stage of the value chain, the different wine blends are created. Critical to this process is the fermentation stage in which the wine is kept in stainless steel tanks. Later, the wine is stored (in steel, ceramic or oak containers) until there is a purchase order, after which bottling takes place. However, premium wines are commonly stored and aged in bottles.

17.3.3. *Post-production: Distribution and Marketing and Sales*

Distribution is a key segment for the wine industry, although the distribution models differ widely by market. The winery studied follows three distribution schemes, all of which include intermediaries, with no direct sales to the consumer. The first model includes an importer that sells the wine to distributors who in turn sell the wine to the final outlets. The second model is very similar to the first one; the only difference is that the distributor is a monopoly in the economy and also owns retail outlets. For example in Norway, the government owns 'Vinmonopolet', a state-owned enterprise that sells beverages with alcohol content higher than 4.75 per cent (Vinmonopolet, 2015). This is the only retail outlet permitted to sell wine. The third model excludes the importer and the wine goes directly to distributors. Thus, distributors in leading export destinations largely control access to market and buy wines according to the final consumer preferences.

With the influx of new entrants into the global wine industry and a broad expansion of the consumer base, marketing has become an essential part of firm strategies. Amongst the low-to-mid end of the market, there has been a move away from a traditional supplier driven approach to labelling and branding, towards one focused on a demand driven approach (Cusmano *et al.*, 2010), with emphasis on occasions, or controversial labels to attract younger consumers attention. The cost of building these new brands has led firms to focus on a smaller number of brands, under which they market a variety of different grapes and blends (Ponte and Ewert, 2009).[4] Nowadays marketing activities, especially promotions are critical to sell wines, especially high volume/low value wines. Other marketing activities popular in the industry includes tasting in the point of sales, training to sommeliers, and special discounts.

---

[4] In some cases, marketers simply own the brand and are not engaged in any way in the production operations.

17.4. SERVICES ALONG THE VALUE CHAIN

A total of 70 services can be identified in this value chain; 33 of which are more directly related to the wine production phase while 23 relate to the agriculture phase; an additional 14 services are transversal operations which support all stages of the value chain. These can be categorised according to the various stages within the chain (see Fig. 17.3): (i) vineyard preparation, (ii) services during grape production, (iii) services during winemaking, (iv) distribution and marketing services and (v) Back office services.

17.4.1. *Pre-production*: *Vineyard Preparation and Grape Production* (*Viticulture*)

Pre-production services include activities in the vineyard preparation and grape production. These services are carried out by in-house personnel as well as third-party providers.

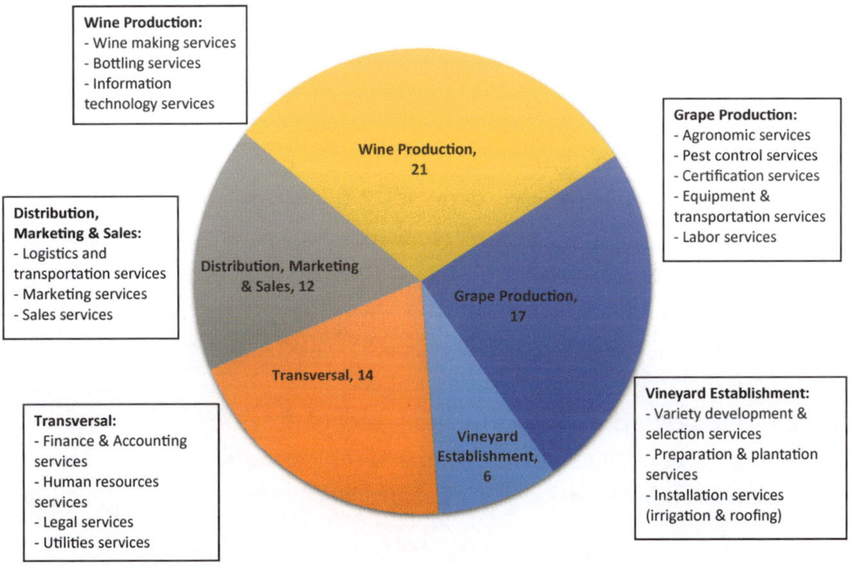

Figure 17.3: Breakdown of Services by Stage and Examples of Key Services.

Source: Authors.

Vineyard preparation is mainly outsourced to experts with respective specialised skills in variety development, irrigation systems, soil testing and vineyard construction. R&D in variety development is typically carried out by local and international research institutions with expertise in grape production and performance. The winery outsources soil studies, soil preparation, installation of drip irrigation, installation of electrical circuits, well construction and recruitment of external personnel to plant of new vines. In addition, the winery hires external technical experts to provide critical information regarding the varieties to be planted and the cultivation technique to be followed.

In grape production operations, the company generally performs key services in-house. Its production team includes one internal agronomist and three technicians that manage both internal and outgrower grape production. However, there are a number of services that are supplemented by external personnel including pesticide application in which the firm hires specialised service providers that apply these chemicals by land and air.[5] During peak periods, the winery also supplements certain activities with external companies to perform services such as trimming and harvesting. The training of personnel is handled through a combination of internal on-the-job training and specialised external courses. In addition, the firm also trains outgrowers to ensure the quality of the grapes to be purchased, especially in the most expensive type of grapes. The winery usually has a dual system for equipment handling at this stage of the chain; it owns basic equipment and outsources repairs and maintenance. The firms also contracts equipment and operators during peak periods. For example it contracts mechanised harvesting services as well as transportation services to move the grapes to the winery.[6] Regarding information technology (IT) services, software development is provided by specialised firms that create customised programs for precision

[5] These external services represent around 25 per cent of the total pesticide application in their plantations.

[6] Harvesting and transportation services are not bundled.

agriculture, including climate, soil and disease monitoring and production tracking.

17.4.2. *Production*: *Wine-making* (*Viniculture*)

While many services in this stage of the value chain can be outsourced, the firm generally undertakes these activities in-house. These services include processing grapes and transforming them into wine. This process is handled by an internal team of employees including an operations manager supported by oenologists, technicians and workers. The company handles the entire process from grinding, fermentation, pressing, storage and bottling. The firm also stores its wine in tanks and barrels. In addition, the company bottles its wine and offers bottling services to other wineries. Testing services are typically outsourced to experts, especially laboratories in local universities. The firm also works with these external laboratories to obtain certification of origin of denomination that is granted by the National Agriculture and Livestock Agency (SAG). The company usually hires experts the first time it applies for a new market certification to make sure they follow each step correctly. Once it obtains the certification, the renewal process is handled internally. It is also important to note that at certain times, such as the creation of new wines, the company may hire an external oenologist. Nonetheless, an internal oenologist works in the winery all year round. In addition, the firm has installed an IT system developed by an expert company, but maintains an internal IT team to manage production and inventory.

17.4.3. *Post-production: Distribution and Marketing and Sales*

In this stage of the value chain, distribution is carried out by external factors including importers and distributors that facilitate entry in destinations markets. Transportation of the wine from the winery to the port is outsourced, as is international freight — the majority of which is by sea. This is the main transportation cost incurred by the firm since the wine is sold free on board and the importer is in charge of all international transportation costs.

Finally, the marketing and sales operations are handled in-house. The company has an internal marketing team that prepares an annual marketing plan. These activities include: penetration of new markets, promotions, training of sommeliers and participation in trade fairs. In addition, specific marketing initiatives are undertaken to ensure participation in exclusive wine rankings, such as Wine Spectator. At this stage of the chain, the firm outsources brand design and other graphics to position the company in the domestic and worldwide market. At the same time, the winery hires an external agency to develop a media marketing campaign in Chile. Sales activities are managed in-house and they provide sales incentives to the salespersons hired by the distributors.

17.4.4. *Back Office Services and Workers Transversal Services*

The back office services are transversal to the chain activities and they support all firm operations. These activities include finance and accounting, human resources, IT, legal services and security. The company carries out some of these services in-house, including, finance and accounting and human resources, occasionally drawing on the support of external advisors. The remaining activities, including legal services, security and software development, are outsourced to specialised firms.

17.4.5. *Outsourcing, Bundling and other Aspects of Services Supply*

Among the 70 different services identified in this value chain, our analysis estimates that 20 are supplied in house, 21 are partially outsourced, while 29 are fully outsourced (see Table 17.1). This shows a general balance between in-house and outsourced provision of services; however, in the analysis of services provision by segment of the value chain certain trends can be observed.

Table 17.1: Breakdown of Services Provision, by Value Chain Stage.

Segments of value chain	In-house	Partially outsourced	Fully outsourced
Vineyard preparation	1	0	5
Grape production	0	10	7
Wine making	6	7	8
Distribution, marketing and sales	7	1	4
Transversal services	6	3	5
Total	20	21	29

Source: Authors.

For example the winery supplies more services in-house in the later stages of the chain, while it follows a more aggressive outsourcing strategy at the beginning of the chain.

The firm's provision of services is concentrated in its core competencies, wine production and distribution, marketing and sales. Those segments of the chain are crucial for its business, especially the wine making stage. In this stage, all the critical activities, such as crushing, fermentation, blending, bottling and storage, are performed in-house. The company outsources other activities to concentrate on what they consider their core strengths. In addition, the Chilean wine sector has developed considerably over the past decade which has allowed the emergence of a critical mass of specialised services providers. With almost 200 vineyards, and US$2 billion in exports, these providers are able to provide specific services — such as harvesting services, laboratory testing and bottling services — to a range of clients that demand the same services.

The establishment of the vineyard is principally fully outsourced since this is a one-time operation. In the case of the grape production, the company has internal staff that are usually supplemented by external companies. In this case, the winery maximises its resources by outsourcing services during peak times instead of maintaining idle staff during slower periods of the year. The company is also moving towards outsourcing transversal services as these are not core firm activities.

The main reasons for outsourcing services are: (1) cost efficiencies and economies of scales (e.g. hiring agricultural services during peak times); (2) services provision in a collective manner (e.g. meteorological services provided by Wines of Chile); (3) access to specialised equipment and skills (e.g. renting the harvesting machine and operator to collect the grapes); (4) government support services (e.g. certification for origin denomination); (5) network economies, such as recruitment services and (6) lack of feasibility in supplying services in-house, such as utilities services (Tables A.1–A.5).

17.5. POLICIES AFFECTING THE VALUE CHAIN

Since the 1990s, the Chilean wine industry has been supported by a range of industrial policy measures, with specific emphasis on exports. These policies have encouraged industry development in different areas including facilitating trade, incentivising employee training and research and development and international promotion among others (see Table A.6). This set of policies has had a significant impact due to very strong industry collaboration in which all stakeholders in the sector are closely aligned and work together on a number of core issues. Some of the most important stakeholders include government agencies such as CORFO, National Agriculture and Livestock Agency (SAG) and ProChile, Chile's export promotion agency; the industry association, Wines of Chile, and local research universities. These policies have supported strong growth in the industry and helped to reduce costs in production and export; a key challenge for policy makers and firms alike that remains today is to reposition Chile has a higher quality wine in the global market.

17.5.1 *Supportive Export-oriented Policies*

Chile's wine exports have benefited to a great degree from the consistent export-oriented trade policies that have been implemented over the past few decades. By 2013, Chile had 22 free trade agreements with 60 economies, providing access to 60 per cent of the world's consumers (Murray, 2013). ProChile and the General Directorate of International Economic Relations (DIRECOM) have been vital in terms of leading negotiations in reduction of quotas and tariffs. By 2015, import tariffs for Chilean wine to China had been reduced to zero, compared to 14 per cent from other economies with Most Favoured Nation status (ProChile, 2011). ProChile is a key factor in identifying new international clients, maintaining and sharing a database of distributors and retailers in all major markets. In Asia, specifically, they financed the establishment of an international office for Wines of Chile, helping to position the Chilean wines in this important market. China is now the fourth-largest importer of Chilean wine. SAG officials constantly visit wineries to control plagues and certify that the pesticide applications are done in time. In addition, SAG periodically visits wineries to quantify the wine stock to support the credibility of the certification of origin initiatives.

17.5.2 *Efficient Logistics: Competitive Transportation, Modern Ports and Streamlined Customs Operations*

Despite considerable distance from its major markets, logistics costs are lower in Chile than in many competing economies. High import and export volumes and a liberalised transportation sector have fostered competition amongst logistics providers, which have helped to reduce costs. The modernisation of the Chilean ports following their privatisation, which began in the late 1990s, has significantly contributed to the economy's capacity to rapidly export large quantities of products. There are numerous ports along the economy's coast, reducing costs and distances for land transportation. Online processing with the implementation of the single-window system has improved the efficiency of customs clearance, reducing unnecessary delays at the ports. This has helped to circumnavigate issues such as strikes of customs staff at the ports; the firm is able to complete all of the customs forms online and deliver the wine cases to the international transport company. This helps to prevent disruptions in the on-time delivery of wine to foreign distributors.

17.5.3. *Human Capital Development and Labour Regulations*[7]

Efforts in terms of human capital development and labour regulation have been key in facilitating growth. This is particularly important in the pre-production stages, because of the shortage of agricultural workers. Training efforts have thus been focused on driving productivity, led on one hand by tax incentives through the National Training and Employment Agency (SENCE), and on the other by the ChileValora program — a joint initiative between the Ministries of Economy, Education, Labour and Social Security and the private sector. SENCE provisions incentivise firms to individually and collectively develop training programs specially to meet the needs of the industry. This company uses all these benefits up to the limit amount established by the government. Inclusively, they have developed their own curricula about efficient operations in the winery. The ChileValora program is a competencies-based certification system launched at the end of the 1990s that certifies every job profile required for the industry (Fernandez-Stark *et al.*, 2011). In addition to increased productivity, flexible labour regulation, which permits subcontracting of labour, has allowed the industry to maintain its competitiveness by relying on temporary contracts to meet peak periods of demand during the year.

17.5.4. *R&D and Variety Development*

Several competitive grant funds are available to carry out research for the wine sector, many of which are financed through the *Fondo de Inversion para la Competitividad* (InnovaChile). Both local and foreign universities (see Box 17.2) play a central role in applied research through the provision of services to the industry. In 2012, a consortium

---

[7]For a discussion of how workforce development initiatives have contributed to the upgrading of the fruit industry in Chile, see Fernandez-Stark *et al.* (2010).

Box 17.1. Chile National Labour Skills Certification System (NLSCS).

This program was created in 1998 by Fundación Chile. This system aims to provide a framework for the recognition of competencies, regardless of how these were acquired with the goal to improve industry competitiveness. The program has involved both public and private stakeholders from 15 different industries. In these 15 industries, more than 500 occupational standards have been established and 10,000 workers have been certified annually. These skills standards have been transferred to vocational training institutions to refine their curricula. At the same time, the workers that have not passed the evaluation are offered training courses to fill the skills gaps and later receive the certification of skills diploma.

This system has identified the skill profiles required for the majority of jobs performed in the wine industry. The certification of skills of the wine sector workers started in 2010. The industry association, Wines of Chile, and the Chilean government havebeen working for more than 15 years to develop this program. To date, more than 4,500 workers have certified their skills in the wine sector.

The main advantages of this system is to facilitate the portability of skills, decrease uncertainty in the hiring process making it more efficient, grant credibility of the workers abilities and most importantly create a proud sentiment among workers certified helping to establish a culture of lifelong learning and development. While this system is still in a nascent stage, in the medium term it will be expanded to include the entire agro sector.

Source: Araneda (2010), Chilealimentos (2010), ChileValora (2013), Kis and Field (2009) and López (2009).

for wine R&D was created by the Chilean Government through InnovaChile and the industry association, Wines of Chile. The total cost of the project is approximately US$30 million for a 10-year period, with each organisation providing 50 per cent of the financing. The goal is to strengthen collaboration with local universities to generate useful knowledge for the wine sector (Vinos de Chile, 2015).

Box 17.2. Chile and University California - Davis Partnership.

Chile and University of California — Davis (UC Davis) have collaborated for over 40 years. In the early days of this agreement, more than 50 Chileans went to the university to study agriculture. They were named the 'Davis boys' and they revolutionised and supported the success of the Chilean export fruit industry.

In 2010, the original agreement was revitalised with the signature of two memorandi of understanding. The first focuses on training and R&D in the area of breeding and genetics for new grape crop varieties. Various Chilean institutions participate in this program including the National Association of Seed Producers, the Institute of Research on Agricultural and Livestock, University of Chile, Pontificia Catholic University of Chile (PUC), Pontificia Catholic University of Valparaiso, University of Talca and University Austral of Chile. The objective of the second agreement is the development of the wine industry through R&D, training and knowledge transfer in the latest technologies. The Chilean participants include Vinnova, a consortium enterprise formed through the partnership of Wines of Chile, PUC and the University of Concepción, with the support of Innova Chile.

In 2015, this partnership was reinforced with the opening of the UC Davis-Chile Life and Science Innovation Centre in Santiago. With a US$12 million initial budget, the centre is focused on R&D to support the agriculture industry including plant breeding, postharvest and climate-change technologies.

Source: UC Davis (2008, 2015).

17.5.5. *Institutionalisation*

As mentioned earlier, strong industry institutionalisation has helped the development of the sector. Wines of Chile, the industry association, representing 90 per cent of all wine exports, has played a key role in the industry success. Members pay a quota proportional to their exports to support a range of programs and activities. Some of these initiatives include: meteorological stations along the economy to provide climate information, scientific studies that analyse the key topics for the sector, and determining areas of appellation of origin for the Chilean wine among many others. The organisation works closely with both public and private sector actors to resolve issues affecting the industry. They usually are involved in joint programs with the government and local universities. This strong coordination and collaboration underlies the development of the industry. This industry association also works closely with ProChile, the Chilean government agency that promotes the export and services around the world. ProChile has a range of programs to support the Chilean wine exporters that range from training to funds for international promotion. A key challenge that the industry faces with respect to its global exports is improving the perception of the quality of its wine. Chile's success as a good value for money wine producer has made it very difficult for individual vineyards to position themselves internationally with higher priced, superior quality wines. As this is a collective problem, ProChile and Wines of Chile must take the lead in promoting an international campaign to reposition Chilean wine in higher value segments.

APPENDIX

Services in The Wine Value Chain

Table A.1: Pre-production: Vineyard Establishment Phase.

Service	Central product classification (CPC) Ver. 2 code	Supplied in-house	Outsourced to affiliated companies and reasons	Outsourced to third-party suppliers/ government and reasons	Bundled
1 Genetic development services	81140 — Research and experimental development services in agricultural sciences	No	No	Yes; comparative lack of expertise	
2 Variety selection	81140 — Research and experimental development services in agricultural sciences	Yes; in-field testing	No	No	
3 Land preparation and plantation of rootstock	86119 — Other support services to crop production	No	No	Yes; efficiencies	
4 Electric installation	54611 — Electrical wiring and fitting services	No	No	Yes; efficiencies	
5 Construction of infrastructure (roofing, etc)	86119 — Other support services to crop production	No	No	Yes; efficiencies	
6 Installation of irrigation equipment	54234 — General construction services of irrigation and flood control waterworks	No	No	Yes; efficiencies and specialised skills	

Source: Authors.

Table A.2: Pre-production: Grape Production (Viticulture).

Service	CPC Ver. 2 code	Supplied in-house	Outsourced to affiliated companies and reasons	Outsourced to third-party suppliers/ government and reasons	Bundled
7 Production management	83115 — Operations management consulting services	Yes	No	Yes; access to specialised knowledge	No
8 Disease and plague control	86119 — Other support services to crop production	Yes; application	No	Yes; efficiencies, specialised knowledge; internal capacity	
9 Pruning of vines	86119 — Other support services to crop production	Yes; supervisory team and trainers	No	Yes; efficiencies labour-intensive for short periods	Bundled with labour sub-contracting
10 Equipment rental for harvesting	86119 — Other support services to crop production	No	No	Yes; efficiencies	Bundled with repair and maintenance
11 Harvesting services	86119 — Other support services to crop production	Yes	No	Yes; efficiencies, specialised knowledge; internal capacity	Bundled with labour sub-contracting
12 Land transport of grapes to processing plants (wineries)	65119 — Other road transport services of freight	No	No	Yes; efficiencies — economies of scale	

(*Continued*)

Table A.2: (*Continued*)

Service	CPC Ver. 2 code	Supplied in-house	Outsourced to affiliated companies and reasons	Outsourced to third-party suppliers/ government and reasons	Bundled
13 IT services production management	8313 — IT consulting and support services	Yes	No	Yes; for specific tasks	
14 Repair and maintenance of equipment	87156 — Maintenance and repair services of commercial and industrial machinery	Yes; for minimal fleet	No	Yes; for specific maintenance	
15 Recruitment of temporary/seasonal workers	8512 — Labour supply services	Yes	No	Yes; network efficiencies	
16 Transportation services for workers to and from orchards	64114 — Local special-purpose scheduled road transport services of passengers	No	No	Yes; economies of scale	
17 Training of temporary/seasonal workers	92919 — Other education and training services, n.e.c.	Yes	No	Yes; specialised skills	
18 Catering services for workers	63393 — Other contract food services	No	No	Yes	

19	Medical services	93121 — General medical services	No	No	Yes	
20	Social insurance for seasonal workers	91320 — Administrative services related to government employee pension schemes; old-age disability or survivors' benefit schemes, other than for government employees	Yes; for internal personnel (required by law)	No	Yes	Bundled with recruitment of workers
21	Post-harvest quality control for all production	83441 — Composition and purity testing and analysis services	No	No	Yes; efficiency, specialised skills	
22	Weather analytical services for all growers	83430 — Weather forecasting and meteorological services	No	No	Yes	
23	Agronomic extension services for outgrowers	83115 — Operations management consulting services	Yes; firm provides this service to their outgrowers	No	Yes; specialised skills	

Source: Authors.

Table A.3: Production: Wine-making (Viniculture).

Service	CPC Ver. 2 code	Supplied in-house	Outsourced to affiliated companies and reasons	Outsourced to third-party suppliers/ government and reasons	Bundled
24 Production management	83115 — Operations management consulting services	Yes	No	No	
25 Architectural services: Design and construction of packing plants	83213 — Architectural services for non-residential building projects	No	No	Yes; efficiency, specialised skills	Bundled with engineering services for plant construction
26 Engineering services: Design and construction of packing plants	83310 — Engineering advisory services	No	No	Yes; efficiency specialised skills	Bundled with architectural services for plant construction
27 Wine makers	83115 — Operations management consulting services	Yes	No	Yes	
28 Crush grape services	88182 — Wine manufacturing services	Yes	No	No	
29 Fermentation. Barrel Storage	67220 — Bulk liquid or gas storage services	Yes	No	No	

30 Bottling services	85400 — Packaging services	Yes	No	No
31 Certification of Operations for Private Standards	83115 — Operations management consulting services	Yes; for recertification	No	Yes; efficiency specialised skills
32 Certification for Public Sanitary and Phytosanitary Standards	91122 — Public administrative services related to health care	No	No	Yes; government services
33 Warehousing services for final goods	67290 — Other storage and warehousing services	Yes	No	No
34 Testing/lab analysis	83441 — Composition and purity testing and analysis services	No	No	Yes; lack of expertise
35 Vineyard and wine production software services: IT services for vineyard and inventory management	Class of 8313 —IT consulting and support services Class of 8314 —IT design and development services Class of 8316 — IT infrastructure and network management services	Yes	No	Yes; specialised services

(Continued)

Table A.3: *(Continued)*

Service	CPC Ver. 2 code	Supplied in-house	Outsourced to affiliated companies and reasons	Outsourced to third-party suppliers/government and reasons	Bundled
36 Training Services	92919 — Other education and training services, n.e.c.	Yes	No	Yes; specialised services	
37 Cleaning services of Operations	85330 — General cleaning services	Yes	No	No	
38 Repair and maintenance of equipment	87156 — Maintenance and repair services of commercial and industrial machinery	Yes	No	Yes	
39 Security guards for factory and warehouses	85250 — Guard services	No	No	Yes	
40 Government inspections on fire prevention, health hazards, environmental protection and other aspects	91290 — Public administrative services related to other public order and safety affairs	No	No	Yes; government services	
41 Recruitment of temporary/seasonal workers	8512 — Labour supply services	Yes; permanent workers recruited in-house	No	Yes; efficiency and network efficiencies	Bundled with social insurance for workers

42 Social insurance for workers	91320 — Administrative services related to government employee pension schemes; old-age disability or survivors' benefit schemes, other than for government employees 91330 — Administrative services related to unemployment compensation benefit schemes	Yes; for internal personnel (required by law)	No	Yes; subcontractors contribute individually
43 Catering services for workers	63393 — Other contract food services	No	No	Yes; efficiency
44 Medical services	93121 — General medical services	No	No	Yes; efficiency

Source: Authors.

Table A.4: Distribution, Marketing and Sales.

Service	CPC Ver. 2 code	Supplied in-house	Outsourced to affiliated companies and reasons	Outsourced to third-party suppliers/government and reasons	Bundled
45 Land transport of wine to the port	65119 — Other road transport services of freight	No	No	Yes; efficiencies — economies of scale	
46 Customs-related services	85999 — Other support services n.e.c.	Yes	No	No	
47 Importer	61221 — Wholesale trade services on a fee or contract basis, of fruit and vegetables	No	No	Yes; lack of expertise, market specific	
48 Design of logo and design of packaging	83919 — Other specialty design services	No	No	Yes; efficiency, specialised skills	
49 Marketing services	83114 — Marketing management consulting services	Yes	No	No	
50 Market research	83700 — Market research and public opinion polling services	Yes	No	Yes; collective action	
51 Market analysis and Intelligence	71591 — Financial consultancy services	Yes	No	No	
52 Market entry	91138 — Public administrative services related to general economic, commercial and labour affairs	No	No	Yes; government services	

53	Sales services	61221 — Wholesale trade services on a fee or contract basis, of fruit and vegetables	Yes	No	No
54	Retail trade services By store	621 — Non-specialised store retail trade services	Yes	No	No
55	Retail services on a fee or contract bases	625 — Retail trade services on a fee or contract basis	Yes	No	No
56	Training services	92919 — Other education and train ing services, n.e.c.	Yes	No	No

Source: Authors.

Table A.5: Back Office Services.

Service	CPC Ver. 2 code	Supplied in-house	Outsourced to affiliated companies and reasons	Outsourced to third-party suppliers/ government and reasons	Bundled
57 Company registration and licensing services (obtaining permit to operate)	91138 — Public administrative services related to general economic, commercial and labour affairs	No	No	Yes; government services	
58 Headquarter services	83118 — Head office services	Yes	No	No	
59 Management services	83111 — Strategic management consulting services	Yes	No	No	
60 Finance andaccounting	Group of 822 — Accounting, auditing and bookkeeping services	Yes	No	Yes; independent audit required by law	
61 Insurance services	7131 — Life insurance and pension services 7132 — Accident and health insurance services 71334 — Other property insurance services 71335 — General liability insurance services	No	No	Yes; not possible to supply in-house	
62 Corporate communications, marketing and public relationship	83121 — Public relations services	Yes	No	No	

63	Legal services, including contract negotiations with foreign buyers	82120 — Legal advisory and representation services concerning other fields of law 82130 — Legal documentation and certification services	No	No	Yes; efficiency, specialised skills
64	Human resources management (Permanent staff)	Class of 8511 — Personnel search and referral services	Yes	No	No
65	Training services	92919 — Other education and training services, n.e.c.	Yes	No	Yes; specialised skills
66	Social insurance for staffs	91320 — Administrative services related to government employee pension schemes; old-age disability or survivors' benefit schemes, other than for government employees 91330 — Administrative services related to unemployment compensation benefit schemes	Yes	No	No
67	IT services	Class of 8313 —IT consulting and support services Class of 8314 —IT design and development services Class of 8316 — IT infrastructure and network management services	Yes; internal IT department supports day to day operations	No	Yes; efficiency, specialised skills

(*Continued*)

Table A.5: (*Continued*)

Service	CPC Ver. 2 code	Supplied in-house	Outsourced to affiliated companies and reasons	Outsourced to third-party suppliers/ government and reasons	Bundled
68 Utilities: Electricity, gas, water and telecommunications services supply andwaste removal	Class of 8631 — Support services to electricity transmission and distribution 86320 — Gas distribution services through mains (on a fee or contract basis) 86330 — Water distribution services through mains (on a fee or contract basis) 94239 — General waste collection services, other Group: 841 — Telephony and other telecommunications services	No	No	Yes; not possible to supply in-house	
69 Security services	85230 — Security systems services 85250 — Guard services	No	No	Yes; specialised skills	
70 Cleaning services	85330 — General cleaning services	Yes	No	No	

Source: Authors.

Table A.6: Policies Affecting Services in the Value Chain.

Government and private policies/services	Authority(ies) in charge	Details	How the policy affects services in the value chain
Incentives for R&D (Tax Incentives and Competitive Grant Funds)	CORFO (Economic Development Agency), CONICYT (National Commission for Scientific and Technological Research)	CORFO offers a series of grant and incentive benefits for the research and development of the wine sector. These instruments are designed to foster collaboration between local and international universities, public research institutions and the private sector	It provides inputs into improved wine production
Competitive Grant Funds for Exports in Agroforestry	PROCHILE (Export Promotion Agency)	The program provides market research and undertakes promotional activities to support the wine exports abroad. In addition, the program provides funding for commercial missions abroad, office rentals abroad for promotion, participation in international trade fairs, short-term training abroad amongst others	It provides exporters with detailed information regarding potential export markets as well as promoting wines abroad in key international events. In addition, it offers financing for other marketing and sales services in the value chain
Modernisation of Customs Services and Single-Window System	Servicio Nacional de Aduanas (National Customs Office)	The online single-window system requires exporters to submit all export related documentation for approval in one place. It is open 24 hours a day, 7 days a week provides immediate processing of documents that do not require further evaluation.	It enables governments to electronically process information, documents and fees both faster and more accurately. In doing so, it increases the speed of customs handling and processing reducing the time products is in transit for international markets

(*Continued*)

Table A.6: (*Continued*)

Government and private policies/services	Authority(ies) in charge	Details	How the policy affects services in the value chain
Tax Credit for Education and Training	Servicio Nacional de Capacitación y Empleo (SENSE) (National Training and Employment Agency)	This incentive allows firms to deduct the cost of training workers in specific workplace competencies, subject to a cap of 1% of total compensation costs. To qualify for this credit, employers must pay social security costs for participating employees. Courses must be approved by SENCE to receive this benefit and can vary from short term courses (<8 hours) to longer courses with academic equivalence (>200 hours)	It facilitates training for employees in the sector, directly contributing to productivity
NLSCS	Joint initiative between the Ministries of Economy, Education, Labour and Social Security	Skills and competencies for every job in the sector were profiled and compiled with the help of the private sector. These results were shared with technical training institutions to ensure that educational programs met industry needs. In addition, practical examinations were made available for workers already in the industry to certify their competencies regardless of how these were acquired	It facilitates sector-wide competency improvements for labour and increases labour mobility contributing to productivity and quality improvements and standards compliance
Flexible Labour Regulations for the Fruit Export Sector	Direccion de Trabajo (Ministry of Labour)	The law allows for short-term contracts for temporary workers in the fruit sector as well as permitting variations in working hours	It facilitates the provision of labour in peak periods (pruning and harvesting) in the value chain at competitive prices.

Source: Authors.

REFERENCES

Araneda, H (2010). Human capital and the national strategy for competitiveness: The case of Chile: Fundación Chile- Innovum. http://www.slideshare.net/innovum/human-capital-and-the-national-innovation-strategy-for-competitiveness-the-case-of-chile.

Chilealimentos (2010). Estándares de Competencias Laborales, Sector Alimentos. http://www.chilealimentos.com/link.cgi/Servicios/capacitacion/estandar/?pos=20&j [7 September 2010].

ChileValora (2013). Competencias Laborales Sector Vitivinicola.http://www.chilevalora.cl/index.php/noticias-index/114-noticias-2013/731-714-trabajadores-vitivinicolas-son-reconocidos-por-hacer-bien-su-trabajo [4 August 2015].

Cusmano, L, A Morrison and R Rabellotti (2010). Catching up trajectories in the wine sector: A comparative study of Chile, Italy, and South Africa. *World Development*, 38(11), 1588–1602.

Farinelli, F (2012). *Natural Resources, Innovation and Export Growth: The Wine Industry in Chile and Argentina*. Maastricht: Maastricht University.

Fernandez-Stark, K, P Bamber and G Gereffi (2010). Workforce development in the fruit and vegetable global value chain. In *Skills for Upgrading: Workforce Development and Global Value Chains in Developing Countries*, G Gereffi, K Fernandez-Stark and P Psilos (eds.). Durham: Center on Globalization Governance & Competitiveness and RTI International.

Fernandez-Stark, K, P Bamber and G Gereffi (2011). *Workforce Development in the Fruit and Vegetable Global Value Chain*. Durham: Center on Globalization Governance & Competitiveness and RTI International.

Kis, V and S Field (2009). Learning for jobs. OECD Reviews of Vocational Education and Training. In *Chile: A First Report*. OECD. http://www.oecd.org/dataoecd/33/13/44167258.pdf [15 March 2015].

López, R (2009). El Sector Frutícola Chileno y las Competencias Laborales: AGROCAP. http://www.fdf.cl/files/Rodrigo L%C3%B3pez (OTIC Agrocap) — Seminario Competencias Laborales FDF (18 Marzo CasaPiedra).pdf [18 March 2015].

Murray, N (2013). Free to trade. *Food News*. June. http://www.direcon.gob.cl/wp-content/uploads/2013/12/Oct5_DG-Food-News_Chilealimentos_Anuga.pdf [15 March 2015].

Ponte, S and J Ewert (2009). Which way is "up" in upgrading? Trajectories of change in the value chain for South African wine. *World Development*, 37(10), 1637–1650.

ProChile (2011). Estudio de Mercado Vino Embotellado en China. Santiago de Chile: Prochile. December. http://www.chilealimentos.com/medios/Servicios/noticiero/EstudioMercadoCoyuntura_2012/Jugos/Vino_China_diciembre_2011.pdf [15 March 2015].

Todovinos.cl (2014). Chile: US$1.879 Millones en Exportaciones de Vinos. http://todovinos.cl/noticias/5083-chile-us1879-millones-en-exportaciones-de-vinos [3 August 2015].

UC Davis (2008). Chilean president and Schwarzenegger preside over signing of Landmark Agreements at UC Davis.http://news.ucdavis.edu/search/printable_news.lasso?id=8691&table=news [12 May 2015].

_______. (2015). UC Davis opens life sciences innovation center in Chile. http://news.ucdavis.edu/search/news_detail.lasso?id=11209 [12 May 2015].

UN Comtrade (2015). *United Nations Commodity Trade Statistics Database*. In U. N. S. Division (ed.).

Vinmonopolet (2015). Vinmonopolet. http://www.vinmonopolet.no [4 August 2015].

Vinos de Chile (2015). Consorcio I+D Vinos de Chile. http://www.vinosdechile.cl/contenidos/investigacion/investigacion-y-desarrollo/ [3 August 2015].

Visser, E (2004). A Chilean wine cluster? Governance and upgrading in the phase of internationalization: Economic Commission for Latin America and the Carribean (September 2004).

Chapter 18

INTEGRATED LOGISTICS SOLUTIONS PROVIDER IN MEXICO

Andre Wirjo and Gloria O. Pasadilla

18.1. INTRODUCTION

This case study is not about a manufacturing firm but a logistics service provider. The analysis focuses on the increasing number of value-added services that third-party logistics service providers now contribute to manufacturing, taking over many tasks that were, in the past, part of manufacturing activities. The case study provides a "flip-side" view of manufacturing-related services, that is, the view from a service provider.

That more services are now outsourced by manufacturing firms to service suppliers have been a result of competitive pressures from a more commoditised manufacturing sector. To improve efficiency and profits, manufacturing firms have increasingly opted to focus on their core competencies and outsource many non-core activities. In the process, service providers, particularly logistics service providers, have expanded the scope of their service offerings. Not only do they provide transportation or customs facilitation services, but they now also offer inventory and warehouse management services, packaging services, and in some cases, even repair and maintenance services. The firm in this case study illustrates the range and the importance of services it provides to manufacturing firms.

18.2. BACKGROUND INFORMATION ON THE FIRM[1]

The firm is a provider of integrated logistics solutions that was formally established in year 2000 to respond to the changing competitive landscape in Mexico. The signing of the North American Free Trade Agreement (NAFTA) in 1994 ushered expectations of more foreign competition across various sectors including logistics services. In response, eight customs brokerage companies decided to join together as partners to establish one logistics firm. The eight partners rotate the presidency of the firm every two years and for the purpose of its operations, divide Mexico into eight different regions. Starting from the provision of customs brokerage services, the firm eventually ventured into domestic trucking and delivery services. It then joined a network of freight forwarders in order to qualify as air and ocean freights agent. It increasingly added more service packages such that today, the firm is an integrated logistics solutions provider offering services ranging from consolidation to packaging and warehousing services to inventory management.

It remains a major customs brokerage firm with its own supply chain consultancy department which provides advice on how to comply with Mexican customs rules and regulations. Its comparative advantage rests on its local knowledge, but it has also since strengthened its own technology and internal procedures to compete with international logistics players such as FedEx and UPS. To supplement its capacity, it uses existing networks to improve service provision, such as participating in online logistics platform whereby customers can choose and book various services such as customs clearance service, trucks and others. Though it competes with international players for the logistics business, it also provides customs clearance and brokerage services for them, thanks in part to the existing law that reserves customs broker licenses to Mexicans.

The firm currently has 45 offices that are divided into three types (Table 18.1). Border offices are located at the US–Mexico border and focused on customs brokering services for both sides of

---

[1]All information about the firm in this case study is from its corporate website as well as from the interview.

Table 18.1: Examples of Firm's Different Offices.

Border offices	Maritime offices	Internal offices
Nogales, Arizona	Altamira, Tamaulipas	Guadalajara, Jalisco
Calexico, California	Tampico, Tamaulipas	Leon, Guanajuato
San Diego, Otay, California	Manzanillo, Colima	Mexico DF
Brownsville, Texas	Puerto Vallarta, Jalisco	Monterrey, Nuevo Leon
El Paso, Texas	Veracruz, Veracruz	Puebla, Puebla
Laredo, Texas	Progreso, Yucatan	Queretaro, Queretaro
McAllen, Texas	Ciudad del Carmen, Campeche	Hermosillo, Sonora
Mexicali, Baja California	Lazaro Cardenas, Michoacán	Merida, Yucatan
Ciudad Juarez, Chihuahua	Guaymas, Sonora	Ciudad Hidalgo, Chiapas
Colombia, Nuevo Leon		Aguascalientes, Aguascalientes
Nogales, Sonora		
Matamoros, Tamaulipas		
Nuevo Laredo, Tamaulipas		
Reynosa, Tamaulipas		

Source: Courtesy of the firm.

the border. Maritime offices are located in various Mexican ports and deal primarily with customs clearance of imports and exports. Internal offices are located in the state's largest cities and act more like sales offices as well as back office support. It also has 18 distribution centres at the US–Mexico border, eight of which are located at the US side.

18.3. LOGISTICS VALUE-ADDED SERVICES IN MANUFACTURING VALUE CHAIN

The firm provides a range of services to manufacturing firms at various stages of their value chains, starting from pre-manufacturing and manufacturing to post-manufacturing and even after-sales. While in theory, the firm can supply these various services for the entire value chain of a manufacturing firm, instead, the firm provides value-added services to different manufacturers and at different stages of the chain. To one firm, it provides only pre-manufacturing services such as transport or logistics arrangements; to another, it provides post-manufacturing services like packaging and deliveries. Thus, the following discussion of services in the value chain should be seen as the spectrum of value-added services that the case study firm offers, though not in an integrated way to one single firm. While the firm provides services, it is likewise a consumer of services. Thus, services utilised by the firm are also included in the analysis.

For the purpose of this case study, the manufacturing value chain begins with the provision of services at pre-manufacturing stage and ends at the after-sales stage (see Fig. 18.1).

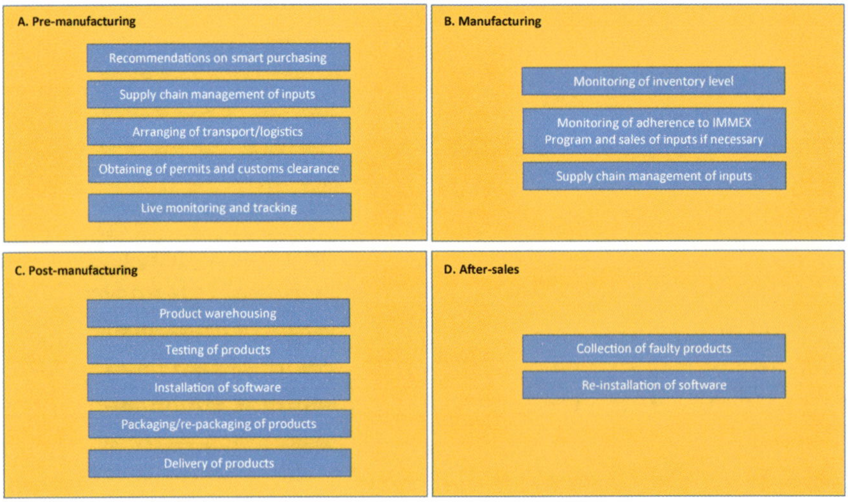

Figure 18.1: Examples of Services Provided by the Case Study Firm at Each Stage of the Manufacturing Value Chain.

Note: All the activities in the value chain are optional and depend on firm's customer on the services that it would like the firm to provide.

Source: Authors.

18.3.1. *Pre-manufacturing: Design, Materials Sourcing and Logistics*

The firm assists its customer in making smart purchasing by advising an optimal procurement and logistics strategy, taking into account the locations and movements of freight. For example it can advise on how to minimise the number of containers through improved packing of items. Smart purchasing also ensures that its customer only imports what it needs for production and avoids overstocking of production materials. If requested by customers, the firm can also involve itself in material inputs sourcing. The firm may also track and manage its customer's purchase orders. This can include scheduling the truck pick up of inputs from various locations as well as its quantity.

Brokering and logistics services like obtaining all the requisite permits for imported goods and arranging all transports such as ocean carriers, air freighters and trucks likewise remain a traditional part of its business model. On a typical pick-up day in the United States, the firm's trucks collect the ordered quantity of inputs, then bring them to different consolidation warehouses for repacking, if necessary, before being moved to the firm's warehouses located near the US–Mexico border for further consolidation. Eight out of 18 firm-owned warehouses near the border are located in the US side. Once the goods are at any of the warehouses, the firm's customs offices from both the US and Mexico sides coordinate with each other to facilitate the border crossing of the goods. If all the required documents are in order, the goods are transported into Mexico, usually by simply changing the US-licensed truck head to a Mexico-licensed one while keeping the trailer containing the goods untouched (see Fig. 18.2). This is somewhat different from the experience in the Association of Southeast Asian Nations (ASEAN) economies where the goods have to be physically removed from one ASEAN economy's truck and reloaded to another carrier that is licensed in the importing economy (see, for example, Haines, Chapter 3, this volume). In the Mexico–US case, at least, reloading need not take place. However, in specific border crossings such as Laredo, Texas, the firm may be required to use a specific border transfer

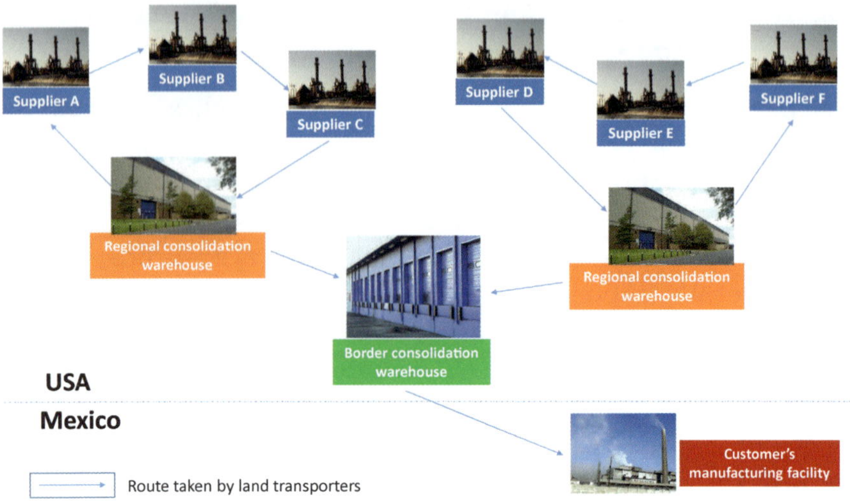

Figure 18.2: Simplified Diagram of How Smart Purchasing Services are Provided by Firm.

Source: Authors.

service from the US border to the Mexico border. This means that instead of changing the truck head only once, the change takes place twice. First, the US truck head has to be changed to the transfer service truck head to move the trailer across the customs area into the other side. Then, once at the Mexican side of the border, the transfer service truck head has to be changed again to the Mexican truck head before driving to different Mexican destinations. In terms of maximum capacity per truck, 20 tonnes are allowed on both sidesand fines are levied if truck exceeds this capacity.

For the entire logistics process, the firm uses a monitoring system — its own, that of the freight service providers or a customer-provided system. The firm may also provide its customers with the tracking system if it wants to monitor its own goods' trail themselves.

18.3.2. *Manufacturing Process*

During the manufacturing stage, the firm can provide value-added services such as inventory management, ensuring that there are always sufficient inventory to work on, and renewing orders the moment the materials inventory falls to a specific threshold level.

Importantly, the inventory management service keeps tab of the duration of storage of the inputs. The Mexican government allows for a maximum of 18 months for imported inputs to be used in the exported goods if it were to qualify for import duty exemption and other incentives.[2] The tax saved by exporting companies through this value-added service by the firm should be considerable if the manufacturing company were to outsource this service.

18.3.3. *Post-manufacturing*

Value-added services offered by the firm in the post-manufacturing stage includes storage of finished products at its warehouses (either before export or immediately after import) as well as product quality testing using various product standards.

---

[2]The Mexican government published the Decree for the Promotion of the Manufacturing, Maquila and Export Services Industry (IMMEX Decree) in November 2006 to increase the competitiveness of its export sector by simplifying the compliance regime and reducing firms' operational costs. The IMMEX Program allows firms to temporarily import material inputs, machinery and equipment without paying the general import tax, value-added tax (VAT) and in some cases, countervailing duties if these inputs are intended for use in the production, transformation and repair of goods for exports. However, a maximum limit of 18 months applies (PwC Mexico, 2014), meaning that the exemption is applicable only if the imported goods are used to make the finished products and exported within 18 months. More details of the IMMEX Program can be obtained at: http://www.economia.gob.mx/industry/foreign-trade-instruments/immex. It is noted that the 2013/2014 tax reform has led to a revision of the IMMEX Program such that VAT and excise tax exemptions on the temporary import of goods have been abolished. This means that maquiladora now needs to pay VAT rate of 16 percent first at point of entry. It then has to obtain certification from the Mexican tax authorities. If it is certified, the VAT that is technically imposed on the imported goods will be eliminated by a full tax credit. If it cannot obtain certification, satisfying liability for VAT/excise tax on imported goods can be done by placing a security via a bond issued by an authorised entity (Deloitte, 2014). However, regardless of the change, it is understood that there is still a maximum period of 18 months where imported goods have to be used to make the finished products and exported to be eligible for the incentive in one form or another.

Another important service provided by the firm at this stage is re-packaging services, for example, providing labels or instruction manuals in Spanish, or changing plugs of electrical equipment to fit the Mexican power source. In one customer case, the firm provided the software installation services for imported computers from Chinese Taipei and the service was done at its own warehouse.

When products are ready, the firm handles the distribution logistics regardless of whether they are to be exported or for the domestic market. Information on where and who they should be delivered to are provided by the firm's customer.

18.3.4. *After-sales Services*

Firm's involvement in after-sales services usually revolves around collection of faulty products from end users to the manufacturing firm's repair stations. For one computer manufacturer customer, the case study firm provided the software re-installation service for the manufacturing firm's computers because it has no affiliate presence in Mexico.

18.4. SERVICES ALONG THE VALUE CHAIN

18.4.1 *Services Identification and Value Contribution*

A total of 40 services (see Fig. 18.3), which are composed of at least 71 services (if classified by UN Central Product Classification, Ver. 2) are identified in the value chain and they are categorised according to the various stages within the chain: (i) pre-manufacturing services, (ii) services during manufacturing, (iii) post-manufacturing services, (iv) after-sales services and (v) back office services.

Of the 71 services, the firm supplies 28 services and utilises 44 services. One service — training services, particularly for tracking/ monitoring systems at the pre-manufacturing stage — is both provided and utilised.[3] With the exception of call centre services, all

[3]Training services at the pre-manufacturing stage is counted only once even though it is both supplied and utilised.

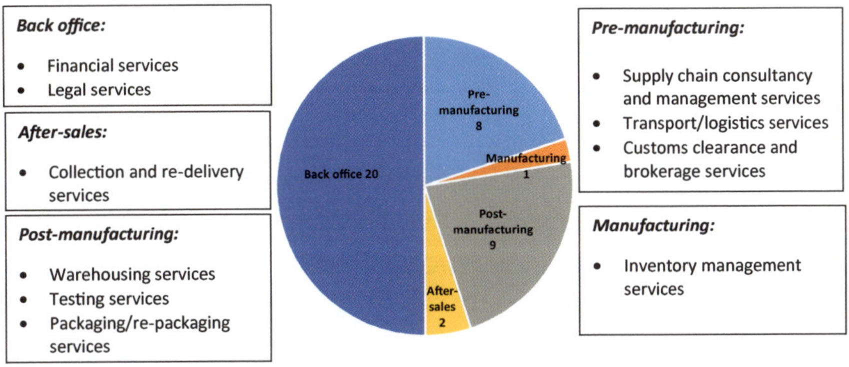

Figure 18.3: Breakdown of Services by Stage and Examples of Key Services.

Source: Authors.

the provided services are found at the pre-manufacturing to after-sales stage. Examples of these services are supply chain consultancy and management services; customs clearance and brokerage services; inventory management services; and packaging/re-packaging services. This is very much in line with firm's strategy and objective of providing services to manufacturing firms at these stages of their value chains.

All the utilised services except training services are found at the back office stage and this is intuitive because just like its manufacturing counterparts, services firms themselves need these services to ensure the smooth functioning of their operations. Examples of such services are financial services, legal services, human resources services and telecommunications services.

18.4.2. *Outsourcing, Bundling and other Aspects of Services Supply*

Among the 71 services classified by UN Central Product Classification, Ver. 2, 18 services are supplied in-house, 19 are partially outsourced while 34 are fully outsourced (see Appendix A). Dividing these services into whether they are supplied or utilised by the firm prior to analysing its outsourcing decisions reveals interesting observations. Of the 28 services supplied by the firm, 14 are provided solely in-house, 6 are partially outsourced while 8 are fully outsourced. On the contrary, of the 44 services utilised by the firm, 5 are provided in-house, 12 are partially outsourced while 27 are fully outsourced.

The fact that the share of in-house services is higher in supplied relative to utilised services (50 vs. 11 per cent) indicates that if possible, firm would keep supplied services, where it has comparative advantage and serve to differentiate itself relative to other logistics firms, in-house. These supplied services include customs clearance and brokerage services, the primary service provided by the firm, as well as value-added services such as supply chain consultancy services, testing services and re-packaging services. However, as the statistics show, this does not mean that all supplied services are provided in-house. Indeed, around half of supplied services are either fully or partially outsourced, for reasons such as: (i) firm's strategy not to be asset-heavy which requires the purchase of a large fleet of trucks, ocean freighters and aircrafts for logistics services; (ii) absence of warehouses in certain cities/towns for warehousing services; (iii) strategy of not hiring many headcounts who would only be needed for one-off tasks and for a short period such as labelling certain products; and (iv) lack of expertise such as freight insurance services.

On utilised services, services are generally outsourced for the following reasons: (i) government services, such as inspections pertaining to environmental, health and safety (EHS); (ii) required by laws and regulations, such as external auditing services by third-party providers; (iii) lack of expertise or specialisation in-house to provide certain services, such as IT and legal services; (iv) lack of feasibility to supply service in-house, such as banking and utilities services; (v) need for strong relationships with government agencies, such as company registration and licensing services; (vi) economies of scale such as property management services; and (vii) network economies, such as recruitment services.

18.5. POLICIES AFFECTING THE VALUE CHAIN

An essential part of this study is the analysis of how policies affect the value chain (see Appendix B).

18.5.1. *Security Concerns*

The integrated logistics solutions provided by the firm are heavily reliant on the smooth movement of goods from its origin to intended destination. Inputs and products need to be transported in a timely manner and with no or little damage or loss. While there are many reasons why goods could be delayed and/or lost during transport, security is one of the main contributing factors in Mexico.[4] Indeed, in a survey conducted by World Bank for the Logistics Performance Index (2014), the experience of criminal activities (e.g., stolen cargo) is reported as one of the major logistics problems in Mexico. Other problems pointed out for shipment delays are pre-shipment inspection and solicitation of informal payments.

Firm representatives estimate that security concerns add about 6 per cent of freight value to the cost of logistics. The increase comes in the form of extra insurance and use of additional measures such as live tracking of trucks and freights or, if necessary, armed escorts and armoured trucks. To address the security issue, the firm has strict rules and regulations concerning confidentiality whereby only the staff who is handling a specific customer know exactly the goods that are being transported and the specific logistical arrangement, that is the time and route of the truck. The firm noted that in many cases, crimes happened because of information provided by the staff of logistics providers.

However, it should be acknowledged that security is more of an issue for the transport of finished goods and less for raw materials and intermediate goods because the former can be easily resold in some black market somewhere while raw materials and

---

[4]More details on the state of security in these states can be obtained at: http://travel.state.gov/content/passports/english/alertswarnings/mexico-travel-warning.html.

intermediate inputs are harder to dispose of. In addition, security is a concern only for certain states and/or routes. The usual routes taken to transport goods from Mexico to the United States and vice versa are generally safe. The same can be said for routes leading to Mexican ports.

18.5.2. *Infrastructure and Customs*

Relative to the situation 20 years ago, Mexico has come a long way in terms of logistics improvements. Mexico enlarged its ports and allowed management of ports by foreigners. Mexican ports, in fact, now serve as good substitutes for saturated US ports such as Long Beach, California and Houston, Texas.

However, the Logistics Performance Index (LPI) released by World Bank shows that Mexico still ranks at the bottom six among Asia-Pacific Economic Cooperation (APEC) economies based on overall score. This single score takes into account six main components,viz.: (i) the efficiency of customs and border management clearance; (ii) the quality of trade and transport infrastructure; (iii) the ease of arranging competitively priced shipments; (iv) the competence and quality of logistics services such as trucking, forwarding and customs brokerage; (v) the ability to track and trace consignments; and (vi) the frequency with which shipments reach consignees within scheduled or expected delivery times. Specifically on (i) customs and (ii) infrastructure, Mexico ranks at the bottom four and six, respectively. In some ports, it still takes up to three days to unload a ship. The firm considers port services in Mexico expensive.

18.5.2.1. *Single windows system*

The Mexican government released a decree on the establishment of single windows system in January 2011. The system, which is called Ventanilla Digital Mexicana de Comercio Exterior or Ventanilla Única became effective in June 2012.[5] This electronic

[5]For the original decree establishing the single windows system in Spanish, please refer to: http://www.economia.gob.mx/files/marco_normativo/D135.pdf.

system serves as a platform for foreign trade operators including customs brokers to electronically file all foreign trade information required by the relevant authorities. It is expected to lead to faster customs clearance and reduced customs costs because it automates and standardises customs processing.[6]

The issue of single windows is an interesting one from the viewpoint of the case study firm. As providers of customs brokerage services, the firm should not be supportive of single windows initiative because by facilitating customs clearance processes, it could take business away from the firm. However, as a provider of integrated logistics solutions, the firm is very supportive of the initiative because the loss of business from customs brokerage services is compensated by the additional businesses spawned by efficient customs systems.

Despite the implementation of single windows, however, customs brokerage services still contribute more than 50 per cent of the case study firm's revenue. This is one evidence that the single windows system has not supplanted the need for custom brokerage services in Mexico, presumably because importers do not want to take the risk of delays in their goods' release due to inadvertent mistakes in the customs declaration or incomplete information or whatever tiny wrongs that customs authorities can find. Customs procedures expertise, not to mention network and connection, apparently fill the gap that the single window system cannot solve.

18.5.2.2. *Transfer services*

In certain border crossing such as Laredo (Texas, United States) — Nuevo Laredo (Tamaulipas, Mexico), transfer services are mandatory to move goods from the Mexican side of the border to the US side and vice versa at a rate of USD120 per truck. Historically, the long wait experienced by truck drivers (up to 8 hours) to clear the customs motivated the establishment of this service. Mexican or American drivers would hand their cargo (through the change in

[6]For more information on the single windows system, please refer to: http://export.gov/mexico/static/VentanillaUnica_Latest_eg_mx_050491.pdf.

truck heads described previously) over to the service providers, who would then wait in line until the trucks are on the other side of the border. With improvements in infrastructure, the wait is now relatively shorter and there is supposedly no more economic reason to use transfer services. However, the transfer service providers union, understandably, prefer the status quo in which they have monopoly moving trucks within the customs area, thus creating an additional logistics cost.

18.5.3. *The Way Forward*

In contrast to the other manufacturing case studies, this case study is about the perspectives of a services firm that supplies services to manufacturing value chains. The complexity of many global value chains as well as the preoccupation about minimising cost have created opportunities for services companies that can provide cheaper value-added services alternative to manufacturing in-house provision of many services. The integrated logistics company in this case acts almost like a "jack of all trades", providing services ranging from the traditional services such as transportation and customs brokering, to less traditional services like inventory management, re-packaging and re-labelling, repairs and ad-hoc services like software installations.

The firm has anticipated the liberalising trend in Mexico by partnering with erstwhile competitors to be able to compete better with global logistics companies that would eventually have freer rein in the Mexican market. Indeed, policies like port management liberalisation had impacted it for the good; at the same time, some policies (like the mandatory use of transfer services in certain customs area) help make their cost higher. Moreover, the fact that the case study firm is a domestic firm operating in its own economy validates the view that improved policies and better business environment will not only benefit foreign firms but also, and in the first place, domestic firms.

APPENDIX A

Table A.1 Pre-manufacturing Stage Including Sourcing and Importation of Raw Materials.

	Service	Corresponding CPC ver. 2 code	Utilised or provided services*	In-house	Outsourced to affiliated companies	Outsourced to third-party suppliers/ government and reasons	Bundled
1	Supply chain consultancy services	83116 — Supply chain and other management consulting services	Provided	Yes	No	No	
2	Supply chain management services	83116 — Supply chain and other management consulting services	Provided	Yes	No	No	
3	Procurement services	83116 — Supply chain and other management consulting services	Provided	Yes	No	No	

(*Continued*)

Table A.1 (*Continued*)

Service	Corresponding CPC ver. 2 code	Utilised or provided services*	In-house	Outsourced to affiliated companies	Outsourced to third-party suppliers/ government and reasons	Bundled
4 Transport services of raw materials	651 — Land transport services of freight	Provided	Yes	No	Yes, efficiency; economies of scale	Bundled as part of the logistics agreement
	652 — Water transport services of freight	Provided	No	No	Yes, efficiency; economies of scale	Bundled as part of the logistics agreement
	6531 — Air transport services of freight	Provided	No	No	Yes, efficiency; economies of scale	Bundled as part of the logistics agreement
	67910 — Freight transport agency services and other freight transport services	Provided	No	No	Yes, efficiency; economies of scale	Bundled as part of the logistics agreement

5	Freight insurance of raw materials	71333 — Freight insurance services	Provided	No	No	Yes, efficiency; economies of scale
6	Training services	92919 — Other education and training services, n.e.c.	Provided/ Utilised	Yes	No	Yes, lack of expertise
7	Storage and warehousing services of raw materials	67290 — Other storage and warehousing services	Provided	Yes	No	Yes, absence in certain cities
8	Customs clearance services and logistics of raw materials	67110 — Container handling services	Provided	Yes	No	No
		85999 — Other support services n.e.c.	Provided	Yes	No	No

*In this table and subsequent ones, "Provided" means that the case study firm supplies the service; "Utilised" means that the firm consumes or uses the service for its own operations.

Source: Authors.

Table A.2: Manufacturing Stage.

Service	Corresponding CPC ver. 2 code	Utilised or provided services	In-house	Outsourced to affiliated companies	Outsourced to third-party suppliers/ government and reasons	Bundled
9 Inventory management services	83116 — Supply chain and other management consulting services	Provided	Yes	No	No	

Source: Authors.

Table A.3: Post-manufacturing Stage.

Service	Corresponding CPC ver. 2 code	Utilised or provided services	In-house	Outsourced to affiliated companies	Outsourced to third-party suppliers/ government and reasons	Bundled
10 Warehousing services for products	67290 — Other storage and warehousing services	Provided	Yes	No	Yes, absence in certain cities	
11 Product testing services	8344 — Technical testing and analysis services	Provided	Yes	No	No	
12 Labelling services	89121 — Printing services	Provided	Yes	No	Yes, efficiency; lack of expertise	
13 Training services	92919 — Other education and training services, n.e.c.	Utilised	Yes	No	Yes, lack of expertise	

(Continued)

Table A.3: *(Continued)*

Service	Corresponding CPC ver. 2 code	Utilised or provided services	In-house	Outsourced to affiliated companies	Outsourced to third-party suppliers/ government and reasons	Bundled
14 Software installation services	87332 — Installation services of personal computers and peripheral equipment	Provided	Yes	No	No	
15 Packaging/ re-packaging services	85400 — Packaging services	Provided	Yes	No	No	
16 Customs clearance services and logistics of parts and products	67110 — Container handling services	Provided	Yes	No	No	
	85999 — Other support services n.e.c.	Provided	Yes	No	No	

17	Transport services of parts and products	651 — Land transport services of freight	Provided	Yes	No	Yes, efficiency; economies of scale	Bundled as part of the logistics agreement
		652 — Water transport services of freight	Provided	No	No	Yes, efficiency; economies of scale	Bundled as part of the logistics agreement
		6531 — Air transport services of freight	Provided	No	No	Yes, efficiency; economies of scale	Bundled as part of the logistics agreement
		67910 — Freight transport agency services and other freight transport services	Provided	No	No	Yes, efficiency; economies of scale	Bundled as part of the logistics agreement
18	Freight insurance of products	71333 — Freight insurance services	Provided	No	No	Yes, efficiency; economies of scale	

Source: Authors.

Table A.4: After-sales Services.

Service	Corresponding CPC ver. 2 code	Utilised or provided services	In-house	Outsourced to affiliated companies	Outsourced to third-party suppliers/ government and reasons	Bundled
19 Collection and re-delivery services	651 — Land transport services of freight	Provided	Yes	No	Yes, efficiency; economies of scale	Bundled as part of the logistics agreement
20 Software installation services	87332 — Installation services of personal computers and peripheral equipment	Provided	Yes	No	No	

Source: Authors.

Table A.5: Business Processes (Back Office Support).

Service	Corresponding CPC ver. 2 code	Utilised or provided services	In-house	Outsourced to affiliated companies	Outsourced to third-party suppliers/ government and reasons	Bundled
21 Company registration and licensing services (obtaining permit to operate)	91138 — Public administrative services related to general economic, commercial and labour affairs	Utilised	Yes	No	Yes, lack of expertise; strong relationship with government agencies; government services	
22 Government licensing and inspections on fire prevention, health hazards, environmental protection and other aspects	91134 — Public administrative services related to transport and communications	Utilised	Yes	No	Yes, lack of expertise; strong relationship with government agencies; government services	

(*Continued*)

Table A.5: (*Continued*)

Service	Corresponding CPC ver. 2 code	Utilised or provided services	In-house	Outsourced to affiliated companies	Outsourced to third-party suppliers/ government and reasons	Bundled
	91290 — Public administrative services related to other public order and safety affairs	Utilised	Yes	No	Yes, lack of expertise; strong relationship with government agencies; government services	
23 Headquarter services	83118 — Head office services	Utilised	Yes	Yes	No	
24 Management services	83111 — Strategic management consulting services	Utilised	Yes	No	No	
25 Accounting, auditing and bookkeeping services	82210 — Financial auditing services	Utilised	No	No	Yes, required by laws and regulations	
	8222 — Accounting and bookkeeping services	Utilised	Yes	No	No	

26	Financial services	71121 — Deposit services to corporate and institutional depositors	Utilised	No	No	Yes, not possible to supply in-house
		71140 — Financial leasing services	Utilised	No	No	Yes, not possible to supply in-house
		71313 — Group pension services	Utilised	No	No	Yes, not possible to supply in-house
		83112 — Financial management consulting services	Utilised	No	No	Yes, not possible to supply in-house
27	Legal services	82120 — Legal advisory and representation services concerning other fields of law	Utilised	No	No	Yes, lack of expertise
		82130 — Legal documentation and certification services	Utilised	No	No	Yes, lack of expertise

(*Continued*)

Table A.5: (*Continued*)

Service	Corresponding CPC ver. 2 code	Utilised or provided services	In-house	Outsourced to affiliated companies	Outsourced to third-party suppliers/ government and reasons	Bundled
28 Insurance services (commercial life and accident/ health insurance, property insurance for the factory compound, product quality insurance, management liability insurance)	7131 — Life insurance and pension services	Utilised	No	No	Yes, better portfolio; not possible to supply in-house	
	7132 — Accident and health insurance services	Utilised	No	No	Yes, better portfolio; not possible to supply in-house	
	71331 — Motor vehicle insurance services	Utilised	No	No	Yes, better portfolio; not possible to supply in-house	
	71332 — Marine, aviation and other transport insurance services	Utilised	No	No	Yes, better portfolio; not possible to supply in-house	
	71334 — Other property insurance services	Utilised	No	No	Yes, better portfolio; not possible to supply in-house	

		71335 — General liability insurance services	Utilised	No	No	Yes, better portfolio; not possible to supply in-house
29	Human resources services	91320 — Administrative services related to government employee pension schemes; old-age disability or survivors' benefit schemes, other than for government employees	Utilised	Yes	No	Yes, lack of expertise
		8511 — Personnel search and referral services	Utilised	Yes	No	Yes, efficiency; network economies
		8512 — Labour supply services	Utilised	Yes	No	Yes, efficiency; network economies

(*Continued*)

Table A.5: (*Continued*)

Service	Corresponding CPC ver. 2 code	Utilised or provided services	In-house	Outsourced to affiliated companies	Outsourced to third-party suppliers/ government and reasons	Bundled
	83113 — Human resources management consulting services	Utilised	Yes	No	No	
30 Corporate communications and public relationship	83121 — Public relations services	Utilised	Yes	No	No	
31 Courier, postal and local delivery services	681 — Postal and courier services	Utilised	No	No	Yes, not possible to supply in-house	
32 Information technology services including for remote monitoring and visibility of freights and stocks	8313 — Information technology (IT) consulting and support services	Utilised	Yes	No	Yes, lack of expertise	
	8314 — Information technology (IT) design and development services	Utilised	Yes	No	Yes, lack of expertise	

		83151 — Website hosting services	Utilised	Yes	No	Yes, lack of expertise	
		8316 — IT infrastructure and network management services	Utilised	Yes	No	Yes, lack of expertise	
33	Telephone-based support services	85931 — Telephone call centre services	Provided	Yes	No	No	
34	Telecommunication services	841 — Telephony and other telecommunications services	Utilised	No	No	Yes, not possible to supply in-house	
		84210 — Internet backbone services	Utilised	No	No	Yes, not possible to supply in-house	
		8422 — Internet access services	Utilised	No	No	Yes, not possible to supply in-house	
35	Maintenance and repair of transport vehicles	8714 — Maintenance and repair of transport machinery and equipment	Utilised	Yes	No	Yes, efficiency; economies of scale	Bundled as part of the logistics agreement

(Continued)

Table A.5: (*Continued*)

Service	Corresponding CPC ver. 2 code	Utilised or provided services	In-house	Outsourced to affiliated companies	Outsourced to third-party suppliers/ government and reasons	Bundled
36 Property management services for firm	72212 — Non-residential property management services on a fee or contract basis	Utilised	No	No	Yes, efficiency; economies of scale	
37 Leasing services	72112 — Rental or leasing services involving own or leased non-residential property	Utilised	No	No	Yes, efficiency; economies of scale	
	660 — Rental services of transport vehicles with operators	Utilised	No	No	Yes, efficiency; economies of scale	
	7311 — Leasing or rental services concerning transport equipment without operator	Utilised	No	No	Yes, efficiency; economies of scale	

38	Security services	85230 — Security systems services	Utilised	No	No	Yes, lack of expertise
		85250 — Guard services	Utilised	No	No	Yes, lack of expertise
39	Cleaning services	853 — Cleaning services	Utilised	No	No	Yes, lack of expertise
40	Utilities (electricity, gas and water supply)	691 — Electricity and gas distribution (on own account)	Utilised	No	No	Yes, not possible to supply in-house
		692 — Water distribution (on own account)	Utilised	No	No	Yes, not possible to supply in-house

Source: Authors.

APPENDIX B

Table B.1: Policies Affecting Services in the Value Chain.

Government policies/services	Authority(ies) in charge	Details	How the policy affects services in the value chain
State of security in parts of Mexico	Various law enforcement agencies	The security situations in certain parts of Mexico raise the probability that trucks may be attacked and goods/freights stolen during transportation.	Firm representatives shared that concerns over security could add to the cost of logistics by approximately 6 percent of freight value. The increase in cost is attributed to extra insurance and additional measures in the form of live tracking of trucks and freights and if necessary, armed escorts and armoured trucks.
Single windows	Servicio de Administración Tributaria (SAT) Secretaría de Economía	Single windows initiative facilitates customs clearance process by establishing a single-stop shop where users could obtain all the required permits for imports. However, despite the implementation, users of single windows face challenges such as getting their goods held at the customs because of wrong declaration, incomplete submission of forms and so on.	Although an alternative to firms, operationalisation of single windows indicate that it is still not a good substitute for customs brokerage services. It may end up increasing the cost and time to clear goods at the customs because firms would begin by using single windows only to obtain the services of customs brokerage halfway.

Transfer services in certain US–Mexico border	Transfer services are provided to move goods from the Mexican side of the border to the US side and vice versa at a rate of USD120 per truck. These transfer services were originally established with the intent to facilitate the long wait at the border, which could stretch up to 8 h. Truck drivers would hand over their trucks over to the service providers, which would then wait in line until the trucks were safely on the other side of the border. However, improvements in infrastructure have made the transfer services obsolete and only serve to extract economic rents from users.	The transfer services raise the logistics cost to the firm.

Source: Authors.

REFERENCES

Deloitte (2014). Taxation and Investment in Mexico 2014. Mexico City: Deloitte. http://www2.deloitte.com/content/dam/Deloitte/global/Documents/Tax/dttl-tax-mexicoguide-2014.pdf [8 July 2015].

Haines, W (2016). Industrial welding services in Thailand. In *Services in Global Value Chains: Manufacturing-Related Services*. Singapore: APEC Secretariat.

Mexico Travel Warning. Bureau of Consular Affairs, U.S. Department of State. http://travel.state.gov/content/passports/english/alerts warnings/mexico-travel-warning.html [8 July 2015].

PwC Mexico (2014).*Doing Business in Mexico — Automotive Industry.* Mexico City: PwC. http://www.pwc.de/de/internationale-maerkte/assets/doing-business-mexico-automotive.pdf. [8 July 2015].

Velazquez, M (2012). Requirement for Exporting to Mexico: Single Window for Trade/Ventanilla Unica para el Comercio Exterior (VUCEM). U.S. Commercial Service, May 2012. http://export.gov/mexico/static/VentanillaUnica_Latest_eg_mx_050491.pdf [17 July 2015].

World Bank Logistics Performance Index (2014). http://lpi.worldbank.org/ [17 July 2015].

Chapter 19

REMANUFACTURING SERVICES IN THE CONSTRUCTION MACHINERY VALUE CHAIN

Katherine Tait and Gary Gereffi

19.1. FIRM OVERVIEW

19.1.1. *History of the Firm*

The firm central to this study began in 1925 as a tractor company based in San Leandro, California, and moved its headquarters to East Peoria, Illinois in 1930. In the 1950s, while its competitors focused on agricultural equipment for the domestic market, the case study firm focused its manufacturing on earth-moving construction equipment and was open to trading internationally. Its remanufacturing production began in 1972, making the company among the first major original equipment manufacturers (OEMs) in the heavy-duty truck and off-roading (HDOR) equipment sector to become involved in remanufacturing. By the late 1980s, the company offered a broad range of earth-moving construction equipment, which included highway trucks, power generation equipment including gas and diesel turbines, as well as engine manufacturing.

19.1.2. *Situation within Construction Machinery Industry*

Today, the firm is the global leader in the manufacture of construction and mining equipment, diesel and natural gas engines, industrial gas turbines, diesel-electric locomotives and mobile electricity generating equipment including drilling rigs and electricity for cruise ships. Its

2014 sales and revenues totalled $55.2 billion Securities and Exchange Commision (SEC), 2014; Filings 10K).[1] The company is structured into seven segments: Construction Industries; Resource Industries; Energy and Transportation; Financial Products; Corporate Services; Customer and Dealer Support; and the Enterprise System Group (Company source, 2014a, 2015f).[2] The company currently employs 114,233 full-time employees globally (Company source, 2015b; SEC, 2014; Filing 10K).

The case study firm is among the three largest remanufacturers of HDOR equipment in the United States.[3] The company describes its current business model with the expression "Seed, Grow, Harvest", which refers to its aims of delivering equipment to clients in the field ("seed"), expanding its market share ("grow"), and maintaining its end-products through on-going client service and the production of remanufactured goods ("harvest"). The latter point is important for this study of manufacturing-related services because the goal of "harvesting" refers to the firm ethic of producing high quality, high value capital goods, which can be maintained through extensive service and maintenance options (Company source, 2015e).[4]

19.1.3. *Remanufacturing at the Firm*

The industrial process of remanufacturing restores end-of-life manufactured goods to OEM specifications. Remanufacturing

---

[1]The firm was ranked first in its industry and in the top 100 overall in the 2015 *Fortune* 500 list (Fortune, 2015).

[2]Its three main product segments are: Construction; Resource and Energy; and Transportation. Additional key segments include Customer & Dealer Support, Corporate Services, the Enterprise System Group and Law & Public Policy (Company source, 2015f). Remanufacturing, Components and Work Tools falls under the Customer and Dealer Support segment, and its remanufacturing division emerged as a distinct division in 2005.

[3]HDOR equipment includes "trucks, bulldozers, excavators, backhoes, asphalt pavers and rollers, farm tractors and combines, rock cutters and tunneling machinery, and oil and gas drilling machinery" (USITC, 2012: 83).

[4]The firm continues to carry parts for any of its own equipment that is still in operation in the world.

involves high levels of design, engineering and technical skill, and its sophistication distinguishes it from refurbishment, rebuilding, recycling and reuse. According to the 2012 US International Trade Commission (USITC) study, the United States is the largest remanufacturer globally, and the value of this type of production is steadily increasing, having reached at least $43 billion and supporting over 180,000 full-time jobs by 2011 (USITC, 2012: i). Heavy-duty and off-road (HDOR) equipment is one of the major remanufacturing sectors in the United States.[5]

Remanufacturing is particularly important for manufacturers of large and expensive construction and mining equipment. By closing the loop on the manufacturing process and introducing reverse logistics, remanufacturing enables the firm to integrate all of the manufacturing-related services along the value chain. It also enables the firm to retain the value of its products, stabilise demand for replacement components, and leverage its exclusive dealer network to provide information and after-sales services to customers. The company is a leader in remanufacturing for the earth-moving equipment industry (Stahel, 1995).

19.1.3.1. *The remanufacturing process*

The dedicated remanufacturing line of the case study firm, which emerged as its own division in 2005, is consistent with its "Seed, Grow, Harvest" business model (Company source, 2015e; Fiksel, 2006). As shown in Fig. 19.1, the remanufacturing division employs over 4000 workers across 17 dedicated remanufacturing facilities (Company source, 2015g). Located in China, France, Mexico, Singapore, the United Kingdom and the United States, these facilities are specifically outfitted to receive, inspect and remanufacture core components (Asia-Pacific Economic Cooperation and United States Agency for International Development [APEC and USAID], 2013; Fiksel, 2012).

---

[5] The majority of US remanufacturing occurs in "aerospace, consumer products, electrical apparatus, heavy-duty and off-road equipment, information technology (IT) products, locomotives, machinery, medical devices, motor vehicle parts, office furniture, restaurant equipment, and retreaded tires" (USITC, 2012: i).

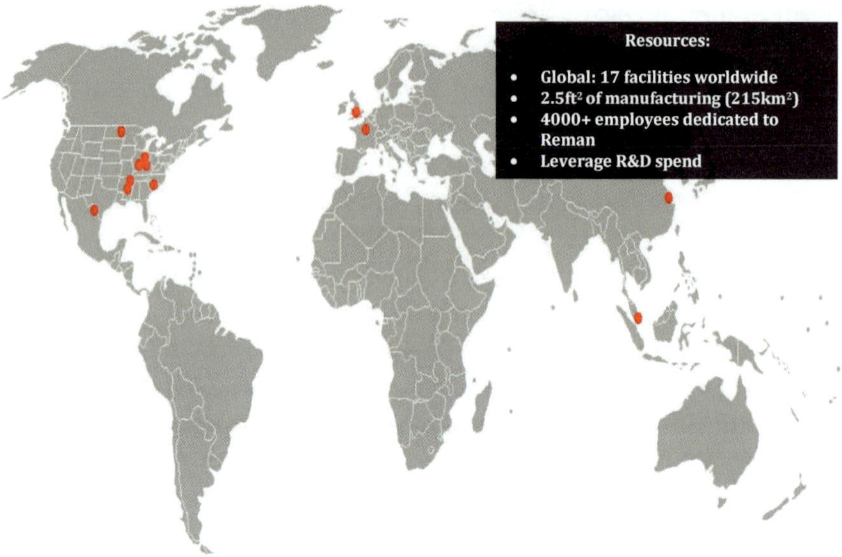

Figure 19.1: Dedicated Remanufacturing Facility Locations.
Source: Company source (2015g).

Remanufacturing earned $1 billion in sales in 2005 and earnings have steadily increased each year since (Fiksel, 2006; Hindo and Arndt, 2006).[6] The case study firm has steadily expanded its remanufacturing capacity to allow for the restoration of components across its product line of over 7200 products (Company source, 2015g).

Figure 19.2 shows how manufacturing operations at the firm are fully integrated and linked to the remanufacturing program for its entire product line. For clients, purchasing a remanufactured product means receiving the same form, fit, function and warranty as a new product at about half the cost (Company source, 2015e). As depicted in Fig. 19.2, the firm describes this as the "lowest cost life cycle" available for their high-quality goods: remanufactured products are expected to perform as well as new products, and also be equally reliable and durable. The remanufacturing process at the firm is centred on core components of finished products, such as engines.

[6]Financial accounting at the firm is performed on an end-to-end basis. As a result, data are not available detailing the percentage of overall business that remanufactured products comprise (Company source, 2015e).

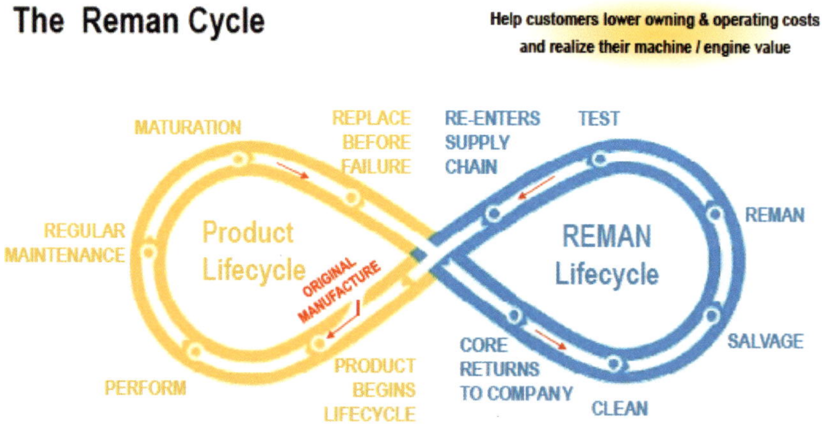

Figure 19.2: Product Life Cycle: Integrating Manufacturing and Reman.
Source: Company source (2015g).

The firm describes its remanufacturing process as follows:

> Our remanufacturing operations are primarily focused on the remanufacture of engines and components and rail-related products. In this business, used engines and related components (core) are inspected, cleaned and remanufactured. In connection with the sale of most of our remanufactured product, we collect a deposit from the dealer that is repaid if the dealer returns an acceptable core within a specified time period. [The firm] owns and has title to the cores when they are returned from dealers. The rebuilt engine or component (the core plus any new content) is then sold as a remanufactured product to dealers and customers (SEC Company Form10-K, 2014: A-12).

The core deposit thus renders end-of-life or worn components to be valuable if dealers find them eligible for remanufacture. This, in turn, gives customers financial incentive to return products that need to be replaced. Once the dealer inspects and accepts the old product, the dealer ships the core to a core-receiving facility, which confirms the dealer's inspection and reimburses the dealer for the deposit. Next, the core is shipped to a remanufacturing facility, where it is disassembled into its individual components, down to each nut and bolt. Qualifying components are then restored in the remanufacturing process, while components that fail inspections are recycled.

Later stages of the process include reassembly, product testing, and preparation for re-sale and after-sales services. The use of its own core components for remanufacturing enables the firm to be vertically integrated throughout its reproduction processes. The role of each phase of this circuit in the value chain will be discussed in the next section.

19.2. MANUFACTURING SERVICES VALUE CHAIN

19.2.1. *Global Value Chain Analysis of the Firm's Services*

The global value chain (GVC) framework is an analytical tool used to examine the global expansion and organisation of supply chains. GVC analysis is particularly useful for better understanding how value creation and capture take place within supply chains, and also for identifying the causes and effects of the reconfiguration taking place among organisations in global industries (Gereffi, 2015: 1, 5). Through its emphasis on how "new patterns of international trade, production, and employment shape prospects for development and competitiveness", the GVC framework highlights the international expansion and geographic fragmentation of contemporary supply chains, and focuses on issues of industry (re)organisation, coordination, governance and power in the chain (Gereffi, 2015: 1; Gereffi and Lee, 2012). In addition, the framework employs the central concepts of "governance" and "upgrading" to demonstrate these dynamics (Gereffi, 2015: 1; Gereffi and Lee, 2012).[7] The concept of

[7]Governance and upgrading represent two converse but complementary ways to view the global expansion of supply chains. *Governance* represents the "top-down" perspective that concerns lead firms, the organisation of global industries and how power influences the distribution of both risk and profits in an industry (Gereffi, 2015: 5). *Upgrading*, by contrast, represents the "bottom-up" perspective and concerns local, regional and national actors interested in strengthening

governance is particularly relevant when applying GVC analysis to lead firms, given that the closed-loop production process that remanufacturing introduces carries with it important opportunities for economic, social and environmental upgrading.

Manufacturing- and remanufacturing-related services are not segmented or discrete; both cut across the entire value chain. Equipment manufacturers typically conduct substantial and highly successful business through services, which can generate up to half of total company revenues (Boston Consulting Group (BCG), 2014: 3). A recent analysis of service performance within the machinery sector has shown that services perform better than machinery sales by having faster growth rates, higher profitability, greater predictability and more resilience during economic downturns (BCG, 2014: 3–4). The degree of value that services provide to these manufacturers varies by equipment, competition in the sector, and the regulatory conditions affecting trade, but generally a higher investment in manufacturing-related services correlates with competitive advantage for machinery companies (BCG, 2014: 3).

The company is a vertically integrated lead firm positioned at the apex of the construction machinery value chain, and, with the exception of its network of certified dealers, it is vertically integrated through ownership across all of its affiliates. Its production process encompasses the full scope of backward and forward linkages to manufacturing, from mining and extracting raw materials through to the point of retail sale. The value chain, however, extends well beyond the point of sale, through manufacturing-related services that offer financing for leasing and purchasing equipment, monitoring support for equipment, technical and information services for equipment use, as well as maintenance, repair and replacement services.

In addition to enhancing value and lowering costs for products used by its customers, remanufacturing also has a significant

---

linkages and enhancing access to markets in order to move up to higher value activities in the chain, ultimately in order to improve their positions in the global economy (Gereffi, 2015: 5; 2011: 39–40).

environmental impact through savings in materials, energy and deferring transportation costs. This can be understood in terms of the closed-loop life cycle as depicted in Fig. 19.2. The remanufacturing process is particularly relevant to value chain analysis not only because remanufacturing is a relatively new mode of industrial production, but also because the majority of remanufactured goods require services in order to be exported (UNCTAD, 2013: 22). All of the services integral to remanufacturing add value to the remanufacturing product line, and the majority of these services involve significant training and skills.

19.2.2. *Value Chain Diagram of Firm Services*

A value chain diagram of manufacturing-related services identifies the value adding stages in the manufacturing process in which services are both integral and add value to the finished product. The value chain diagram in Fig. 19.3, outlines the services associated with both manufacturing and remanufacturing at the firm, and emphasises (in yellow) those phases that add particularly high value.[8] Figure 19.3 also demonstrates how remanufacturing-related services are superimposed on the manufacturing value chain.

19.2.2.1. *Value addition of manufacturing-related services*

19.2.2.1.1. Product development

The manufacturing-related services value chain shown in Fig. 19.3 begins with several product development services required before manufacturing can take place. Of these services, the product

[8]GVC diagrams illustrate the relationship between lead firms and all of the economic activities and companies in a supply chain, and therefore provide both the "top-down" and "bottom-up" views of an industry. By showing how the governance and upgrading sides of an industry interrelate, both lead firms and key local, regional and national actors can focus upon those leverage points needed to facilitate trade and/or improve one's position within the sector.

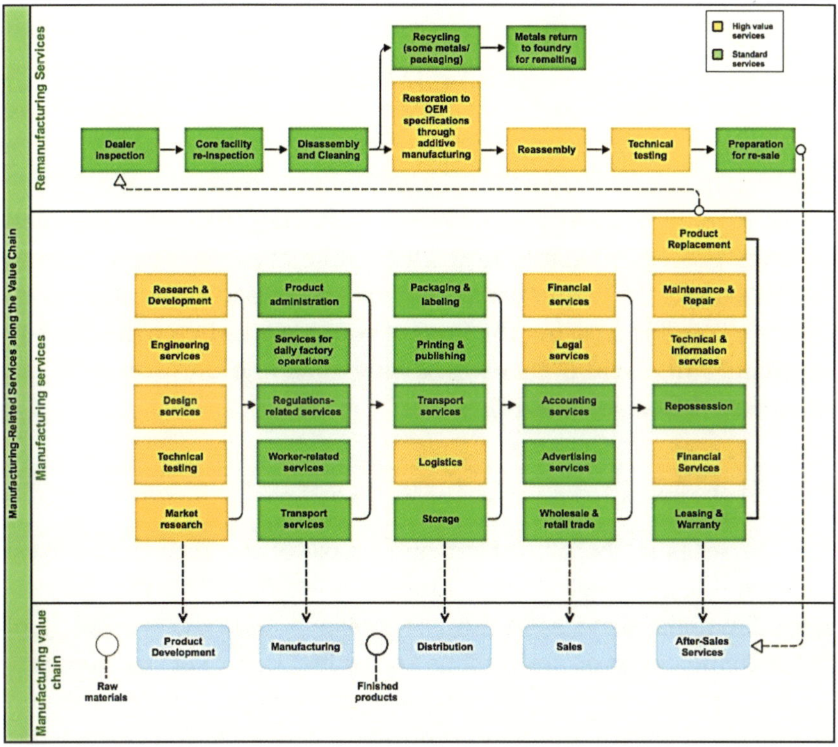

Figure 19.3: Value Chain of Manufacturing and Remanufacturing Services.
Source: Adapted from the firm.

development phase of the value chain, which includes research and development (R&D), engineering, and design services, generates important value addition for both manufacturing and remanufacturing. Through innovating new products and technologies, these phases have the capacity for eventual additive remanufacturing, enabling the extension of the traditional after-sales life cycle of products. Continued investment in R&D, equipment design and engineering will enable the company to sustain its competitive advantage, allowing it to "offer clients better and more competitive machine features and performance and also to differentiate their service capabilities from other competitors" (BCG, 2014: 10).

19.2.2.1.2. Distribution

Logistics services, during the distribution phase of the services value chain, also deliver crucial value-addition for the firm. These services include streamlining the transportation of products to the dealer network, as well as the timely delivery of replacement parts (as well as maintenance and repair services). Well-implemented logistics processes enable lead OEMs like the case study firm to "both reduce the capital invested in spare parts and speed up delivery", and increasing the flow of components helps the parts to maintain value while also supporting machine performance and efficiency (BCG, 2014: 5, 10).

19.2.2.1.3. Sales

Sales services in the manufacturing phase of the firm value chain that provide relatively high value include financial and legal services. Financial services in the sale phase include the arrangement of financing for product leasing and sales. Legal services include patent protection for proprietary designs, technologies and products, as well as the preparation and provision of contracts for leasing and sales.

19.2.2.1.4. After-sales

The after-sales services phase of the value chain is crucial for the firm. Because of the large scale and substantial expense of heavy-duty construction equipment, warranties are essential for customers. Financial services in the after-sales phase pertain to refinancing, which can be especially relevant when clients purchase remanufactured products as replacements. Information and technical services are high-value-adding after-sales services as well. These include the acquisition of information on condition and performance of core components already in the field, as well as the development of equipment manuals, training clients to use equipment, managing equipment fleets and other on-going client support.

19.2.2.1.4.1. Maintenance and repair

Maintenance and repair services include on-going equipment support services so that clients reduce equipment downtime and prolong the life and efficiency of equipment. Customer service agreements typically offer customers predictable operation costs, fewer unexpected operating problems and reduced down time for servicing, and better service levels; all of these contribute to long-term relationships between the firm, dealers and customers (BCG, 2014: 6, 9). By investing in remanufacturing, the case study firm has demonstrated an innovative maintenance strategy that simultaneously enables to firm to retain the value of its worn components, maintain demand for restored components and continue on-going relationships with customers.

19.2.2.1.4.2. Product replacement

Product replacement is the crucial link to remanufacturing, and this phase closes the loop and reverses the logistics for manufacturing at the company. By offering remanufactured components to fulfil product replacement needs, the firm reduces cost and downtime for customers while maintaining high service levels through dealer familiarity with OEM specifications, inspections, and the product line more broadly.

19.2.2.2. *Value addition of remanufacturing-related services*

Remanufacturing is a much more refined, sophisticated and developed process than recycling, refurbishment and rebuilding. As such, it is the highest level of re-use for industrial goods. Expert knowledge of firm specifications is integral to the remanufacturing process (Company source, 2015e). Remanufacturing-related services add value to every stage of the production process, and many of the services are inherently high value. Most services involve a high degree of technical, engineering and design skill, a deep understanding of proprietary specifications, extensive training, as well as technological development. Since the exchange of remanufactured products involves core components, the company is able to re-capture the value of its previously manufactured goods at the very beginning of the Reman value chain (Fig. 19.2). Thus, the firm has reduced the resource and material intensity of the remanufacturing process. The retrieval of core components also enables the firm to offer customers lower prices on replacement components.

19.2.2.2.1. Disassembly, cleaning and inspections

Both dealers and dedicated remanufacturing specialists conduct inspections of retrieved core components to determine whether they qualify for remanufacturing. This stage eliminates wait time that would otherwise be required if only the core receiving facility conducted inspections. Once dealers have approved core components for remanufacturing, they are shipped to the remanufacturing facility for re-inspection, where the components are completely disassembled and cleaned using a patented cleaning process. The second inspection then takes place using detailed remanufacturing criteria to determine if the components can be remanufactured to meet original design and engineering specifications.

19.2.2.2.2. Restoration to OEM specifications

The high-value adding engineering and design phase of the remanufacturing process involves the application of advanced additive-manufacturing technology that restores components to original specifications and good-as-new quality. Many of the technologies applied during this phase are developed by the dedicated remanufacturing division of the firm.

19.2.2.2.3. Reassembly, technical testing and preparation for re-sale

Reassembly involves a combination of remanufactured and new parts, and includes engineering updates. Reassembled products are then tested to meet manufacturer specifications.[9] Finally, in

[9]These specifications include knowledge of the exact capacity for a part, heat treatment, and the internal dimensions of a part (Company source, 2015e).

preparation for re-sale, new serial numbers (where applicable) are issued to the products, which are painted and prepared for sale as remanufactured products. The preparation for re-sale phase also creates value for remanufacturing through innovations such as metal spray, laser welding and other technologies that can restore tolerances to products, rendering them identical or even superior to original parts.

19.2.2.2.4. After-sales services

After-sales services are as essential for remanufacturing products as for new products, because the firm assumes responsibility for the core components.

19.2.2.3. *Service inputs across production phases*

The value chain analysis in this case study encompasses a minimum of 66 manufacturing-related services pertaining to the establishment, pre-production, manufacturing, post-manufacturing, back office and post-sales phases of production, as shown in Tables A.1–A.6.[10] The tables provide detailed descriptions of 66 services that enter into the services value chain across each of these manufacturing phases. The third column of the tables refers to the Central Product Classification (CPC) codes associated with each service. Where 3-digit group codes and 4-digit class codes are listed, the service can be further disaggregated into approximately 80 categories, demonstrating that no fewer than 66 services are involved in the services value chain for this firm. The fourth column of the tables indicates whether the service is provided in-house by the firm, by an in-group company or is outsourced to a third-party provider. Approximately 38 per cent of the services in Tables A.1–A.6 are provided in-house, 8 per cent in-group, 18 per cent are outsourced and 36 per cent are provided by a combination of in-house, in-group or third-party agents.[11]

---

[10]Author approximations based on analysis of case study firm.

[11]Over two-thirds (70.8 per cent) of services involving coordination of more than one party have an in-house component.

Of the 66 services, manufacturing services (Table A.3) comprise 29 per cent of the total, while raw materials, input and pre-production services (Table A.2) comprise 20 per cent. Delivery, sales and post-manufacturing services (Table A.4) and back office, utilities and general services (Table A.5) each comprise 15 per cent of the total services in the value chain. Establishment (Table A.1) and post-sales services (Table A.6) each comprise 10 per cent of the total. While these proportions do not represent costs or value, our analysis of the services value chain suggests that design, R&D, and engineering services, as well as the remanufacturing-related inspections of core components, are the services that add the most value to the pre-production phase. Remanufacturing services, including disassembly and restoration of core components through remanufacturing, re-assembly and technical testing, also add significant value to the manufacturing phase. In addition, our research showed that the IT services in the back office phase are particularly crucial for the firm, as these facilitate product ordering and core processing required for the remanufacturing business model.

19.2.2.4. *Key role of firm dealer network*

A notable aspect of the remanufacturing reproduction process at the company, and a key contributor to its success, is the extent to which it is driven by strong personal relationships between the company, its dealers and their shared clients. The strong and exclusive relationships between the firm and its dealers have enabled the company to secure an extremely high core-return rate of 93 per cent. This translates into a high level of customer loyalty through on-going relationships established in channels such as equipment rental, new product and parts sales, preventive maintenance monitoring and the provision of warrantees for a range of firm equipment including construction machinery, trucks, engines and electric power generation equipment (Company source, 2015e, 2015g; Rental Equipment Register (RER) Magazine, 2015: 1).

Dealers are crucial to its business because they exclusively sell and support company products and do not distribute competitive products. This means that dealers recommend the firm-brand replacement components and create a stable flow of demand for these components.[12] Their understanding of the import and export restrictions relevant to exchange and international trade is another way dealers contribute to the remanufacturing business model (Company source, 2015e).[13] Dealers can be particularly useful when trade policy for remanufactured goods and core components lacks clarity.

Dealers occupy a crucial role in the remanufacturing circuit by providing irreplaceable liaison services between clients and the firm. Their client-facing services include detailed knowledge of products and permission to carry the brand exclusively, their ability to conduct inspections, as well as the authorisation to conduct certified re-builds, which can be highly profitable for dealers. Dealers also provide firm-facing services, such as fluency in the local language of economies where dealerships are based, understanding how to represent the brand to clients, as well as familiarity with the exchange process for core components and international trade restrictions (Company source, 2015e).

19.2.2.5. *Environmental impact of remanufacturing*

Remanufacturing and its related services generate important and far-reaching environmental benefits, including reduced waste from industrial production, decreased demand for raw materials, and

[12]Dealers can independently provide "certified rebuild" services to specific firm products, which are distinct from remanufacturing because they involve the complete machine. These certified machine rebuilds will generally use remanufactured products and components in the process, enabling the customer to benefit from the value added engineering updates, technological and design inputs inherent in the firm's remanufacturing process. These dealer provided certified rebuilds offer service revenue for the dealer and extended life of machines for the customer.

[13]The company's network of local dealers is also noteworthy because it is highly stable: the firm requires dealers to have a succession plan, and dealerships and client bases rarely change. As a result, dealers have two to three generations of experience representing the brand and have established strong client relationships.

the creation of a "closed-loop reverse logistics process" that recovers end-of-life products (Company source, 2013; Fiksel, 2012; Opresnik and Taisch, 2015).[14] As shown in Fig. 19.2, the remanufacturing cycle helps the firm to meet its sustainability goals through extending the time those resource-intensive goods are in use and reducing the materials and energy used in the production process.

The remanufacturing process at the company delivers many benefits of dematerialisation that Fiksel has outlined in his research on sustainable industrial processes, including "increasing material efficiency in operations; designing products with reduced mass, packaging, or life cycle energy requirements; reducing transportation requirements in the supply chain, thus reducing fuel and vehicle utilisation; and substitution of electronic services for material-intensive services" (2012: 140). These issues are discussed in more depth in the triple value (3V) model developed by Joseph Fiksel, as well as in the circular economy literature. Both the 3V and the circular economy models are "closed-loop" models that understand the conservation of capital to positively affect the economic, social and environmental systems over time. An example of how the sustainable practices implemented through remanufacturing fit into the circular economy model is available in Fig. 19.4.

19.3. THE IMPACT OF TRADE POLICY AND THE REGULATORY ENVIRONMENT ON THE FIRM'S SERVICES VALUE CHAIN

Trade policy and the regulatory environment affect the majority of services along the manufacturing and remanufacturing value chains. Based on the current regulatory environment affecting remanufacturing, the firm has highlighted specific policy priorities that support its overarching goal of strengthening regional economic integration. Currently, its highest priorities are to establish a shared legal

[14]For more information, see the Ellen MacArthur Foundation (2012) and Fiksel (2012).

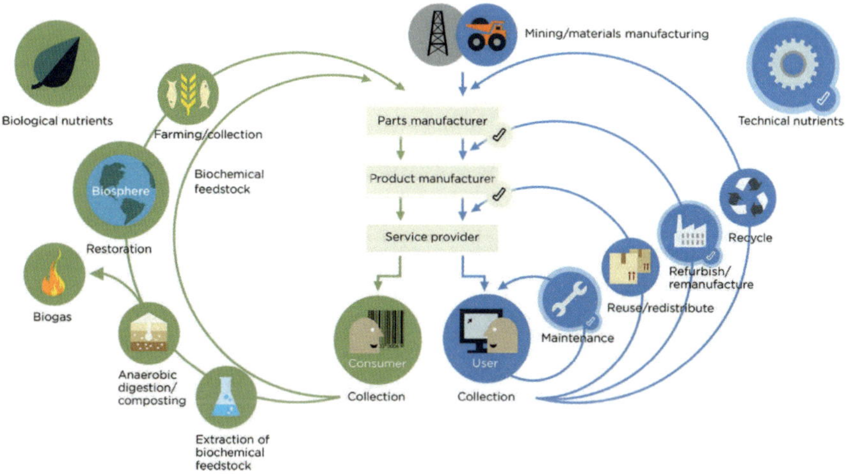

Figure 19.4: Case Study Firm in the Circular Economy.

Source: Ellen MacArthur Foundation.

definition of "remanufactured", support movement toward the Trans Pacific Partnership (TPP), and practice sustainable industrial production. These priorities shape the policy discussion in the following.

19.3.1. *Trade Facilitation of Remanufactured Goods*

The most important issue for the remanufacturing division of the case study firm is the establishment of a shared international definition of remanufactured products. Currently, the absence of such a definition contributes to regulatory bans, import restrictions and confusion among customs officials about the origins and content of remanufactured goods. Without this clarification, economies such as China, Indonesia and Turkey can maintain their existing importation restrictions on remanufactured goods while simultaneously supporting initiatives to bolster domestic remanufacturing, for example through inviting foreign direct investment in the establishment of remanufacturing facilities at home. If such facilities are established, there are not restrictions for foreign firms in hiring local labor. In addition, the lack of clarity about the content of remanufactured goods means that customs officials in many markets will continue to categorise the goods as used or waste. This complicates international trade by limiting both the importation of remanufactured goods and the exchange of core components.

A related issue is that the definition of "core" also needs to be established legally in order to help remanufacturers obtain exception and permission to export non-functioning mechanical devices. Core components have financial value, involve a regulated chain of command, and are properly packaged for re-sale on the market. Extant trade barriers rooted in these definitional issues constrain growth of the remanufacturing industry, for example, by preventing the firm from providing equipment for large-scale infrastructural projects, many of which are currently underway in China.[15]

19.3.1.1. *The APEC Pathfinder Initiative*

The APEC Pathfinder Initiative, launched in 2011, advocates the facilitation of trade in remanufactured goods (APEC, 2011). The initiative notes that exported manufactured goods often face "trade-restrictive non-tariff measures" including documentation requirements, import licensing requirements or prohibitions against importing used goods. Many economies already treat remanufactured goods as new, including Australia, Canada, Chile, Chinese Taipei, Japan, Korea, Mexico, New Zealand, Papua New Guinea, Singapore and the United States (APEC, 2011). Still, APEC continues to work toward increasing economy participation.

The Pathfinder Initiative supports arguments set forth by the firm that "the treatment of these high-quality, lower-cost goods as

[15]In the past, the corporation has faced obstacles to participation in large-scale foreign infrastructure projects. A key example is the Three Gorges Dam Project on the Yangtze River in Hubei Province, China, whose construction began in December 1994. While the firm had secured the required export credit insurance and deal financing from ExIm Bank to become the equipment provider for this project, the largest earth-moving project of that decade globally, it lost the business when ExIm Bank coverage was withdrawn by the US government. In turn, a key competitor gained the advantage of providing the equipment (Company source, 2015e).

used goods can impede their trade, limit the opportunities for remanufacturing industries to grow in developing and developed economies, and cause the APEC region to miss out on the significant environmental benefits from more robust trade in remanufactured goods" (APEC, 2011).[16] As a result, the Pathfinder Initiative recommends:

- The application of import-related measures only to used goods, as remanufactured goods are *not* used goods — remanufactured goods should not face import limitations
- The release of import prohibitions against remanufactured goods, in specific sectors and across sectors
- Tariff measures and border charges should treat remanufactured goods as new goods
- The same technical regulations, conformity assessment procedures, documentation and import licensing requirements that apply to new goods should also apply to remanufactured goods.

Trade facilitation measures are needed to lift tariff measures, border charges, assessment procedures and licensing requirements that currently treat remanufactured goods as used goods. A recent joint report by the World Trade Organization (WTO) and the United Nations Conference on Trade and Development (UNCTAD) has underscored the importance of trade facilitation measures, asserting that they are crucial to the successful participation in GVCs (WTO/UNCTAD, 2013: 5).

---

[16]This is a key point because even subsequent trade conventions such as the Basel Convention in 2014, organised to promote the control of the trans-boundary movement of hazardous wastes and their disposal, explicitly explore the concept of remanufactured goods as waste. While the Basel Convention states that business and industry "contribute to support and promote environmentally sound management of hazardous wastes and other wastes," the twelfth meeting of the convention in 2014 considered including remanufactured goods (Secretariat of the Basel Convention, 2011). This would be damaging to the firm and other remanufacturers (Company source, 2015e).

19.3.1.1.1. Trade policy and the location of remanufacturing facilities

Restrictive trade policies on both the export and import of remanufactured goods can limit the scope of locations in which remanufacturing facilities and industries can establish themselves. The location of remanufacturing facilities is relevant to GVC analysis for several reasons. First, facilities are located in markets where firms have permission to export their goods. Second, they are often located near markets with increasing demand for remanufactured goods. Third, remanufacturing facilities can be located near emerging markets and developing economies; this proximity enables linkages for that economy to the manufacturing-related services value chain in the corresponding industry. Such linkages are crucial for industrial clustering, the development of regional value chains, and eventual regional integration. Fourth, as mentioned by the USITC report on remanufacturing, "some of the largest U.S.-based HDOR equipment remanufacturers, have remanufacturing operations located in or near markets experiencing growth in demand for aftermarket parts, including in Brazil, China, and India" (USITC, 2012: 2–24).

Dedicated remanufacturing facilities may strategically locate near major infrastructural development projects, or near other lead firms in order to provide exclusive remanufacturing services for third parties. For example, the company's remanufacturing division is the exclusive remanufacturing partner with Honeywell Garret for turbochargers and Weir Minerals for industrial pumps and components (PR Newswire, 2005, 2011; Bloomberg, 2011). Thus, in terms of policy, remanufacturing industries need trade partnerships and new markets in order to expand their sales. Without this access, the linkages between small and medium size firms are constrained.[17]

[17]For instance, the firm may face obstacles to providing equipment and services for major infrastructural projects taking place in developing and emerging economies. With an expanded definition of remanufactured goods, government policymakers can also support the domestic establishment and growth of the

19.3.2. *Sustainable Development*

In its recent 2014 Sustainability Report, the firm communicated its sustainability goals and outlines the metrics it uses to measure its environmental impact. By 2020, the firm aims to reduce energy intensity, greenhouse gas emissions, water consumption intensity and by-product materials intensity by half, relative to its previous 2006 levels (Company source, 2014b: 2). Additional operations goals include using alternative or renewable sources to meet 20 per cent of energy needs, designing new construction to meet green building criteria, and reducing both workplace injuries and lost-time case rates (Company source, 2014b: 2).

In their analysis of the long-term economic and environmental benefits of the closed-loop supply chain, Li *et al* (2014: 6314). emphasise the importance of government policy intervention in supporting the development of the remanufacturing industry. Specifically, they suggest that government subsidies are one useful intervention that can promote domestic remanufacturing (Li *et al.*, 2014: 6322).[18] Beyond the clear environmental benefits of sustainable industrial practices, however, by having advanced logistical and technological capabilities to remanufacture, the firm is well-positioned to further regional integration and enhance its access to new markets globally.

19.4. CONCLUSIONS

The company at the centre of this study is the lead firm in the construction machinery industry. With the exception of its network of dealers, it is vertically integrated through the ownership of its entire corporate supply chain. Its production process encompasses

---

remanufacturing industry in their own economies. This, in turn, can help to provide the long-term environmental benefits and the growth of skilled professions that accompany remanufacturing in their economies.

[18]Since technological innovation and the skill levels of workers are integral to the development of this industry, these aspects of the remanufacturing industry are also policy priorities.

the full scope of backward and forward linkages to manufacturing, from the use of raw materials from mining and extraction through to the point of retail sale. The firm's services value chain is not segmented; rather, its services span the entire value chain. The 66 services spanning the production and reproduction processes at the company contribute significant value to its services value chain. Of the 66 total services, manufacturing services comprise 29 per cent, while raw materials, input and pre-production services comprise 20 per cent. Post-manufacturing and back office services each comprise 15 per cent of total services in the value chain, while establishment and post-sales services each comprise 10 per cent of the total.

Manufacturing-related services support equipment from the initial purchase of equipment through its life cycle. Financing and leasing agreements support client purchases. Business, financial and consulting services also support customers over the duration of their relationship with the firm. These services are integral to the industrial process of remanufacturing. In addition to enhancing the value and lowering costs for customers, remanufactured products offered through the dedicated remanufacturing division introduce significant positive environmental benefits through savings in materials, energy and deferred transportation costs.

Construction and mining equipment manufacturers generate a large share of their business through offering services. In fact, sales of services typically perform better than machinery sales by having faster growth rates, higher profitability, greater predictability and more resilience during economic downturns (BCG, 2014: 3–4). Although the value of these services varies by sector, type of machinery, and the regulatory conditions affecting trade, a higher investment in manufacturing-related services typically correlates with competitive advantage for machinery companies (BCG, 2014: 3).

The remanufacturing process is particularly relevant to value chain analysis because the majority of remanufactured goods require services in order to be exported (UNCTAD, 2013: 22). The key policy issue that the case study firm currently faces is the need for a shared international understanding and definition of remanufactured products as having the same quality, durability and performance guarantees as newly manufactured products. A revised definition will facilitate lifting restrictions on imports and exports of remanufactured goods, particularly core components. In turn, this will loosen constraints on the expansion of the remanufacturing industry. By supporting a clear definition of remanufactured goods, government policymakers can also support the establishment and growth of the remanufacturing industry in their own economies.

APPENDIX

Manufacturing-related Services Relevant to the Construction Machinery Value Chain at the Firm

Table A.1: Establishment Stage.

Stage/category	Service	CPC ver. 2 code	Service provided in-house, by an in-group company, or outsourced
1. Government services (licensing etc.)	1. Business liaison services (e.g., services from trade development centres or SME centres)	Class of 9113 — Public administrative services related to the more efficient operation of business	Outsourced
	2. Company registration and licensing services	Class of 9113 — Public administrative services related to the more efficient operation of business	Outsourced (government)
	3. Information and statistical services	91138 — Public administrative services related to general economic, commercial and labour affairs	Outsourced
	4. Visa and immigration services for foreign investors/employees	91290 — Public administrative services related to other public order and safety affairs	In-house in coordination with outsourced local agents

2. Other services (professional etc.)	5. Business consultant services	Class of 8311 — Management consulting and management services	Outsourced
	6. Banking and finance services	71135 — Non-mortgage loan services for business purposes	In-house and outsourced
	7. Legal services	82130 — Legal documentation and certification services	In-house and outsourced

Source: Authors

Table A.2: Raw Materials, Input/Pre-production Stage.

Stage/category	Service	CPC ver. 2 code	Service provided in-house, by an in-group company, or outsourced
1. Procurement of raw materials	8. Procurement agent for raw material sourcing	85999 — Other support services n.e.c.	In-house and outsourced
	9. Customs-related services for raw materials imported	85999 — Other support services n.e.c.	Outsourced
	10. Quality assurance services (of raw materials)	83441 — Composition and purity testing and analysis services	Likely in-group
2. Logistics	11. Freight transportation services (of raw materials) by road, rail, sea or air	Division: 65 — Freight transport services	Likely in-group, coordinated with a third party
3. Storage	12. Storage of raw materials — general storage	67290 — Other storage and warehousing services	In-house
4. Product design	13. Conception and design of product	83920 Design originals	In-house
	14. Industrial design	83912 Industrial design services	In-house
	15. Patent acquisition[18]	83960 Trademarks and franchises	In-house

[18] The value chain diagram for this firm categorises patent acquisition under "Legal Services" in both the Sales & Preparation for Resale phases; patent acquisition is important for both manufactured and remanufactured goods.

5. Product development — R&D	16. and 17. Conception and design of product	81129 — Research and experimental development services in other engineering and technology	In-house
			In-house
	18. Engineering services	81400 — Research and development originals — Innovation of new products and technologies, including those used for additive reman.	In-house
		83310 — Engineering advisory services — Development of OEM specifications and fabrication of prototypes	
6. Other services: Core component inspections	19. Dealer inspection	End-of-life components inspected for remanufacture eligibility	In-house
	20. Core facility re-inspection	End-of-life components inspected for remanufacture eligibility	In-house

Source: Authors

Table A.3: Manufacturing Stage.

Stage/category	Service	CPC ver. 2 code	Service provided in-house, by an in-group company, or outsourced
1. Production administration	21. Production Administration — Production Management	83115 — Operations management consulting services	In-house
	22. Production Administration — Repair and maintenance of factory equipment	87156 — Maintenance and repair services of commercial and industrial machinery	In-house
2. Services supporting factory daily operations	23. Cleaning services of factory	85330 — General cleaning services	In-house
	24. Engineering Services	Class: 8332 — Engineering services for specific projects	In-house
	25. Specialized cleaning services for machines and equipment	85340 — Specialized cleaning services	In-house
	26. Repair and maintenance services of machines and equipment	87156 — Maintenance and repair services of commercial and industrial machinery	In-house
	27. and 28. Waste collection and recycling services	9421 — Collection services of hazardous waste	Outsourced
		94229 — Collection services of non-hazardous recyclable materials, other	Outsourced

3. Services from government regulation requirements	29. and 30. Government inspections on fire prevention, health hazards, environmental protection and other aspects	91133 — Public administrative services related to mining and mineral resources, manufacturing and construction 91290 — Public administrative services related to other public order and safety affairs	Outsourced (government) Outsourced (government)
4. Worker-related services	31. Catering services for workers	63393 — Other contract food services	In-group company
	32. Medical services	93121 — General medical services	In-group company
	33. and 34. Social insurance for factory workers	91320 — Administrative services related to government employee pension schemes; old-age disability or survivors' benefit schemes, other than for government employees 91330 — Administrative services related to unemployment compensation benefit schemes	In-house and outsourced In-house and outsourced
5. Others (please specify): Worker-related	35. Wellness services		In-group company
6. Others: Core disassembly	36. Core disassembled into individual core components using advanced qualification technology		In-house

(*Continued*)

Table A.3: (*Continued*)

Stage/category	Service	CPC ver. 2 code	Service provided in-house, by an in-group company, or outsourced
7. Others: Restoration to OEM specifications	37. Remanufacture core components using advanced salvaging and additive manufacturing technology		In-house
8. Others: Re-assembly	38. Recombination of new and remanufactured components, with addition of engineering updates, for eventual sale as remanufactured product		In-house
9. Others: Technical testing	39. Validation that reassembled, remanufactured product meets standards for reliability, durability and quality, to OEM specifications		In-house

Source: Authors.

Table A.4: Delivery and Sales/Post-manufacturing Stage.

Stage/category	Service	CPC ver. 2 code	Service provided in-house, by an in-group company, or outsourced
Packaging and labeling	40. Design of packages	83919 — Other specialty design services	In-house
	41. Packaging services	85400 — Packaging services	In-house
Delivery to wholesaler/ retailer	42. Customs-related services[19]	85999 — Other support services n.e.c.	Outsourced — customs brokers
	43. Land transport of goods (from warehouse to domestic port)	65112 — Road transport services of freight by tank trucks or semi-trailers	In-group and outsourced
Sales services	44. Retail trade services — by distributors	Group of 621 — Non-specialized store retail trade services[20]	In-group company (dealer network) and outsourced
	45. Retail services on a fee or contract bases	Group of 625 — Retail trade services on a fee or contract basis	In-group company (dealer network) and outsourced
	46. and 47. Retail Administration — Operations management	83115 — Operations management consulting services	In-group company and outsourced
		83116 — Supply chain and other management consulting services	In-group company and outsourced

[19]The firm considers customs authorizations from third party customs-broker services to belong in "Other Professional Services" during the Back office phase.

[20]The network of dealers is key for this phase.

(Continued)

Table A.4: (*Continued*)

Stage/category	Service	CPC ver. 2 code	Service provided in-house, by an in-group company, or outsourced
Leasing/Hire purchase services	48. Hire purchase services of machinery	71140 Financial leasing services	In-house, in-group and outsourced
Others (please specify): Printing and Publishing	49. Production of equipment manuals for training and equipment use		In-house

Source: Authors

Table A.5: Back Office, Utilities and General Services.

Stage/category	Service	CPC ver. 2 code	Service provided in-house, by an in-group company, or outsourced
1. Finance	50. Auditing on financial accounts	Group of 822 — Accounting, auditing and bookkeeping services	Outsourced, with assistance from in-house finance division.
	51. and 52. Financial services	71121 — Deposit services to corporate and institutional depositors	In-group.
		71313 — Group pension services	Outsourced.
	53. Internal auditing (including audits of financial accounts and corporate governance)	Group of 822 — Accounting, auditing and bookkeeping services	In-house and in-group.
	54. Insurance services for machinery	71334 — Other property insurance services	In-group and outsourced.
2. General management	55. and 57. IT and information system management, consulting and support, with webpage development	Class of 8313 — IT consulting and support services	In-house and outsourced.
		Class of 8314 — IT design and development services	In-house and outsourced.
		Class of 8316 — IT infrastructure and network management services	In-house and outsourced.

(*Continued*)

Table A.5: (*Continued*)

Stage/category	Service	CPC ver. 2 code	Service provided in-house, by an in-group company, or outsourced
3. Legal	58. and 59. Legal services[21]	82120 — Legal advisory and representation services concerning other fields of law	In-house and outsourced.
		82130 — Legal documentation and certification	In-house and outsourced.

[21] The global value chain analysis presented in the earlier discussion considers Legal Services to belong in the Sales phase of the value chain for manufacturing and remanufacturing.

Source: Authors.

Table A.6: Post-sales Services.

Stage/category	Service	CPC ver. 2 code	Service provided in-house, by an in-group company, or outsourced
1. Sales/after-sales Customer services	60. Customer services — Loyalty program	95999 — Other services provided by membership organisations n.e.c.	In-house.
	61. Customer services — Technical support	85931 — Telephone call centre services	Outsourced.
	62. Customer services — warranty and repair services	Group of 872 — Repair services of other goods	In-house.
	63. Replacement/ re-possession services for leased equipment	Group of 731 — Leasing or rental services concerning machinery and equipment without operator	In-house.
		Group of 732 — Leasing or rental services concerning other goods	In-house.
		71140 Financial leasing services	In-house.
2. Repair and maintenance	64. Repair and maintenance of machines for customers	87156 Maintenance and repair services of commercial and industrial machinery	In-house and in-group (dealer network).

(*Continued*)

Table A.6: (*Continued*)

Stage/category	Service	CPC ver. 2 code	Service provided in-house, by an in-group company, or outsourced
3. Remanufacturing services	65. Recovery of core components: buy-back services	62587 — Retail trade services on a fee or contract basis, of mining, construction and civil engineering equipment and components	In-house, in-group and outsourced (in cooperation with customs regulations and trade policy).
	66. Remanufacturing services *(Note: for related services, see rows 6–9 in Table A.3.)*	Class of 4446 — Parts for the goods of classes 4441 to 4444 (Machinery and equipment for special purposes)	In-house and in-group (dealer network).

Source: Authors.

REFERENCES

Adler, DP (2004). Comparing energy and other measures of environmental performance in the manufacturing and remanufacturing of engine components. M.S. Thesis, Department of Mechanical Engineering, Michigan Technological University.

Asia-Pacific Economic Cooperation (2011). Annex D: Pathfinder on facilitating trade in remanufactured goods. Appendix 12 from The 23rd APEC Ministers Meeting, Honolulu, HI (11 November 2011). http://www.apec.org/Meeting-Papers/Ministerial-Statements/Annual/2011/2011_amm/annex-d.aspx [8 June 2015].

_____ (2013). Annex D: Pathfinder on facilitating trade in remanufactured goods. [24 June 2013].

_____ (2014). Manufacturing related services in supply chains/value chains. http://goo.gl/P1Jkpu [12 December 2014].

Asia-Pacific Economic Cooperation and United States Agency for International Development (2013). Remanufacturing resource guide. Prepared for APEC Technical Assistance Training Faculty (TATF) Program, managed by USAID. http://www.apec.org/~/media/Files/Groups/MAG/20131120_APEC-RemanResourceGuide_Sep2013.pdf [10 May 2015].

Barrientos, S, G Gereffi and A Rossi (2011). Economic and social upgrading in global production networks: A new paradigm for a changing world. *International Labour Review*, 150(3–4), 319–340.

Bernard, S (2011). Remanufacturing. *Journal of Environmental Economics and Management*, 62(3), 337–351.

Bloomberg (2011). [Remanufacturing] Reaches commercial agreement with Weir minerals-North America (25 May 2011). http://www.bloomberg.com/apps/news?pid=newsarchive&sid=alZ7MHdQrg8Q [7 June 2015].

Boston Consulting Group (BCG) (2001). E-Services: Redefining the aftermarket opportunity. BCG Report on Industrial Products & Processes and Automotive Sector by Tom Erixon and Dan Johnson. https://goo.gl/9oSWEL [7 June 2015].

_____ (2014). Creating value for machinery companies through services. BCG Perspectives Report on Industrial Products & Processes and Operations by Victor Du, Kelly Howe, Rahul Jain, Rafael Rilo, Patrick Staudacher, Pekka Vanne and Francisco Salmerón (26 May 2014). https://goo.gl/xcLj7F [7 June 2015].

_____ (2015). Battling for global leadership in construction equipment. BCG Perspectives Report on Globalization and Automotive Sector by Nikolaus Lang, Bob Zhai, Brian Collie, and Peter Ullrich (24 June 2015). https://goo.gl/LF3SFd [18 July 2015].

Company source (2008). Remanufacturing worldwide locations. [23 June 2015].

_____ (2010a). [Firm] Remanufacturing process (17 November 2010). Video. [3 June 2015].

_____ (2010b). [Firm] Values in action: Worldwide code of conduct. [23 June 2015].

_____ (2012). History of remanufacturing at [the firm]. [May 2015].

_____ (2013). [Firm] Celebrates five years in West Fargo with grand opening of facility expansion. [25 June 2015].

_____ (2014a). Corporate overview presentation. [23 June 2015].

_____ (2014b). Sustainability report fact sheet. [5 June 5 2015].

_____ (2014c). Year in review. Annual Report. [5 June 2015].

_____ (2015a). Employment site. [10 June 2015].

_____ (2015b). First quarter (1Q) fact sheet. [23 June 2015].

_____ (2015c). Global footprint — China. [18 July 2015].

_____ (2015d). Governmental affairs news. U.S. Ambassador Visits [Remanufacturing] Facility in Singapore. [7 July 2015].

_____ (2015e). Executive interview at firm headquarters. [12 June 2015].

_____ (2015f). Organizational chart (1 May 2015). [23 June 2015].

_____ (2015g). Presentation on firm remanufacturing at the 2015 world remanufacturing summit. Amsterdam, NL. [16 June 2015].

_____ (2015h). Remanufacturing page — Corporate site. [3 June 2015].

Cheung, D and A Wirjo (2016). Production of precision die and machine parts in Thailand. In *Services in Global Value Chains: Manufacturing-Related Services*. Singapore: APEC Secretariat.

Colvin, G (2011). [Firm] is absolutely crushing it. *Fortune*, 163(7), 54. [8 June 2015].

Ellen MacArthur Foundation (2012). [Firm] case study. [9 June 2015].

Elliott, S (2014). [Remanufacturing]: Hungry for more. Duxes (Shanghai) Business Consulting, Inc. [8 June 2015].

Fabick Cat (2015). [Dealers] have joined forces (1 July 2015). [18 July 2015].

Fiksel, J (2012). A systems view of sustainability: The triple value model. *Environmental Development*, 2, 138–141.

Fortune 500 (2015). http://fortune.com/fortune500/ [19 July 2015].

Fuller, K (2012). RIT's Nabil Nasr touts remanufacturing in testimony to U.S. International Trade Commission. Rochester Institute of Technology University News (5 March 2012). http://www.rit.edu/news/story.php?id=48934 [10 June 2015].

Gereffi, G (2005). The global economy: Organization, governance and development. In *Handbook of Economic Sociology*, NJ Smelser and R Swedberg (eds.), 2nd ed., pp. 160–182. Princeton: Princeton University Press/Russel Sage Foundation.

_____ (2011). Global value chains and international competition. *Antitrust Bulletin*, 56(1), 37–56.

_____ (2015). Global value chains and development. Report written for the United Nations Industrial Development Organization (UNIDO). [6 April 2015].

Gereffi, G, A Goger, A Abdulsamad, D Hamrick, G Ahmed, A Guinn, J Daly and J Morgan (2014). *Shifting Sources of Wheat Supply for MENA Countries: The Rise of the Black Sea*. Durham, NC: Duke Center on Globalization, Governance & Competitiveness.

Gereffi, G and J Lee (2012). Why the world suddenly cares about global supply chains. *Journal of Supply Chain Management*, 48(3), 24–32.

Hauser, W and RT Lund (2003). *The Remanufacturing Industry: Anatomy of a Giant*. Boston: Department of Manufacturing Engineering, Boston University.

_____ (2008). *Remanufacturing: Operating Practices and Strategies*. Boston: Boston University.

Hindo, B and M Arndt (2006). Everything old is new again. *Bloomberg.com*. [8 June 2015].

Japke, O (2009). Practice guidelines: Development of a framework for assessing the economic benefits of remanufacturing. MA Thesis, Cranfield University, Cranfield UK. www.remanufacturing.org.uk [9 June 2015].

Low, P (2013). The role of services in global value chains (4 December 2013). Working Paper. Fung Global Institute. http://www.fungglobalinstitute.org/en/working-paper-the-role-of-services-in-global-value-chains/ [10 May 2015].

_____ (2014). A loaf of bread's well-serviced journey (20 August 2014). *South China Morning Post*. http://www.scmp.com/comment/insight-opinion/article/1577631/loaf-breads-well-serviced-journey [22 June 2015].

Lund, RT (1984). Remanufacturing. *Technology Review*, 87(2), 18.

_____ (2012). *The Database of Remanufacturers*. Boston: Boston University.

Lund, RT and FD Skeels (1983). *Guidelines for an Original Equipment Manufacturer Starting a Remanufacturing Operation*. No. DOE/CS/40192-T2. Cambridge: Massachusetts Institute of Technology, Center for Policy Alternatives.

Nasr, N and M Thurston (2006). *Remanufacturing: A Key Enabler to Sustainable Product Systems*. Rochester Institute of Technology.

Opresnik, D, C Zanetti and M Taisch (2013). Servitization of the manufacturer's value chain. In *Advances in Production Management Systems (APMS 2013: PART II, IFIP AICT 415)*, V Prabhu, M Taisch, D Kiritsis (eds.), pp. 234–241. Berlin, Heidelberg: IFIP Advances in Information and Communication Technology, Springer.

Opresnik, D and M Taisch (2015). The manufacturer's value chain as a service — The case of remanufacturing. *Journal of Remanufacturing*, 5(1), 1–23.

Parker, D and P Butler (2007). *An Introduction to Remanufacturing*. Aylesbury: Centre for Remanufacturing and Reuse.

PR Newswire (2005). [Firm] remanufacturing services and honeywell announce North American turbocharger remanufacturing alliance (5 October 2005). [7 June 2015].

_____ (2011). [Firm] and vestas wind systems reach agreement on remanufacturing wind turbine components (7 November 2011). [7 June 2015].

Remanufacturing Industries Council (2015). Definition and importance of remanufacturing. http://remancouncil.org/ [5 June 2015].

Rental Equipment Register Magazine (2015). [Firm] dealers to merge (18 February 2015). [18 July 2015].

Stahel, W (1995). [Firm] remanufactured products group. The Product-Life Institute, Geneva. Research for the Geneva Environment Meetings in 1995. [9 June 2015].

Trefis, T (2014). [Firm] intensifies focus on dealer performance to drive results in a tough macro (11 March 2014). [9 July 2015].

United Nations Conference on Trade and Development (UNCTAD) (2013). World investment report 2013 — Global value chains: Investment and trade for development. New York and Geneva.

United States International Trade Commission (USITC) (2012). Remanufactured goods: An overview of the U.S. and global industries, markets and trade. *USITC Publication*, 4356, 332–525.

United States Securities and Exchange Commission (SEC) (2014). Company form 10-K. http://www.sec.gov [22 June 2015].

Wirjo, A and G Pasadilla (2016). Manufacturing of oil and gas industry equipment in Singapore. In *Services in Global Value Chains: Manufacturing-Related Services*. Singapore: APEC Secretariat.

World Trade Organization (WTO) (2014). World trade report 2014. In *The Rise of Global Value Chains*, pp. 78–127. https://goo.gl/s3Y18c [25 June 2015].

Chapter 20

MANUFACTURING OF CONSUMER ELECTRONIC APPLIANCES IN INDONESIA

Emmanuel A. San Andres

20.1. THE CONSUMER ELECTRONICS INDUSTRY IN INDONESIA

The electronics and electrical appliance industry in Indonesia had a gross output of IDR 170.4 trillion (USD 12.8 billion) in 2013, an increase of 5.3 per cent over the IDR 161.8 trillion (USD 12.2 billion) recorded in 2012. In 2008–2013, the industry has seen steady growth, growing at an annual average rate of 10 per cent during the period (Fig. 20.1). On the other hand, employment in the industry had a sharp fall in 2013, employing 216,550 workers during the year compared to 274,194 workers employed in 2012.

According to data from Badan Pusat Statistik (BPS), up to 78.2 per cent of input costs in the electronics and appliance industry are from raw materials (i.e., parts, components or cost of input goods). This reflects the general cost structure of manufacturing firms where 85.3 per cent of input costs are from raw materials. As may be expected, the electronics and appliance industry is more energy intensive than manufacturing firms in general. While manufacturing firms generally spend 7.4 per cent of total input costs on electricity and gas, the electronics and appliance industry allocates 14.8 per cent of input costs to energy (Fig. 20.2).

The electronics industry can be categorised into three segments: consumer electronics, industrial electronics and components electronics. In Indonesia, the consumer electronics segment is the most

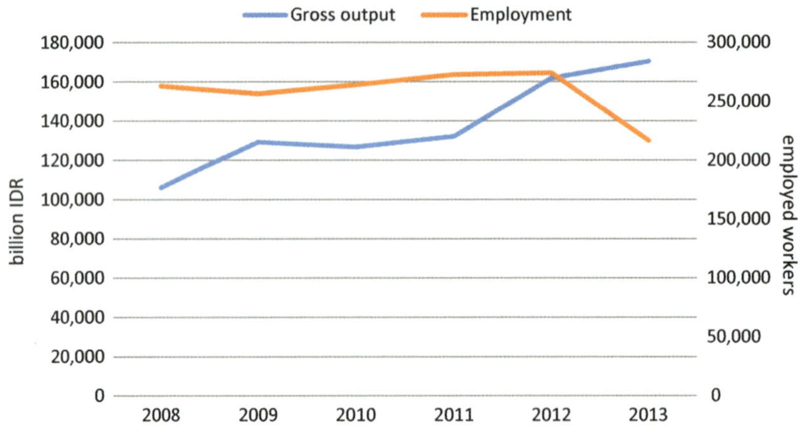

Figure 20.1: Electronics and Appliance Industry Indicators, 2008–2013.

Note: Figures represent gross output and employment of medium and large manufacturing enterprises in the subsectors of "Computer, Electronic, and Optical Products" and "Electrical Equipment".

Source: Author's computation based on BPS data.

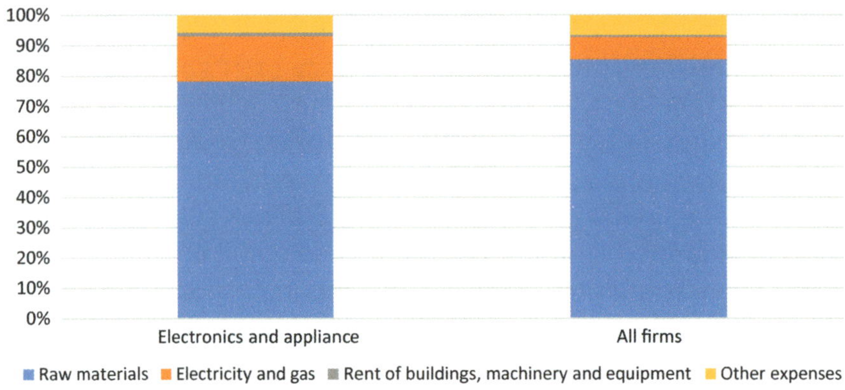

Figure 20.2: Input Cost Structure, 2012.

Note: Figures for "electronics and appliance" represent cost inputs of medium and large manufacturing enterprises in the subsectors of "office, accounting, and computing machinery", "electrical machinery and apparatus n.e.c.", "radio, television and communication equipment and apparatus", and "medical, precision and optical instruments, watches and clocks".

Source: Author's computation based on BPS data.

developed owing to a large domestic market of 62 million households, while the industrial electronics segment (e.g., office equipment, telecommunications, data servers) has been developing due to the expanding telecommunications sector. The components electronics segment (e.g., microchips, motherboards, transistors), on the other hand, is relatively weak, with the electronics sector highly dependent on imported components to assemble.

The electronics industry in Indonesia started in the 1950s with the establishment of the PT Transistor Radio Thayeb Mfg. Co. as the first producer of "Tjawang" transistor radios in the economy.[1] In the 1970s, the government encouraged joint ventures between domestic firms and foreign electronics companies in a bid to spur technology transfer. In the 1970s, the electronics sector shifted from assembling imported components to producing components in Indonesia. In the mid-1980s, the introduction of several deregulation measures shifted policy in the electronics industry from import substitution to export-orientation, encouraging the establishment of more electronics firms as well as IT companies. However, the 1997 Asian financial crisis and the resulting decline in household purchasing power led to the closure of many electronics firms in Indonesia. Although electronics exports recovered quickly after the crisis due to favourable terms of trade (i.e., depreciation of the Indonesian rupiah), the electronics parts and components industry found itself unable to compete with cheap imports from China. Likewise, local electronics producers found it more profitable to import products from China and market them domestically with their own brands. Moreover, a previous luxury tax on electronics did not lead to the development of locally produced electronics; rather, it encouraged smuggling and the finding of loopholes to gain access to foreign brands.

Although Indonesia had the good idea to encourage the development of the electronics parts and components industry in order to tap into the international electronics global value chains, the unfortunate

---

[1]This discussion on the development of Indonesia's electronics industry is taken from Negara (2010). Fragmentation of electronics and textile industries from Indonesia to CLMV countries. In *A Study on Upgrading Industrial Structure of CLMV Countries*, R Banomyong and M Ishida (eds.), ERIA Research Report 2009, No. 7–3. Jakarta: ERIA

timing of the Asian financial crisis and the unintended consequences of protectionist measures hindered its development. As a result, Indonesia has not been able to integrate its electronics industry into global production networks in the same way Malaysia and Thailand have.

20.2. COMPANY BACKGROUND

The company interviewed for this case study is a leading manufacturer of consumer electronic appliances in Indonesia. This company has had a presence in Indonesia for more than three decades producing refrigerators, air conditioners, washing machines, fans and entertainment systems. It employs more than a thousand employees and had a sales turnover of more than a trillion rupiah in 2014. For this analysis, we focus on the value chain for air conditioners as this appliance contributes the most in terms of sales revenue.

The manufacturing company that we interviewed (Company A) is a joint venture between an Indonesian conglomerate (Company B) and an international consumer electronics appliance

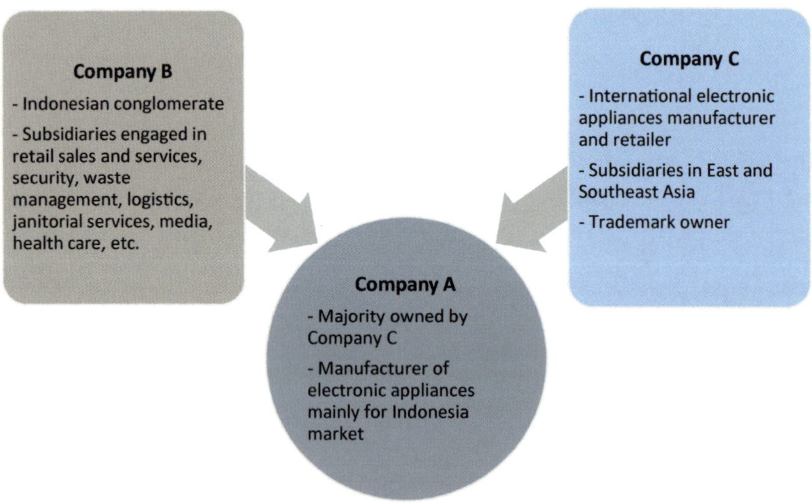

Figure 20.3: Ownership Structure of Company A.

Source: Author.

firm (Company C). Company C owns the electronics brand manufactured by Company A (Fig. 20.3).

20.3. DESCRIPTION OF THE VALUE CHAIN

Company A is in a unique setting due to its age and affiliation with both a large local conglomerate and a major international electronics firm. Hence, most pre- and post-manufacturing processes are handled by affiliate companies, and the company focuses almost solely on the manufacturing aspect of the value chain.

Figure 20.4 shows the value chain for Company A. In the product design stage, Company A receives instructions on technology and design from Company C. Company B, on the other hand, provides inputs on quantities needed for a particular product. Company A studies these specifications and proceeds to the pre-manufacturing stage, where it procures needed raw materials and, if needed, new equipment are installed and production capacity is expanded. It then manufactures products to Company C's specifications and in the quantities required by Company B. In the post-manufacturing stage, Company A packages and delivers the products to Company B, which takes over all post-manufacturing activities.

In the product design stage, most of the activities are carried out by Company C, which owns the brand and most of the technology.

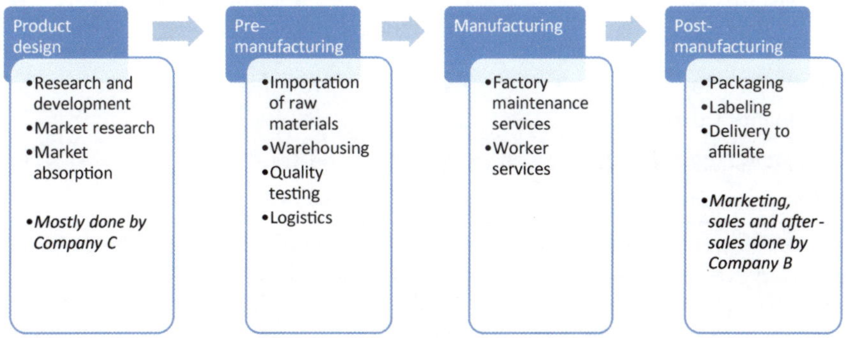

Figure 20.4: Value Chain for Company A.

Source: Author.

Although Company A provides inputs to Company C and its affiliates elsewhere in Asia, final decisions on product design and technology are taken from Company C. On the other hand, most of the activities in the post-manufacturing stage — including marketing, sales, after-sales services and consumer feedback — are handled by affiliates under Company B. Most services required by the firm are also procured from Company B.

20.4. SERVICES ALONG THE VALUE CHAIN

The ownership structure of Company A — that is, a joint venture between two large firms with their own complex value chains — has made its own manufacturing value chain relatively simple. Likewise, its age and long presence in Indonesia is reflected in its procurement of services along its value chain. Whereas newer firms may be more likely to outsource most of their services, in the case of Company A most services are either in-house or with affiliate companies rather than outsourced to third parties.

Appendix provides a listing of the services procured by Company A in the course of its value chain, as well as the sources

Table 20.1: Service Supply along the Value Chain.

Phase	Number of services	In-house	Company B or C	Third party
Establishment	6	5	0	2
Pre-manufacturing	13	10	4	3
Manufacturing	27	15	11	9
Post-manufacturing	10	4	7	2
Office and General Services	31	19	11	11
Entire Value Chain	87	53	33	27

Note: The sum of in-house, affiliate and 3rd party services may not add up to the total number of services as some services may be provided by more than one modality. For example, waste management services can be done in-house (employed janitors), by affiliate companies (janitorial services sister company), and third parties (specialised waste management) depending on the disposal requirements.

Source: Author.

of these services. As can be seen in Table 20.1, there are 87 various services required by Company A in its value chain, mainly in its manufacturing phase; office and general services account for the most number of services as these support the entire value chain. A majority of services are supplied in-house, followed by services provided by affiliate companies. Services outsourced to third parties are a minority in Company A's value chain.

20.4.1. *Establishment and Product Design*

Given Company A's decades-long presence in Indonesia, it does not need to procure services related with setting up a business. However, it still requires annual renewal of business-related licenses and permits according to applicable laws. Likewise, it needs to regularly apply for and renew work visas for some expatriate staff from Company C. For these regulation-related services, Company A has in-house administrative staff to handle paperwork and file them with appropriate government agencies. Company A also has in-house legal and financial departments to handle legal and bank-related issues associated with running the firm.

At the product design stage, most services related to product research and development are handled by Company C. Hence, research into new technologies, industrial design and packaging are implemented by Company C, with Company A providing inputs and feedback by in-house engineers. Patent acquisition, if required, is done in-house by Company A. Most market research is done by Company C, although Company A also procures the services of a third-party market research firm for monthly market data.

At this stage, there is minimal outsourcing of services to third parties. Due to its decades-long presence in Indonesia, Company A has already built up internal support systems for government relations that can handle licenses, permits and visas. On the other hand, the proprietary nature of product development requires that technological and design researches are conducted with affiliates or in-house. Only market research, which requires regular surveys and focused group discussions, can be outsourced to third parties.

20.4.2. *Pre-manufacturing Stage*

All of Company A's raw inputs — such as sheet metal, plastic powder or chemicals — are imported from various economies such as China; Japan; Korea; Malaysia; Singapore; and Thailand. Most paperwork related to the importation of raw materials — for example, procurement transaction and customs coordination — is done in-house, although Company A sometimes hires a third-party translator to communicate with suppliers based in other economies. Freight insurance, if needed, is procured from third parties.

Logistics services for bringing in raw materials are procured from various sources. Sea, air and rail transportation to import materiel are procured from third-party suppliers, but land transportation from the port to Company A's warehouses are procured only from Company B. Storage services for raw materials is done in-house by Company A, which has built its own storage warehouses including facilities for storing chemicals and gases (e.g., refrigerants).

If new equipment needs to be installed for the production of new products, Company A will do the installation in-house for small equipment. However, for major equipment, Company A may require the specialised services of Company C or an external supplier depending on the technical requirements of the equipment.

At the pre-manufacturing stage, the main services that are outsourced to third parties relate to overseas transport — whether sea or air transportation of raw materials. Land transportation, on the other hand, is sourced from affiliates under Company B. The other main aspect of this stage — warehousing — is done in-house by Company A, which over decades has built sufficient infrastructure to meet its storage needs.

20.4.3. *Manufacturing Stage*

Of the 27 various services procured by Company A during the manufacturing stage, 15 are provided in-house, 11 are procured from one of the affiliates under Company B, and nine are procured from external suppliers (see Table A.3 in Appendix for a detailed list of these services).

Most services related to the day-to-day operations of the factory and other manufacturing facilities — including production administration, warehousing, water treatment and equipment maintenance — are provided in-house as these are integral to Company A's operations. Other services for the maintenance of the grounds, such as gardening, security and waste disposal are provided by sister companies under Company B, while Company C sometimes provides engineering services for specific projects. If specialised expertise is needed for some services, such as equipment cleaning or hazardous waste collection, the services of external suppliers are procured. Likewise, engineering services (e.g., machining) are either provided by in-house engineers or obtained from external suppliers depending on required expertise.

On the other hand, most services related to workers' welfare are provided by sister companies under Company B. These services include catering, dormitory facilities, medical services and retail outlets. Company A provides human resource services in-house, as well as recreational (e.g., karaoke) and religious facilities (i.e., mosque). Retirement and health services for workers are provided in-house by Company A, while mandated social benefits are obtained from Badan Penyelenggara Jaminan Sosial (BPJS).

In the manufacturing stage, which is the most crucial stage in Company A's operations, the most essential services relate to engineering, maintenance and workers' welfare services. Most of these services are provided in-house or through affiliates due to intra-firm trust — whenever possible, Company A prefers to work with Company B or C to meet service requirements. Services from third parties are procured only when expertise is not available in-house or with affiliates, or if mandated by law (e.g., BPJS). Moreover, as Company A does not produce its own water supply or electricity, these services are procured from external suppliers: the power distributor for electricity and the water utility company for water.

20.4.4. *Post-manufacturing Stage*

After the products are manufactured, Company A makes them ready for packaging and delivery. Decisions on packaging design are done by Company C as it owns the brand, although Company A also provides inputs on packaging design. Packaging services, on the other hand, are all done in-house by Company A.

Company A produces units on an available-to-promise basis (ATP), which means it produces a specific number of units by a due date for immediate delivery. While it has an in-house facility for final products, it is only for one-day storage for outbound delivery the next day. Delivery services are provided by a transportation firm under Company B.

Orders for Company A's products come from a sister firm under Company B, which handles all sales and after-sales services. Hence, all retail services, advertising, warranties and after-sales services are handled by Company B and are external to Company A's operations. After Company A produces the appliances, it is immediately shipped out to Company B which sells the item. The only involvement of Company A in sales or after-sales services is the provision of spare parts to Company B.

Company A is peculiar in that it practically has no post-manufacturing activities. Product design and branding are handled by Company C, while marketing, sales and after-sales are done by Company C. As such, most post-manufacturing activities such as advertising, customer relations, repairs and retail are not utilised by Company A.

20.4.5. *Office and General Services*

As Company A has had a presence in Indonesia for several decades, it has already built up a considerable set of in-house service structures, which were a necessity before outsourcing became common practice around the world. As such, most office and general services in Company A are provided in-house, including those that are commonly outsourced to third parties by other firms such as financial services (e.g., treasury services), IT support, health care and pensions, and human resource services (e.g., personnel search and headhunting).

However, its affiliation with Company B has also led to some office services to be outsourced, such as security, telecommunications and public relations. On the other hand, Company C continues to provide most research and development services, product design and packaging, and upper management services (e.g., strategic direction). General services outsourced to third parties include utilities, courier services, external auditing, specialised IT, repair or waste management services.

As with the manufacturing stage, Company A prefers to provide office and general services in-house or from affiliates due to trust issues. Being a decades-old company, Company A has had time to develop in-house service structures — such as health care support, treasury services, pensions — that other younger firms may struggle to develop and are outsourced by necessity. Likewise, Company A's age may be leading to path dependence on doing business as it always has: after having developed in-house service structures, it may be more costly to outsource certain services regardless of possible efficiency gains.

20.5. POLICIES AFFECTING THE VALUE CHAIN

Due to its long presence in Indonesia, Company A has adjusted well to many regulatory requirements and policies affecting its value chain. It reports no problems of getting imports through customs inspections and clearance. Policies on foreign equity restrictions are also not a problem as the electronics manufacturing industry allows full foreign ownership, while policies on foreign workers do not affect Company A as the only foreign workers in the firm are a few members of higher management. Likewise, regulations on quality standards are not an issue as Company A is in the forefront of setting standards and is one of the few firms in Indonesia that can provide product testing services.

A policy area that Company A has had difficulty with is the tariff structure in Indonesia, particularly for importing its raw materials. Company A reports that tariffs for completely built air conditioners and refrigerators are lower than those for inputs such as sheet metal, which can be a disincentive to develop local manufacturing capacity.

Based on data from the World Trade Organization (WTO),[2] average ad valorem duties for air conditioners and refrigerators are 10 per cent, while duties for treated or coated flat-rolled iron or non-iron steel are 12.5 per cent. According to Company A, the cause of this issue is that Harmonized System (HS) codes for steel sheets are not differentiated. Although treated or coated steel sheets for automotive and electronic appliances are different products with different standards of treatment or coating, they share the same HS code. Hence, while the government may wish to protect local manufacturers of automotive steel sheets, these protections end up hurting local manufacturers of electronic appliances as well.

Corporate tax rates in Indonesia are another policy aspect that affects Company A's value chain, albeit indirectly. As Company C has a manufacturing presence in various economies in Southeast Asia including Indonesia, higher corporate tax rates in Indonesia can affect investment and manufacturing decisions of its parent firm as taxes ultimately affect the bottom line. Indonesia's corporate tax rate stands at 25 per cent; in comparison, the corporate tax rate in Viet Nam is 22 per cent and set to go down further to 20 per cent in 2016. According to Company A, this distorts the market and makes it difficult to compete with manufactured goods from China or other Southeast Asian economies, making it cheaper to manufacture goods abroad and import them into Indonesia.

Property rights protection and labour laws on compensation are other policy areas that Company A thinks can be improved. Company A reports that trademark infringement has been an ongoing problem with some local competitor firms adopting logos, industrial design and packaging that can be confused with those of Company A. Moreover, patent applications tend to be very lengthy in Indonesia, with the whole process from application to patent certificate taking about one and a half years.[3] On the other hand, rigid

---

[2] WTO Data on MFN Tariff for Indonesia, last updated on 29 April 2015.

[3] According to the World Intellectual Property Organization, the time it takes to obtain a patent can range from a few months up to 18 months, depending on the rigorousness of the examination process. http://www.wipo.int/sme/en/faq/pat_faqs_q4.html

labour laws on compensation reward seniority over performance. Company A suggests that there should be more flexibility on compensation issues in order to promote productivity.

Although not directly identified by Company A as a policy constraint, more can be done at the policy level to upgrade the skills of workers and develop the capabilities of local suppliers. According to Company A, they often find it difficult to hire skilled workers for their manufacturing processes. This is surprising given Indonesia's large pool of young workers, with demographic dividends expected to continue for several decades. However, the constraint is not in finding people willing to work but finding workers with matching skills. Policy can do much to address this skills gap by providing workers with more access to training opportunities, including vocational education and lifelong learning. In order to ensure that these training will impart skills needed by manufacturers such as Company A, efforts to upgrade workers' skills will need to be coordinated with the needs of industries. This could be done with labour market monitoring of skills demand and supply, as well as regular consultation with industry groups. This feedback mechanism — involving both statistical monitoring of the labour market and consultation with stakeholders — should be part of a larger policy to promote manufacturing.

This policy, in turn can be directed towards upgrading the capacity of local manufacturers to provide inputs higher up the value chain. According to Company A, local firms can only provide basic manufacturing services, and there are no local suppliers that are capable of providing them with higher value-added manufactured inputs such as casings or mouldings. At present, they need to ship plastic powder to China or Malaysia where they will be transformed into, say, air-conditioning casings, before importing them back to be assembled in Indonesia. The government can thus explore ways to develop local manufacturing capacity so that firms can produce outputs higher up the value chain. Moreover, development of manufacturing can be synchronised with the skills development of Indonesian workers.

In the past decades, several services that were once provided in-house in Company A have been outsourced to affiliate of third-party firms. These services include cleaning, transportation, security and waste disposal. The decision to outsource these services, however, had more to do with cost efficiency and trust rather than due to policy influences. Of course, some regulations direct Company A to outsource some services, such as external auditing or specialised waste disposal, but these are more the exception rather than the rule.

APPENDIX

Services Used in the Value Chain and Service Supplier

Table A.1: Establishment Phase.

Stage/category	Service	Central product classification (CPC) ver. 2 code	Supplied in-house	Supplied by another company in the group	Outsourced to third parties
1. Government services (licensing, etc.)	Company registration and licensing services	91138 — Public administrative services related to general economic, commercial and labour affairs	X		X
	Information and statistical services	9113 — Public administrative services related to the more efficient operation of business			X: monthly market data
	Visa and immigration services for foreign investors/employees	91290 — Public administrative services related to other public order and safety affairs	X: admin staff		X: airport transfers

(*Continued*)

Table A.1: (*Continued*)

Stage/category	Service	Central product classification (CPC) ver. 2 code	Supplied in-house	Supplied by another company in the group	Outsourced to third parties
2. Other services (professional, etc.)	Banking and finance services	71121 — Deposit services to corporate and institutional depositors 71135 — Non-mortgage loan services for business purposes	X		X: bank
	Legal services	82130 — Legal documentation and certification services	X		

Source: Author.

Table A.2: Raw Materials, Input/Pre-production Stage.

Stage/category	Service	CPC ver. 2 code	Supplied in-house	Supplied by another company in the group	Outsourced to third parties
1. Procurement of raw materials	Procurement agent for raw material sourcing	85999 — Other support services n.e.c.	X		X: translation services for foreign supplier
	Customs-related services for raw materials imported	85999 — Other support services n.e.c.	X: customs relations department		
	Quality assurance services (of raw materials)	83441 — Composition and purity testing and analysis services	X		
2. Logistics	Freight transportation services (of raw materials) by road, rail, sea or air	65 — Freight transport services		X: Company B for land transport	X: sea, air and rail transport
	Freight insurance of raw materials	71333 — Freight insurance services			X: depends on CIF or FOB
3. Storage	Storage of raw materials — general storage	67290 — Other storage and warehousing services	X		

(*Continued*)

Stage/category	Service	CPC ver. 2 code	Supplied in-house	Supplied by another company in the group	Outsourced to third parties
	Storage of raw materials — tank farm	67220 — Bulk liquid or gas storage services	X		
4. Product design	Conception and design of product	83920 — Design originals	X	X: Company C	
	Industrial design	83912 — Industrial design services	X	X: Company C	
	Patent acquisition	83960 — Trademarks and franchises n/a	X		
5. Installation services	Installation services of equipment	873 — Installation services (other than construction)	X: for small equipment	X: Company C for major equipment	X: major equipment
6. Others	Training services for staff	92919 — Other education and training services, n.e.c.	X		

Source: Author.

Table A.3: Manufacturing Stage.

Stage/category	Service	CPC ver. 2 code	Supplied in-house	Supplied by another company in the group	Outsourced to third parties
1. Production administration	Production Administration — Production management	83115 — Operations management consulting services	X		
	Production Administration — Repair and maintenance of factory equipment	87156 — Maintenance and repair services of commercial and industrial machinery	X: regular maintenance		X: for overhaul maintenance of equipment
	Production Administration — Quality assurance and compliance with ISO	83441 — Composition and purity testing and analysis services	X		
	Warehousing services for intermediate goods	67220 — Bulk liquid or gas storage services 67290 — Other storage and warehousing services	X		
2. Services supporting factory daily operations	Cleaning services of factory	85330 — General cleaning services	X: minor cleaning	X: mostly Company B	

(Continued)

Table A.3: (*Continued*)

Stage/category	Service	CPC ver. 2 code	Supplied in-house	Supplied by another company in the group	Outsourced to third parties
	Engineering Services	83310 — Engineering advisory services 8332 — Engineering services for specific projects	X	X: Company C	
	Gardening services for factory lawn	85970 — Landscape care and maintenance services		X: Company B	
	Security guards for factory and warehouses	85250 — Guard services	X: chief security officer	X: Company B	
	Sewage water treatment services	94110 — Sewerage and sewage treatment services	X		
	Specialised cleaning services for machines and equipments	85340 — Specialised cleaning services			X

	Repair and maintenance services of machines and equipment	87156 — Maintenance and repair services of commercial and industrial machinery	X	X	X
	Waste collection and recycling services	9421 — Collection services of hazardous waste 94229 — Collection services of non-hazardous recyclable materials, other 94239 — General waste collection services, other		X: Company B X: Company B	X: need special license
3. Services from government regulation requirements	Government inspections on fire prevention, health hazards, environmental protection and other aspects.	91133 — Public administrative services related to mining and mineral resources, manufacturing and construction 91290 — Public administrative services related to other public order and safety affairs	X		X: government

(*Continued*)

Table A.3: *(Continued)*

Stage/category	Service	CPC ver. 2 code	Supplied in-house	Supplied by another company in the group	Outsourced to third parties
4. Worker-related services	Catering services for workers	63393 — Other contract food services		X: Company B	
	Dormitory for factory workers	63220 — Room or unit accommodation services for workers in workers hostels or camps		X: Company B	
	Medical services	93121 — General medical services		X: Company B	
	Personnel search and referral services — Recruitment of factory workers	85112 — Permanent placement services, other than executive search services	X		

Retail services for workers	6212 — Non-specialised store retail trade services, of food, beverages and tobacco		X: Company B	
Recreational facilities and services for workers	96520 — Sports and recreational sports facility operation services	X		
Social insurance for factory workers	91320 — Administrative services related to government employee pension schemes; old-age disability or survivors' benefit schemes, other than for government employees 91330 — Administrative services related to unemployment compensation benefit schemes	X: retirement and health benefits		X: mandated social benefits (BPJS)

(*Continued*)

Table A.3: (*Continued*)

Stage/category	Service	CPC ver. 2 code	Supplied in-house	Supplied by another company in the group	Outsourced to third parties
5. Utilities	Utilities (electricity, gas and water supply)	691 — Electricity and gas distribution (on own account)			X
		692 — Water distribution (on own account)			X
6. Engineering services	Manufacturing services such as basic machining	886 — Basic metal manufacturing services	X		X: depends on required capabilities
		887 — Fabricated metal product, machinery and equipment manufacturing services	X		X: depends on required capabilities
7. Others	Product testing to obtain certification	8344 — Technical testing and analysis services	X		

Source: Author.

Table A.4: Delivery and Sales/Post-manufacturing stage.

Stage/category	Service	CPC ver. 2 code	Supplied in-house	Supplied by another company in the group	Outsourced to third parties
1. Packaging and labelling	Design of packages	83919 — Other specialty design services	X	X: Company C	
	Packaging services	85400 — Packaging services	X		
2. Delivery to wholesaler/ retailer	Cargo handling services	67110 — Container handling services		X: Company B	X
	Customs-related services	85999 — Other support services n.e.c.		X: Company B	
	Land transport of goods (from warehouse to domestic port)	65112 — Road transport services of freight by tank trucks or semi-trailers		X: Company B	
	Water transport (from domestic port to destination port)	65213 — Coastal and transoceanic water transport services of intermodal containers by container ships			X

(*Continued*)

Table A.4: (*Continued*)

Stage/category	Service	CPC ver. 2 code	Supplied in-house	Supplied by another company in the group	Outsourced to third parties
3. Sales services	Retail trade services — By store	621 — Non-specialised store retail trade services 622 — Specialised store retail trade services		X: Company B	
	Storage and warehousing services for finished goods	67220 — Bulk liquid or gas storage services 67290 — Other storage and warehousing services	X: one-day storage; available-to-promise	X: Company B	
4. Others	Warranty, maintenance, repair and overhaul services	8714 — Maintenance and repair of transport machinery and equipment	X: spare parts provision to Company B		

Source: Author.

Table A.5: Back office, Utilities and General Services.

Stage/category	Service	CPC ver. 2 code	Supplied in-house	Supplied by *another company in the* group	Outsourced to third parties
1. Finance	Auditing on financial accounts	822 — Accounting, auditing and bookkeeping services	X		X: external auditors
	Financial services	71121 — Deposit services to corporate and institutional depositors	X		
		71313 — Group pension services	X		
		71701 — Services of holding equity of subsidiary companies	X		
	Internal auditing (including audits of financial accounts and corporate governance)	822 — Accounting, auditing and bookkeeping services 83118 — Head office services	X	X: Company C	X: external auditors
	Insurance services for machinery	71332 — Marine, aviation and other transport insurance services			X

(*Continued*)

Table A.5: (*Continued*)

Stage/category	Service	CPC ver. 2 code	Supplied in-house	Supplied by *another company in the* group	Outsourced to third parties
2. General management	Corporate communications, marketing and public relationship	83121 — Public relations services	X		
	Courier, postal and local delivery services	681 — Postal and courier services			X
	Estate management	72112 — Rental or leasing services involving own or leased non-residential property	X		
	Human resources management	83113 — Human resources management consulting services	X		
	Human resources management — Personnel search and referral services for back office staff	8511 — Personnel search and referral services	X		

I.T. and information system management, consulting and support, with webpage development	8313 — Information technology (IT) consulting and support services 8314 — Information technology (IT) design and development services 8316 — IT infrastructure and network management services	X		X: depends on capacity/ requirements
Safety and security services	85230 — Security systems services 85250 — Guard services		X: Company B	
Telecommunication services	841 — Telephony and other telecommunications services 84210 — Internet backbone services 84221 — Narrowband Internet access services 84222 — Broadband Internet access services		X: Company B	

(Continued)

Table A.5: (*Continued*)

Stage/category	Service	CPC ver. 2 code	Supplied in-house	Supplied by *another company in the* group	Outsourced to third parties
	Uniform — Design and alternation	83919 — Other specialty design services	X		X: sewing services
3. Legal	Legal services	82120 — Legal advisory and representation services concerning other fields of law 82130 — Legal documentation and certification services	X		
4. Research and development	Product development/R&D	81129 — Research and experimental development services in other engineering and technology 81400 — Research and development originals	X	X: Company C	
5. Utilities	Electricity supply	8631 — Support services to electricity transmission and distribution			X

		Gas supply	86320 — Gas distribution services through mains (on a fee or contract basis)			X
		Water supply	86330 — Water distribution services through mains (on a fee or contract basis)			X
6.	Headquarter services	Headquarter services	83118 — Head office services	X	X: Company C	
		Management services	83111 — Strategic management consulting services	X	X: Company C	

Source: Author.

Chapter 21

FRESH CHERRY INDUSTRY IN CHILE

Penny Bamber and Karina Fernandez-Stark

21.1. INDUSTRY OVERVIEW

The growth of global value chains (GVCs) in the fresh fruit sector over the past three decades has resulted in its increased industrialisation, with sophisticated packing plants and highly efficient logistics operations becoming key tenants of competitiveness. The recent evolution of the cherry sector illustrates these changing dynamics. Although a highly perishable product with a considerable quality premium, distant inter-regional trade of fresh cherries has grown significantly over the past 15 years thanks to the incorporation of new handling and packaging technologies and services.

In 2013, global trade in fresh cherries reached US$1.5 billion, up from US$0.5 billion in 2003 (UN Comtrade, 2015).[1] The Asia Pacific region plays a central role in the global industry, accounting for approximately half of both demand and supply.[2] The two largest cherry exporters are the United States and Chile, accounting for 29 per cent and 27 per cent of world exports respectively.[3] China is the

[1] All UN Comtrade statistics are based on calculations using 2013 import data, using product code HS-080929.

[2] In 2013, APEC member economies accounted for 63 per cent of all imports and 67 per cent of all exports.

[3] Turkey is the world's third largest cherry exporter, with an 11 per cent market share. The majority of its exports are destined to intra-regional trade in the European Union.

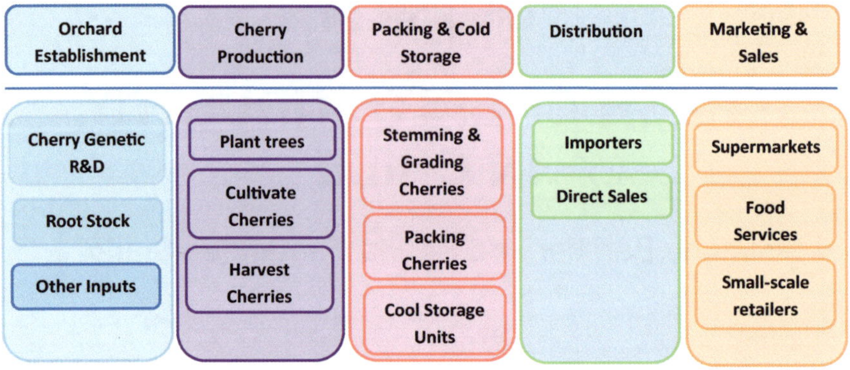

Figure 21.1: Simplified Cherry Industry Value Chain.

Source: Authors.

single largest market, accounting for one-third of all fresh cherry imports. The Asian markets demand not only high volumes, but also high-quality fruit, despite an estimated 30–45 days for container shipment. This places increased emphasis on the need for effective production, packaging and logistics solutions as well as efficient distribution and sales channels. Figure 21.1 illustrates a simplified cherry global value chain.

Chile is one of the few producing economies with primarily export-driven cherry production, with less than one in four cherries being sold in the domestic market (Fresh Fruit Portal, 2013). Chilean cherry export growth has been driven by the sector's ability to rapidly scale up volumes and produce high-quality fruit. The sector began to expand in 2000 in response to opportunities arising from increased demand from China (FreshPlaza, 2015); by 2014, exports had reached 100,000 tonnes, and the value per kilogram of Chilean cherries was almost twice the global average.[4] Exports are projected to increase by a further 50,000 tonnes by 2018 as the result of aggressive expansion of producer orchards

[4] The global average price per kilogram in 2013 and 2014 was US$4.37 and US$4.65 respectively. Chile earned US$6.03 and US$7.27 during these years. New Zealand captured the highest value with US$11.53/kg in 2014 (UN Comtrade, 2015).

(FreshPlaza, 2015). With the economy's orchard expansion outpacing that of rival, United States, this would make Chile the world's largest cherry exporter.

21.2. BACKGROUND INFORMATION ON THE FIRM[5]

The firm represented in this case is one of Chile's top five fresh fruit exporters. In addition to cherries, the company exports a diverse range of fruits, including apples, avocados, blueberries and grapes to over 50 different markets in the Americas, Asia, Europe and the Middle-East, and North Africa. Although its core business is its export operations dominated by large number of packing plants in Chile and Peru, the firm also cultivates fresh fruit through production owned by its business partners, in addition to running large contract grower operations. This model allows the company to guarantee its supply.

With over 30 years of experience, the privately owned Chilean firm has operated under an export-oriented strategy since its inception. It began producing cherries for export in 2008, and had expanded to almost 10 different varieties by 2015. Together with blueberries, cherries are one of the highest value fruits exported by the company and returns can be as much as four times that of the production of other staple exports, such as apples. Two-thirds of the firm's cherry exports are destined to the Chinese market, followed by North America and the United Kingdom.

All of the firm's cherry exports are destined for the fresh cherry market due to its high quality. While cherries can be processed — including frozen, canned and preserved such as the popular maraschino cherry used in cocktails around the world, the firm does not carry out these activities and its products are directed exclusively to the fresh cherry market. Fresh cherries can be sold for 70–100 per cent more than cherries destined for processing and even lower quality fruit, which is sold on the domestic market, nets higher prices than that offered by processing firms.

[5]All information about the firm in this case study is from its corporate website as well as from the interview.

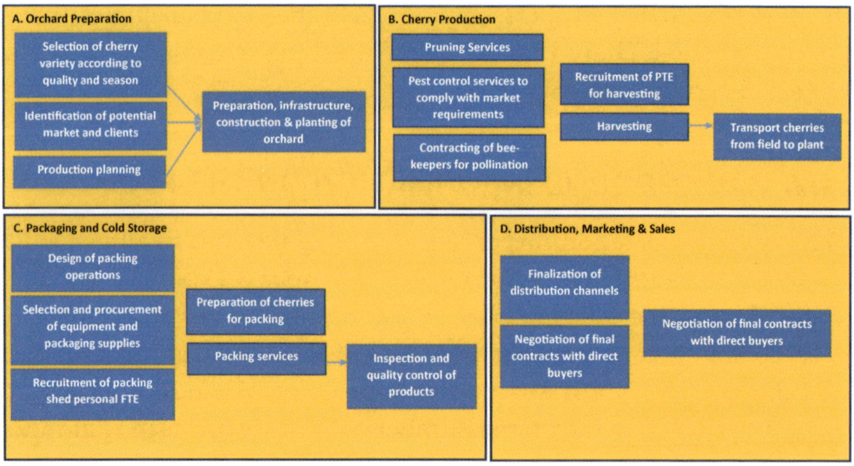

Figure 21.2: Dimension of the Value Chain Covered by Case Study.

Source: Authors.

21.3. DESCRIPTION OF THE VALUE CHAIN

This case study will focus on the firm's Chilean operations, which accounts for the majority of the firm's exports and all cherry exports. The product of focus for this study is fresh cherries, including both sweet and sour varieties. These products are grown over a 700 km stretch, from Santiago in Central Chile to Osorno in the South.

For the purposes of this study, the value chain is considered to begin with research and development (R&D) in both genetics and field performance of different varieties, this is followed by orchard establishment, cherry production, packaging, distribution and marketing (Fig. 21.2).

21.3.1. *R&D and Pre-orchard Establishment*

The first stage in the cherry global value chain is the development and selection of cherry variety. The firm works with several different varieties including Bing, one of the most highly demanded cherry types, and is continuously experimenting with new alternatives to improve production. The development and selection of the right variety of cherry for production underlies two key aspects of competitiveness in international cherry trade — fruit quality and seasonality. Fruit quality in key markets is determined by size, colour, taste and the absence of blemishes. Higher quality fruit nets higher prices. Prices, however, are also determined by when the fruit enters the market, that is, its seasonality. While consumers are often unaware of the differences between the varieties beyond sweet and sour, different cherry varieties perform differently according to the specific climatic and geological conditions in which they are grown. Thus, significant research and testing must be undertaken prior to the plantation of a commercial orchard to ensure quality production and that harvest can be planned to optimise installed packing plant capacity and meet peaks in market demand.

At this stage, the firm provides extensive information on new varieties, their performance characteristics and the firm's projected demand to its outgrowers who then independently determine whether or not to produce these varieties. Once the cherry varieties are selected, the land and rootstock must be prepared and planted and irrigation and roofing infrastructure installed. Due to the impact of rain on the final quality of cherries — which can lead to losses of 30 per cent of the crop, roofing of the cherry trees has become essential. Newly planted orchards then take some four to five years to reach maturity and their average productive lifespan is 30–35 years.

21.3.2. *Cherry Production*

Production has four key annual steps required for orchard optimisation: pruning, pest control, pollination and harvesting. Other ongoing operations include irrigation and fertilisation. Pruning must be undertaken once a year to ensure maximum commercial harvesting heights. Pest control is essential to ensure fruit quality and minimise the spread of disease. Pollination of cherry blooms is actively managed through the presence of beehives in the orchards which can increase production by over 50 per cent. Finally, the cherries are harvested by hand to protect against damage before being transported to the packing plant. While hand-picking the cherries makes this a particularly labour intensive activity, mechanised harvesting has not yet evolved sufficiently to protect the fruit from damage which undermines its arrival quality in long-distance shipping.

21.3.3. *Packing, Cold Chain Management and Logistics*

During the packing stage, fruit is first hydro-cooled and fumigated to slow ripening and kill any spores upon entry from the field. It is then cleaned, stems are checked and the fruit is graded before being packaged for exports. While in early days of exporting, these operations were done by hand, today, the firm's packing operations have been mechanised. Mechanisation of washing and checking of stems, combined with electronic grading has significantly improved the firm's productivity. The shift from manual to mechanised packing plant operations can increase volumes by a factor of 5. This is essential given the increased volume of fruit being produced each year. This equipment represents a particularly high barrier to entry, as it is expensive, requires economies of scale in production and is highly specific to cherries, meaning that it is only used for a few months of the year during the harvest season.

The cherries are first packaged into own-brand bags and placed into 5 kg cardboard boxes which are then loaded into refrigerated containers or trucks for transport to the port. Unlike other fruit such as apples and grapes, cherries cannot be kept in cold storage. Normally, cherries must be consumed between 7 and 10 days following harvest, limiting firm exports to regional markets[6]; however, the specialised "modified atmosphere" packaging used by the firm slows ripening considerably and allows for the fruit to maintain its quality during long-distance shipping. This increases the fruit's shelf life to up to 60 days, and when combined with effective cold chain management, this essentially allows exporters to reach any global market.

[6] Even shipping from Chile to the United States takes approximately 14 days.

Prior to shipping, however, the fruit must be certified to ensure that it meets the sanitary and phytosanitary requirements of the different markets. These requirements are particularly strict in the United States and Europe. As a result of this, over the past decades, the packing plants have become pristine operating environments, akin to clean manufacturing factories in other industries.

21.3.4. *Distribution, Marketing and Sales*

Cherries are sent by road from the packing plants to one of a number of ports in Chile; early season cherries to China, for example, are shipped by air and can net prices as high as US$20 per kilogram; while mid-season volumes are shipped by sea. The majority of the fruit is shipped in refrigerated containers, which has drastically facilitated the export of fruit, by allowing mixed containers of different fruit to be sent to different markets. As a fruit exporter with multiple different fresh products, the firm is thus able to bundle sales of a variety of fruits for different clients, providing it with a definite competitive advantage over single product exporters. Given the high demand for cherries, mixing cherries exports provides the firm with leverage for the sale of commodity products such as Red Globe grapes.

The firm utilises a range of different distribution and marketing channels depending on the market. There are three primary models: In the first model, the firm manages direct sales to supermarkets, this is common for very big buyers such as Walmart; in the second model, the firm ships to independent importers or brokers who arrange commercial agreements with other sales points, from wholesale operations to small greengrocers. This is a common distribution method in China. Under the importer model, the importer charges an 8 per cent fee over the sales value to place fruit in the retail market.[7] In the third model, the firm owns the importer in the destination economy, either exclusively or as part of a joint venture with other fruit exporters. This model is more common in the European Union. As the product is largely a high-value commodity,

[7] Discounting arrival expenses.

minimal branding, or marketing is undertaken to promote the sale of the fruit.

Logistics and distribution plays a fundamental role in ensuring that the firm's fruit earns high prices and meets client needs; a glut of fruit on the market will depress prices, while there are certain high-demand dates for which the firm must ensure correspondingly large volumes are shipped — such as that for the Chinese New Year in early February. As mentioned earlier, this requires integrated planning from orchard locations, through production processes such as pruning to harvesting schedules, but is fundamentally dependent on logistics operations and the speed with which importers or brokers can place the product in the market. Any hold ups in the system, such as customs delays or problems with a broker, can upset this balance and result in millions of dollars of losses.

21.4. SERVICES ALONG THE VALUE CHAIN

21.4.1. *Services Identification and Value Contribution*

The cherry global value chain has a very high incidence of services. A total of 71 services, which have been further disaggregated into 88 separate services (see Fig. 21.3) are identified in this value chain and they are categorised according to the various stages within the chain: (i) orchard establishment services, (ii) services during cherry production, (iii) packing services, (iv) distribution and marketing services and (v) transversal services.

These services are discussed as follows according to the respective stage of the value chain.

21.4.2. *Orchard Establishment Services*

The primary services required in the establishment of a new orchard include variety testing and selection, the plantation of the orchard and the construction of the related infrastructure such as roofing and irrigation. Of these, only variety selection is carried out

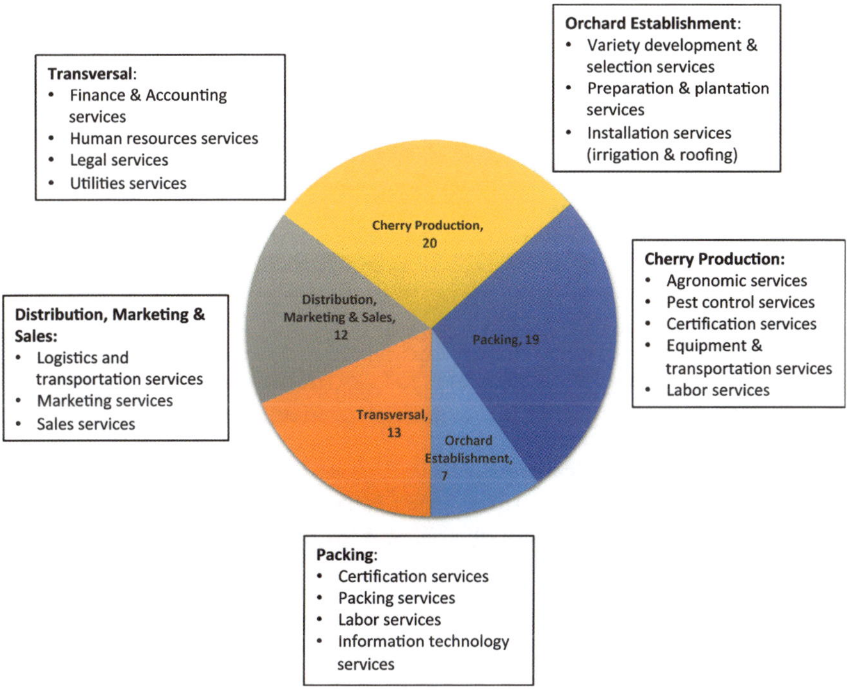

Figure 21.3: Breakdown of Services by Stage and Examples of Key Services.
Source: Authors.

in-house, performed together with affiliated cherry production by the company's owners. The firm's technical department sources and experiments with different varieties in nurseries hosted by these producers to identify combinations that optimise production. Variety development itself is mostly carried out by specialised companies in California, United States and the producers pay royalties according to areas planted and yields. The firm negotiates these royalties on behalf of both their in-house production and their outgrowers.

21.4.3. *Cherry Production Services*

During cherry production, the firm provides technical assistance, and production and harvest supervision to its outgrowers. There are 50 agronomists on staff in charge of these activities,[8] while additional external technical assistance is hired when highly specialised knowledge is needed. The firm also provides financial services to its outgrowers. With years of experience with outgrower operations, it has strong in-house knowledge of outgrowers' potential to deliver quality production, and thus their capacity to repay loans. This allows the firm to provide more competitive lending rates to producers.

Other services hired by producers include consulting services regarding pesticide application and standards compliance; beehive rental for pollination; post-harvest quality analysis; information technology (IT) for precision agriculture; subcontracting of labour and related services such as medical and catering during labour-intensive periods of pruning and harvesting; security services; weather advisories; transport services and renting machinery for harvest amongst others. These services are all outsourced to third-party providers.

21.4.4. *Packing Services*

In the packing stage of the value chain, key services required include packaging selection, certification (public & private), quality and residue testing, IT, equipment maintenance and cleaning, human resources services such as recruiting, training, transportation, medical provision and catering. The firm outsources the majority of these operations to specialised service providers such as laboratories for quality testing and the United States Department of Agriculture (USDA) for approval for exports to the United States.[9] However, it also partially outsources others and maintains in-house capacity to support day-to-day operations. For example, training is provided in-house where possible, but when new certifications are required for new buyers, external trainers are

---

[8] These agronomists work with producers of all fruits, not only cherries.

[9] Indeed, the USDA rents a space within the firm's packing plants to carry out its certification for export.

contracted. Regular equipment maintenance and cleaning is undertaken by an internal maintenance department; however, specialised repairs are outsourced.

While the firm owns two dedicated packing plants for cherries, they also rent a third plant, as well as cold storage facilities during the peak of the season. Interestingly, outsourcing of packing of cherries is also becoming more frequent. Indeed, the firm uses these services to manage peak periods when their in-house packing plants do not have enough capacity. In these cases, the contracted packing plant operates as a turn-key facility — that is, they must have all certifications and quality control in place. The exporter provides the cherries, packaging and specifications.

The Chilean sanitary and phytosanitary agency, *Servicio Agrícola y Granadero* (SAG), must certify all exported produce to ensure it meets regulatory requirements for its different markets. The USDA also has operations in the packing plant and provides certification for produce destined for the US market.

21.4.5. *Distribution, Marketing & Sales Services*

Due to the seasonality of the products, the firm does not have an internal transportation fleet, as this would lie idle for part of the year. Transportation options are widely available in the Chilean market. While the shipping industry in Chile is quite competitive due to the economy's high export volumes, economies of scale are important to manage costs, as these are approximately the same as production costs. Thus, the firm has established a company together with other leading exporters in the industry to negotiate shipping rates. This affiliated operation has developed sophisticated software which allows it to aggregate export volumes across firms and thus to determine the optimum arrangements for shipping companies, routes and methods for fresh produce based on the exported product and destination market. This service is also sold to other exporters in Chile. With their ability to negotiate lower prices at different stages of the logistics operations, combined with online customs clearance procedures, the firm does not use bundled logistics services as these are more expensive.

The principal services in the marketing and sales segments of the value chain are logo and packaging design and market research, combined with commercial operations needed to identify new clients and managing these relationships. The firm outsources market research services, while all commercial operations are carried out in-house. Market research and marketing in Chile's fruit export sector is almost exclusively carried out by the Association of Fruit Exporters (ASOEX) and ProChile, Chile's export promotion agency. This is the result of limited differentiation of the final product.

21.4.6. *Transversal Services*

In addition to services provision in each segment of the value chain, there are a number of transversal services that support the firm's day-to-day operations. These services include finance and accounting, human resource management and are primarily managed in-house. The most important service that is outsourced is the provision of the SAP information systems used for managing production, logistics and inventories. Legal services for negotiating contracts are also outsourced.

21.4.7. *Outsourcing, Bundling and other Aspects of Services Supply*

Among the 88 individual services identified in this value chain, our analysis estimates 12 services are supplied in-house, 32 are partially outsourced, while 44 are fully outsourced (see Table 21.1). This is illustrative of the aggressive outsourcing strategy for services followed in the sector.

Generally, the firm's reasons for outsourcing are driven by its overall strategy to focus on its core competencies and maintain its fixed operations as efficient and knowledge intensive as possible, operating in the highest value segments within the Chilean context. This is essential given relatively low margins, high costs of

Table 21.1: Breakdown of Services Provision, by Value Chain Stage.

Stage of value chain	In-house	Partially outsourced	Fully outsourced
Orchard establishment	2	0	5
Production	1	11	9
Packing	0	14	8
Distribution, marketing and sales	2	2	9
Transversal	7	5	13
Total	**12**	**32**	**44**

Note: Services performed by affiliated companies are included as "in-house" services in this analysis.

Source: Authors.

equipment and a short production season.[10] The sophistication and maturity of the fruit export sector in Chile means that local knowledge and services are widely available at competitive prices either independently or through affiliated firms, thus facilitating a high degree of services outsourcing and allowing the firm and its peers to focus on their core competencies. Furthermore, given the maturity of the fruit export sector, service providers have become increasingly specialised, and there has been a tendency away from bundled service packages. This is highlighted in the firm drawing on a limited number of bundled service operations.

Those services provided in-house are focused principally on the firm's core competencies: sourcing fruit at the right time and right quality, packing and placing the products in their key markets, as well as transversal services. Partially outsourced services are generally those that have a day-to-day aspect in addition to specialised service needs, as well as those with in-house supervision or control of quality and specifications, with external services suppliers for the bulk of the work. The firm has a total of 400 full-time employees. 120 of the full-time staff are in administrative, commercial and

[10] The cost of installation of one equipped packing plant can be as high as US$20 million.

management (15) positions or product specialists (50) providing technical assistance in production operations. The remaining 280 are in operations, including machine and cold room operators. The firm has an estimated 2000 seasonal employees on staff during peak periods. Seasonal employees primarily work on the packing lines in the pack house and many of these are subcontracted.[11]

As noted in Table 21.1, individual decisions to outsource are largely based on (i) efficiencies and economies of scale, such as transport services (ii) access to specialised equipment and skills, (iii) collective provision of services through industry association and public institutions, (iv) independent audits and government regulations (v) network economies, such as recruitment services, and (vi) lack of feasibility in supplying services in-house, such as utilities services.

21.5. POLICIES AFFECTING THE VALUE CHAIN

An important component of this study is the analysis of how policies, both government and private, impact the value chain discussed here, and hopefully, identify areas of improvements (see Appendix). This firm faces few policy challenges, for example, in trade policy (customs and logistics) and even in labour policies. This is due to the fact that Chile's policy environment is well-oriented to fruit exports, thanks to over 30 years of experience in the export sector. Furthermore, as one of Chile's largest fruit exporters, the firm is well positioned to both influence and comply with regulations in the economy.

21.5.1. *Export-oriented Policies*

Chile's fresh fruit exports have benefited to a great degree by the consistent export-oriented trade policies that have been implemented over the past few decades. By 2013, Chile had 22 free trade

[11] Many of these workers also work with the firm in the packing plants for other products during other periods of the year.

agreements (FTAs) with 60 economies, providing access to 60 per cent of the world's consumers (Murray, 2013). ProChile and DIRECOM (*la Dirección General de Relaciones Económicas Internacionales*), together with the National Agriculture and Livestock Agency (SAG) have actively worked to open markets for the economy's produce. ProChile and DIRECOM have been vital in terms of leading negotiations in reduction of quotas and tariffs. SAG is particularly proactive in ensuring compliance with Sanitary Phytosanitary (SPS) standards around the world. Indeed Chile's domestic standards are considered to be world-class, which facilitates access to most global markets.[12] Measures implemented and closely monitored by SAG guard against the entry of plant disease at all borders, and they work actively with all producers and exporters to ensure disease and pest control in the fields and packing plants. This helps both to reduce use of pesticides and ensures import destinations of the disease-free nature of Chilean fruit. As a result of these efforts all major markets are open for all of Chile's fruit — including cherries and blueberries.

21.5.2. *Institutionalisation*

Strong levels of institutionalisation have proven very helpful to ongoing sector growth. The Association of Fruit Exporters (ASOEX) is a particularly strong organisation. Representing 96 per cent of all fruit exports, the organisation provides a range of services from developing inputs for free trade agreement negotiations and managing public good issues such as research in packing techniques, disease, labour issues, certification and SPS problems, to carrying out a range of marketing services such as market research and promotion of Chilean produce in key international shows. Members pay a quota proportional to their exports. The organisation works

---

[12] In addition, SAG collaborated with the USDA to allow Chilean exporters to participate in the Animal and Plant Health Inspection Service - US Agency for International Developement (APHIS-USAID) Pre-Clearance Program which clears Chilean produce for import into the United States before it even leaves the pack-house (SAG, 2015).

closely with both public and private sector actors to resolve issues affecting the industry. The organisation has also been proactive in ensuring Chilean producers can easily meet private standards in key export markets. For example, the organisation was instrumental in the creation of ChileGAP. ChileGAP is considered equivalent to the private GlobalGAP certification required by many large buyers in the EU and the US markets.[13] Allowing producers to receive certification under the ChileGAP regulations, rather than requiring audits by GlobalGAP made certification significantly more affordable for even small producers.

21.5.3. *Efficient Logistics: Competitive Transportation, Modern Ports and Streamlined Customs Operations*

While transportation costs from Chile are invariably high, given the distance from major markets, logistics costs are lower than in many other competitor economies. High import and export volumes and a liberalised transportation sector have fostered competition amongst logistics providers, which have helped to reduce costs. The modernisation of the Chilean ports following their privatisation, which began in the late 1990s, has significantly contributed to the economy's capacity to rapidly export large quantities of perishable products. There are a large number of airports and ports (8–10) in Chile today that are designed to handle these products, from Coquimbo in the North to Chile Chico in the South. This means that no fruit has to travel more than 250 km by land before reaching the port. This helps reduce transportation cost, potential delays and maintain the quality of the fruit. Online processing with the implementation of the single-window system has improved the efficiency of customs clearance, reducing unnecessary delays at the ports. However, strikes amongst customs officials remain a key challenge which affects not only cherries but the export of Chile's wide variety of fresh products from fruit to mussels and salmon. These have occurred on a more regular basis in recent years. These

[13]ChileGAP was validated in 2008 by GlobalGAP. It was the first national standards to be accepted as equivalent to the demanding private standard.

strikes upset the streamlined operations, and can result in losses of millions of dollars in perishable produce (Fresh Fruit Portal, 2013).

21.5.4. *Human Capital Development & Labour Regulations*[14]

Efforts in terms of human capital development and labour regulation have been key in facilitating growth. Training efforts have been focused on driving productivity, led on one hand by tax incentives through the National Training and Employment Agency (SENCE), and on the other by the ChileValora program — a joint initiative between the Ministries of Economy, Education, Labour and Social Security and the private sector. SENCE provisions incentivise firms to individually and collectively develop training programs specially to meet the needs of the industry. The ChileValora program is a competencies-based certification system launched at the end of the 1990s that certifies every job profile required for the industry, from equipment operation to packing of fruit for export (Fernandez-Stark *et al.*, 2011). In addition to increased productivity, flexible labour regulation, which permits subcontracting of labour, has allowed the industry to maintain its competitiveness by relying on temporary contracts to meet peak periods of demand during the year. The introduction of childcare facilities also supports the participation of women in the packing plants.

21.5.5. *R&D and Variety Development*

As the industry has demonstrated success in cherry production, policies have also been put in place to improve productivity and constantly upgrade packing technologies to ensure freshness. Public–private partnerships such as that between Chile's Agricultural Research Institute (INIA, *Instituto de Investigacion Agropecuaria*) and ASOEX co-finance this research. Several competitive grant funds are available to carry out research for the fruit export sector, many of which are financed through the Fondo de Inversion para la

---

[14] For a discussion of how workforce development initiatives have contributed to the upgrading of the fruit industry in Chile, see Fernandez-Stark *et al.* (2010).

Competitividad (InnovaChile). The universities play a central role in applied research through the provision of services to the industry. Despite advances in terms of adjusting varieties to the Chilean context, additional efforts are still required in supporting research in the development of new varieties. Although INIA has made strides in developing new grape varieties, no work is yet being undertaken to develop new cherry varieties.

21.5.6. *Moving Forward*

This case study presents a cherry producer-exporter's perspective on the importance of services in its operations. Policies clearly impact the firm's access to and use of services, as well as its ability to provide them. To date, strong efforts by the Chilean government, combined with significant collaboration by industry actors, from concentrated, economy-wide efforts in promoting Chilean fruit abroad, to trade facilitation and logistics improvements, have supported the growth of the sector. Continued collaboration between the government and private sector and research and educational institutions will further support the sector's consolidation.

In particular, this collaboration over policies affecting services provision is required with respect to labour regulation, innovation and modernisation of government and trade processes. First, labour flexibility is important to the sector's competitiveness and minimising overhead; however, this needs to be assessed against the context of job and social security, together with retention strategies to increase productivity. Second, innovation in the adaptation of foreign varieties to the Chilean context and, in the long term, development of new varieties in Chile will help to position the economy not only as a producer of cherries, but as an exporter of knowledge-intensive research services. This requires continued support for local R&D. Finally, continued streamlining of government protocols regarding licensing and export and import procedures can help support increased volumes of exports, as well as expedite payments through the value chain to the different actors from producers to their wide range of service providers.

APPENDIX

Table A.1: Pre-agro/Orchard Establishment Phase.

	Service	CPC ver. 2 code	Supplied in-house	Outsourced to affiliated companies and reasons	Outsourced to third-party suppliers/ government and reasons	Bundled
1	Variety development	81140 — Research and experimental development services in agricultural sciences	No	No	Yes; comparative lack of expertise	
2	Variety selection	81140 — Research and experimental development services in agricultural sciences	Yes; in-field testing	Yes; in-field testing	No	
3	Identification and purchase of appropriate property for orchards	72240 — Real estate appraisal services on a fee or contract basis	No	No	Yes; efficiencies	
4	Negotiation and payment of royalties for closed varieties	85999 — Other support services n.e.c.	Yes; core strategy	No	No	

(*Continued*)

Table A.1: (*Continued*)

Service	CPC ver. 2 code	Supplied in-house	Outsourced to affiliated companies and reasons	Outsourced to third-party suppliers/ government and reasons	Bundled
5 Land preparation and plantation of rootstock	86119 — Other support services to crop production	No	No	Yes; efficiencies	
6 Construction of orchard infrastructure (roofing, etc.)	86119 — Other support services to crop production	No	No	Yes; efficiencies	
7 Installation of irrigation equipment	54234 — General construction services of irrigation and flood control waterworks	No	No	Yes; efficiencies and specialised skills	

Source: Authors.

Table A.2: Cherry Production.

	Service	CPC ver. 2 code	Supplied in-house	Outsourced to affiliated companies and reasons	Outsourced to third-party suppliers/ government and reasons	Bundled
8	Production management	83115 — Operations management consulting services	Yes	No	Yes; access to specialised knowledge	No
9	Certification of operations for private standards	83115 — Operations management consulting services	Yes; for recertification	No	Yes; for new certifications	No
10	Certification for public SPS	91131 — Public administrative services related to agriculture, forestry, fishing and hunting	No	No	Yes; government services	No
11	Disease and plague control	86119 — Other support services to crop production	Yes; application	No	Yes; efficiencies, specialised knowledge	
12	Pruning of cherry trees	86119 — Other support services to crop production	Yes; supervisory team and trainers	No	Yes; efficiencies labour-intensive for short periods	Bundled with labour sub-contracting

(*Continued*)

Table A.2: *(Continued)*

Service		CPC ver. 2 code	Supplied in-house	Outsourced to affiliated companies and reasons	Outsourced to third-party suppliers/ government and reasons	Bundled
13	Pollination of cherry blooms	86119 — Other support services to crop production	No	No	Yes; efficiencies — economies of scope	
14	Equipment rental for harvesting	86119 — Other support services to crop production	No	No	Yes; efficiencies	Bundled with repair and maintenance
15	Harvesting services	86119 — Other support services to crop production	Yes; supervisory team and trainers	No	Yes; efficiencies	Bundled with labour sub-contracting
16	Land transport of cherries from orchard to packing plant	65119 — Other road transport services of freight	No	No	Yes; efficiencies — economies of scale	
17	Repair and maintenance of equipment	87156 — Maintenance and repair services of commercial and industrial machinery	Yes; for minimal fleet	No	Yes; for specific maintenance	Bundled with equipment rental
18	Recruitment of temporary/ seasonal workers	8512 — Labour supply services	Yes	No	Yes; network efficiencies	

19	Transportation services for workers to and from orchards	64114 — Local special-purpose scheduled road transport services of passengers	No	No	Yes; economies of scale	Bundled with recruitment
20	Training of temporary/seasonal workers	92919 — Other education and training services, n.e.c.	Yes	No	Yes; specialised skills	
21	Catering services for workers	63393 — Other contract food services	No	No	Yes	
22	Medical services	93121 — General medical services	No	No	Yes	
23	Social insurance for seasonal workers	91320 — Administrative services related to government employee pension schemes; old-age disability or survivors' benefit schemes, other than for government employees	Yes; for internal personnel (required by law)	No	Yes,	Bundled with recruitment of workers

(*Continued*)

Table A.2: (*Continued*)

Service	CPC ver. 2 code	Supplied in-house	Outsourced to affiliated companies and reasons	Outsourced to third-party suppliers/ government and reasons	Bundled
	91330 — Administrative services related to unemployment compensation benefit schemes				
24 Post-harvest quality control for all production	83441 — Composition and purity testing and analysis services	No	No	Yes; efficiency, specialised skills	
25 Weather analytical services for all growers	83430 — Weather forecasting and meteorological services	No	No	Yes	
26 Agronomic extension services for outgrowers	83115 — Operations management consulting services	Yes; firm provides this service to their outgrowers	No	Yes; specialised skills	
27 Financing for the production of cherries by outgrowers	7113 — Credit granting services	Yes; specialised knowledge	No	No	

Source: Authors.

Table A.3: Packing.

	Service	CPC ver. 2 code	Supplied in-house	Outsourced to affiliated companies and reasons	Outsourced to third-party suppliers/government and reasons	Bundled
28	Production management	83115 — Operations management consulting services	Yes	No	Yes; specialised skills	
29	Architectural services: design and construction of packing plants	83213 — Architectural services for non-residential building projects	No	No	Yes; efficiency, specialised skills	Bundled with engineering services for plant construction
30	Engineering services: design and construction of packing plants	83310 — Engineering advisory services	No	No	Yes; efficiency specialised skills	Bundled with architectural services for plant construction
31	Certification of operations for private standards	83115 — Operations management consulting services	Yes; for recertification	No	Yes; efficiency specialised skills	

(*Continued*)

Table A.3: (*Continued*)

	Service	CPC ver. 2 code	Supplied in-house	Outsourced to affiliated companies and reasons	Outsourced to third-party suppliers/government and reasons	Bundled
32	Certification for public SPS	91122 — Public administrative services related to health care	No	No	Yes; government services	
33	Quality assurance and compliance with standards	83441 — Composition and purity testing and analysis services	No	No	Yes; efficiency, economies of scale	
34	Software systems for inventory management and packing operations	Class of 8313 —IT consulting and support services	Yes; in-house department for day to day operations	No	Yes; efficiency — specialised skills, economies of scale	
		Class of 8314 —IT design and development services				
		Class of 8316 — IT infrastructure and network management services				

35	Testing and selection of appropriate packaging materials	83442 — Testing and analysis services of physical properties	Yes	No	Yes; efficiency, specialised skills	
36	Rental of additional packing facilities	72112 — Rental or leasing services involving own or leased non-residential property	No	No	Yes; efficiency — extra capacity in peak periods	
37	Packing services	85400 — Packaging services	Yes	No	Yes; efficiency — extra capacity in peak periods	Bundled with cold storage services
38	Temporary cold storage prior to export	67210 — Refrigerated storage services	Yes	No	Yes; efficiency — extra capacity in peak periods	Bundled with packaging services
39	Repair and maintenance of equipment	87156 — Maintenance and repair services of commercial and industrial machinery	Yes	No	Yes	

(*Continued*)

Table A.3: (*Continued*)

	Service	CPC ver. 2 code	Supplied in-house	Outsourced to affiliated companies and reasons	Outsourced to third-party suppliers/government and reasons	Bundled
40	Specialised cleaning services for machines and equipment	85340 — Specialised cleaning services	Yes	No	Yes; specialised services	
41	Pest control	86119 — Other support services to crop production	Yes; for daily operations	No	Yes; efficiency and specialised skills	
42	Government inspections on fire prevention, health hazards, environmental protection and other aspects	91290 — Public administrative services related to other public order and safety affairs	No	No	Yes; government services	
43	Recruitment of temporary/seasonal workers	8512 — Labour supply services	Yes; permanent workers recruited in-house	No	Yes; efficiency and network efficiencies	Bundled with social insurance for workers

44	Social insurance for workers	91320 — Administrative services related to government employee pension schemes; old-age disability or survivors' benefit schemes, other than for government employees	Yes; for internal personnel (required by law)	No	Yes; subcontractors contribute individually	Bundled with labour subcontracting
		91330 — Administrative services related to unemployment compensation benefit schemes	No	No		
45	Catering services for workers	63393 — Other contract food services	No	No	Yes; efficiency	
46	Medical services	93121 — General medical services	No	No	Yes; efficiency	

Source: Authors.

Table A.4: Distribution, Marketing and Sales.

	Service	CPC ver. 2 code	Supplied in-house	Outsourced to affiliated companies and reasons	Outsourced to third-party suppliers/ government and reasons	Bundled
47	Consulting services for optimisation of logistics providers	83116 — Supply chain and other management consulting services	No	Yes; economies of scale	No	
48	Customs-related services	85999 — Other support services n.e.c.	Yes	No	No	
49	Land transport of goods (from packing plant to domestic port)	65111 — Road transport services of freight by refrigerator vehicles	No	No	Yes; efficiency — economies of scale	Bundled international shipping
50	Cargo handling services	67110 — Container handling services	No	No	Yes; efficiency — economies of scale	Bundled international shipping
51	International shipping, by sea and air	65213 — Coastal and transoceanic water transport services of intermodal containers by container ships 65211 — Coastal and transoceanic water transport services of freight by refrigerator vessels	No	No	Yes; efficiency — economies of scale	Bundled cargo handling services

		65319 — Air transport services of other freight			
52	Freight insurance	71333 — Freight insurance services	No	No	Yes; not possible to supply in-house
53	Reception, customs-services and handling at destination	67110 — Container handling services; 85999 — Other support services n.e.c. Customs-related services	No	No	Yes; efficiency — economies of scale
54	Storage and warehousing services for cherries at destination	67210 — Refrigerated storage services	No	No	Yes; efficiency
55	Importer (Concessioned fruit)	61221 — Wholesale trade services on a fee or contract basis, of fruit and vegetables	No	Yes; efficiency, certain markets	Yes; lack of expertise, market specific
56	Design of logo and design of packaging	83919 — Other specialty design services	No	No	Yes; efficiency, specialised skills
57	Market Research	83700 — Market research and public opinion polling services	Yes	No	Yes; collective action

(Continued)

Table A.4: (*Continued*)

Service	CPC ver. 2 code	Supplied in-house	Outsourced to affiliated companies and reasons	Outsourced to third-party suppliers/ government and reasons	Bundled
58 Sales services	61221 — Wholesale trade services on a fee or contract basis, of fruit and vegetables	Yes; core strategy	No	No	

Source: Authors.

Table A.5: Transversal Services.

	Service	CPC ver. 2 code	Supplied in-house	Outsourced to affiliated companies and reasons	Outsourced to third-party suppliers/ government and reasons	Bundled
59	Company registration and licensing services (obtaining permit to operate)	91138 — Public administrative services related to general economic, commercial and labour affairs	No	No	Yes, government services	
60	Headquarter services	83118 — Head office services	Yes	No	No	
61	Management services	83111 — Strategic management consulting services	Yes	No	No	
62	Finance and Accounting	Group of 822 — Accounting, auditing and bookkeeping services	Yes	No	Yes; independent audit required by law	
63	Insurance services (commercial life and accident/health insurance, property insurance for the factory compound, product quality insurance, management liability insurance)	7131 — Life insurance and pension services	No	No	Yes, not possible to supply in-house	

Table A.5: (*Continued*)

Service	CPC ver. 2 code	Supplied in-house	Outsourced to affiliated companies and reasons	Outsourced to third-party suppliers/ government and reasons	Bundled
	7132 — Accident and health insurance services	No	No	Yes, not possible to supply in-house	
	71334 — Other property insurance services	No	No	Yes, not possible to supply in-house	
	71335 — General liability insurance services	No	No	Yes, not possible to supply in-house	
64 Corporate communications, marketing and public relationship	83121 — Public relations services	Yes	No	No	

65	Legal services, including contract negotiations with foreign buyers	82120 — Legal advisory and representation services concerning other fields of law 82130 — Legal documentation and certification services	No	No	Yes; efficiency, specialised skills
66	Human resources management (Permanent staff)	Class of 8511 — Personnel search and referral services	Yes	No	No
		92919 — Other education and training services, n.e.c.	Yes	No	Yes; specialised skills
67	Social insurance for staffs	91320 — Administrative services related to government employee pension schemes; old-age disability or survivors' benefit schemes, other than for government employees	Yes	No	No
		91330 — Administrative services related to unemployment compensation benefit schemes	Yes	No	No

(*Continued*)

Table A.5: (*Continued*)

Service	CPC ver. 2 code	Supplied in-house	Outsourced to affiliated companies and reasons	Outsourced to third-party suppliers/ government and reasons	Bundled
68 IT services	Class of 8313 —IT consulting and support services	Yes; internal IT department supports day to day operations	No	Yes; efficiency, specialised skills	
	Class of 8314 —IT design and development services				
	Class of 8316 — IT infrastructure and network management services				
69 Utilities: Electricity, gas, water and telecommunications services supply and waste removal	Class of 8631 — Support services to electricity transmission and distribution	No	No	Yes; not possible to supply in-house	
	86320 — Gas distribution services through mains (on a fee or contract basis)	No	No	Yes; not possible to supply in-house	

		86330 — Water distribution services through mains (on a fee or contract basis)	No	No	Yes; not possible to supply in-house
		94239 — General waste collection services, other	No	No	Yes; not possible to supply in-house
		Group: 841 — Telephony and other telecommunications services	No	No	Yes; not possible to supply in-house
70	Security services	85230 — Security systems services	No	No	Yes; specialised skills
		85250 — Guard services	No	No	Yes; efficiency
71	Cleaning services	85330 — General cleaning services	Yes	No	No

Source: Authors.

Table A.6: Policies Affecting Services in the Value Chain.

Government and private policies/ services	Authority(ies) in charge	Details	How the policy affects services in the value chain
Incentives for R&D (Tax Incentives and Competitive Grant Funds)	CORFO (Economic Development Agency), CONICYT (National Commission for Scientific and Technological Research)	CORFO offers a series of grant and incentive benefits for the R&D in the fruit sector. These instruments are designed to foster collaboration between local and international universities, public research institutions and the private sector	It provides inputs into improved production, harvesting and post-harvesting techniques, resulting in better quality fruit with higher post-harvest performance
Competitive Grant Funds for Exports in Agroforestry	PROCHILE (Export Promotion Agency)	The program provides market research and undertakes promotional activities to support the growth of cherry exports abroad. In addition, the program provides funding for commercial missions abroad, office rentals abroad for promotion, participation in international trade fairs, short-term training abroad amongst others	It provides exporters with detailed information regarding potential export markets as well as promoting cherries abroad in key international events. In addition, it offers financing for other marketing and sales services in the value chain
Online Certification for Agricultural and Forestry Exports	Servicio Agricola y Granadero (SAG)	Certification that exported products meet the phytosanitary requirements of the destination market. This certification can be obtained online	It facilitates entry of cherries into destination markets

SAG/USDA-APHIS/ASOEX Pre-Clearance Program	SAG, USDA and ASOEX	This public–private model provides jointly managed inspection sites in Chile for pre-clearance primarily for fresh fruit destined to the US market. The joint inspection program has been in place since 1982, and has numerous sites across Chile including all major ports	It streamlines import procedures at US ports allowing Chilean fruit to quickly enter the US market
Modernisation of Customs Services and Single-Window System	Servicio Nacional de Aduanas (National Customs Office)	The online single-window system requires exporters to submit all export related documentation for approval in one place. It is open 24 h a day, 7 days a week provides immediate processing of documents that do not require further evaluation. This system was launched in a pilot phase in 2013; fresh cherries were included in the first phase	It enables governments to electronically process information, documents and fees both faster and more accurately. In doing so, it increases the speed of customs handling and processing reducing the time perishable produce is in transit for international markets
Tax Credit for Education and Training	Servicio Nacional de Capacitación y Empleo SENCE (National Training and Employment Agency)	This incentive allows firms to deduct the cost of training workers in specific workplace competencies, subject to a cap of 1 per cent of total compensation costs. To qualify for this credit, employers must pay social security costs for participating employees. Courses must be approved by SENCE to receive this benefit and can vary from short term courses (<8 h) to longer courses with academic equivalence (>200 h)	It facilitates training for employees in the sector, directly contributing to productivity

(Continued)

Table A.6: (*Continued*)

Government and private policies/ services	Authority(ies) in charge	Details	How the policy affects services in the value chain
National Labour Skills Certification System (NLSCS)	Joint initiative between the Ministries of Economy, Education, Labour and Social Security	Skills and competencies for every job in the sector were profiled and compiled with the help of the private sector. These results were shared with technical training institutions to ensure that educational programs met industry needs. In addition, practical examinations were made available for workers already in the industry to certify their competencies regardless of how these were acquired	It facilitates sector-wide competency improvements for labour and increases labour mobility contributing to productivity and quality improvements and standards compliance.
Flexible Labour Regulations for the Fruit Export Sector	Direccion de Trabajo (Ministry of Labour)	The law allows for short-term contracts for temporary workers in the fruit sector, as well as permitting variations in working hours	It facilitates the provision of labour in peak periods (pruning, harvesting, packing) in the value chain at competitive prices

Source: Authors.

REFERENCES

Fernandez-Stark, K, P Bamber and G Gereffi (2010). Workforce development in the fruit and vegetable global value chain. In *Skills for Upgrading: Workforce Development and Global Value Chains in Developing Countries*, G Gereffi, K Fernandez-Stark and P Psilos (eds.). Durham: Center on Globalization Governance & Competitiveness and RTI International.

Fernandez-Stark, K, P Bamber and G Gereffi (2011). Workforce development in the fruit and vegetable global value chain. In *Skills for Upgrading: Workforce Development and Global Value Chains in Developing Countries*, G Gereffi, K Fernandez-Stark and P Psilos (eds.). Durham: Center on Globalization Governance & Competitiveness and RTI International.

Fresh Fruit Portal (2013). Cherries of the World 2014. http://www.freshfruitportal.com/wp-content/uploads/2013/12/cherries_of_the_world_2014.pdf [June 2015].

FreshPlaza (2015). Chile: More than 100,000 tones of cherries exported. *Fresh Plaza*. April 2. http://www.freshplaza.com/article/134803/Chile-More-than-100,000-tons-of-cherries-exported) [June 2015].

Murray, N (2013). Free to trade. *Food News*. http://www.direcon.gob.cl/wp-content/uploads/2013/12/Oct5_DG-Food-News_Chilealimentos_Anuga.pdf [June].

SAG (2015). Exportaciones: Productos Vegetables a Estados Unidos. http://www.sag.cl/ambitos-de-accion/productos-vegetales-estados-unidos [June 2015].

UN Comtrade (2015). United Nations Statistics Division Trade Statistics. United Nations: New York.